MERCURY RISING

ALSO BY JEFF SHESOL

Supreme Power: Franklin Roosevelt vs. the Supreme Court

Mutual Contempt: Lyndon Johnson, Robert Kennedy, and the Feud That Defined a Decade

MERCURY RISING

John Glenn, John Kennedy, and the New Battleground of the Cold War

JEFF SHESOL

W. W. NORTON & COMPANY
Independent Publishers Since 1923

Printed in the United States of America
First Edition

For information about permission to reproduce selections from this book, write to Permissions, W. W. Norton & Company, Inc., 500 Fifth Avenue, New York, NY 10110

For information about special discounts for bulk purchases, please contact W. W. Norton Special Sales at specialsales@wwnorton.com or 800-233-4830

Manufacturing by LSC Communications, Harrisonburg
Book design by Ellen Cipriano
Production manager: Anna Oler

Library of Congress Cataloging-in-Publication Data

Names: Shesol, Jeff, author.
Title: Mercury rising : John Glenn, John Kennedy, and the new battleground of the Cold War / Jeff Shesol.
Description: First edition. | New York, NY : W. W. Norton & Company, [2021] | Includes bibliographical references and index.
Identifiers: LCCN 2020051050 | ISBN 9781324003243 (hardcover) | ISBN 9781324003250 (epub)
Subjects: LCSH: Project Mercury (U.S.) | Space flights. | Space race—United States. | Friendship 7 (Spacecraft) | Glenn, John, 1921–2016. | Kennedy, John F. (John Fitzgerald), 1917–1963—Influence. | Astronautics and state—United States—History—20th century.
Classification: LCC TL789.8.U6 M4766 2021 | DDC 629.45/4—dc23
LC record available at https://lccn.loc.gov/2020051050

W. W. Norton & Company, Inc., 500 Fifth Avenue, New York, N.Y. 10110
www.wwnorton.com

W. W. Norton & Company Ltd., 15 Carlisle Street, London W1D 3BS

1 2 3 4 5 6 7 8 9 0

To Rebecca

What we are doing here in the United States is having a good hard look at how a great nation faces a great problem.

—JAMES WEBB,
ADMINISTRATOR OF NASA,
APRIL 13, 1961

You lived and died alone, especially in fighters. Fighters. Somehow, despite everything, that word had not become sterile. You slipped into the hollow cockpit and strapped and plugged yourself into the machine. The canopy ground shut and sealed you off. Your oxygen, your very breath, you carried with you into the chilled vacuum, in a steel bottle. If you wanted to speak, you used the radio. You were as isolated as a deep-sea diver, only you went up, into nothing, instead of down. . . . You were alone. At the end, there was no one you could touch.

—JAMES SALTER,
GODS OF TIN

CONTENTS

MERCURY RISING

Introduction

John F. Kennedy spent most of the morning of February 20, 1962, as the rest of America did: in front of a television. When he woke up he turned on the set in his bedroom, so he and his wife, Jacqueline, could watch the countdown to the launch of John Glenn's spacecraft, *Friendship 7,* and when he arrived in the Oval Office he clicked on the television there. Another set was on in the family quarters, where Jackie remained, and another in the third-floor playroom, where four-year-old Caroline and her friends were watching the broadcast.

It was this way almost everywhere. One block north of the White House, a crowd gathered at a construction site to watch the workmen's TV; at a nearby hospital, patients strong enough to leave their beds made their way, on foot or by wheelchair, to watch the set in the lobby. At Grand Central Terminal in New York, where CBS had mounted a twelve-by-sixteen-foot screen in the mezzanine, nearly ten thousand commuters stood side by side, heads angled upward. Ten months earlier, when the Soviet Union had launched the first man into space—Yuri Gagarin, who won the place in history that Glenn and the other Mercury astronauts had sought for themselves and their nation—it had done so in secrecy. But Glenn, as the *New York Times* columnist James Reston wrote, was taking "the whole country along for the ride."

The country had been on edge for months. Americans had watched, with mounting dread, as Glenn's flight was postponed due to heavy cloud cover, postponed due to high seas in the Atlantic, where his capsule was supposed to splash down, postponed due to what NASA, with studied imprecision, called "technical difficulties." Every delay—ten scrubs by mid-February—gave the press more time to ruminate about the many ways an astronaut might die.

In October 1957 the Soviet Union had become the first to send a satellite—Sputnik—into orbit, and then, in short succession, the first to put an animal into orbit, the first to land an unmanned craft on the moon, and the first to record images of the far side of the moon. American rockets—less powerful than Soviet boosters—tended, with alarming frequency, to explode on the launchpad or send their payloads into the sea. By April 1961, when Gagarin made his single orbit around the Earth, the specter of Soviet domination of the heavens had acquired, for many Americans, a grim sort of inevitability. Alan Shepard's ballistic flight three weeks later, which made him the first American in space, did little to lift the pall of self-doubt.

Congress was losing patience. Committees held emergency hearings and demanded a crash campaign—round-the-clock shifts at NASA facilities—to catch up. "Unless we do something very imaginative," one congressman snapped, "America is lost." The next step, many feared, was a Russian nuclear base in orbit, or on the surface of the moon. Some critics said that the space program was moving too fast already, cutting corners, taking cavalier risks. *The Nation* urged the administration to quit playing "Russian roulette" and to call off manned missions indefinitely. "All seven astronauts should be disqualified at once," the magazine argued, "for the perfectly sound reason that they have become too famous to burn in public."

And John Glenn was the most famous of the Mercury Seven—the only one who had been famous before he became an astronaut. In July 1957, he had appeared, smiling and waving on the tarmac in front of his F8U Crusader, on the front page of newspapers across the country; he had just flown from Los Angeles to New York in three hours and twenty-three minutes, a record. At the airfield in Brooklyn, as a band played "Anchors Aweigh," the Marine Corps major had leapt down from the cockpit, hugged his wife, Annie, and their two children, and said that "everything went smooth as silk." SUPERSONIC CHAMPION, the *New York Times* anointed him. But Glenn's appeal went well beyond those heroics, or the numerous medals he had won in World War II and Korea. He was, simply put, everything America wished to see in itself in an age of insecurity: he was cool under pressure, yet warm and good-humored; he spoke of God and country without irony, but also without sanctimony; he brought the self-effacing values of the small town to the fiercest kinds of air combat. All of the Mercury Seven were admired, but Glenn was the one America adored.

This was why, at 9:47 a.m., when Glenn was finally launched into orbit, most Americans were subdued, almost silent—even many of the fifty thousand spectators who spread across the sand dunes, highways, and motel balconies of Cocoa Beach to watch the rocket rise. On the New York City subway, a crackling voice on the loudspeakers implored riders to "please say a little prayer for him." The request was heard every ten minutes for the duration of the flight.

KENNEDY HAD HIS OWN reasons to be anxious. His first year as president had been difficult and perilous—a year in which "we have talked our extinction to death," as the poet Robert Lowell wrote. Mounting crises had, it seemed, drawn the United States and Russia ever closer to mutual destruction: Soviet premier Nikita Khrushchev, convinced he could bully the young president into submission, browbeat him at a summit in Vienna; pledged to support wars of liberation in every part of the world; threatened repeatedly to seize West Berlin; approved the construction, almost overnight, of a concrete wall to divide the city; and resumed nuclear testing—a relentless display, over the plains of central Asia, of apocalyptic power. "There are limits," Kennedy told an adviser toward the end of 1961, "to the number of defeats I can defend in one twelve-month period."

The space race, in that light, was one more theater of war, one more arena in which the United States was on the defensive. Since the launch of Sputnik, Soviet superiority in space had fueled doubts about American resolve. "Satellite pessimism," a political scientist called it, and in the free world it was on the rise. In a Gallup poll, 44 percent of British respondents—and a similar proportion in France and West Germany—said that the Soviets would have the strongest military in ten years' time; only 19 percent chose the United States.

No one, therefore—besides Glenn himself—had more riding on the flight of *Friendship 7* than Kennedy. His fate and Glenn's were bound together. Yet he could do nothing to ensure the flight's success. For all the power of his office, the president was a spectator today. He could only watch the television and hope.

About ninety minutes after liftoff, Kennedy picked up the phone—a direct line connecting the Oval Office to the control center at Cape Canaveral—to exchange a few words with the astronaut. While a NASA technician worked to patch the call through to the capsule, Kennedy held an index card. "Voice Procedures," it read. "First call: 'Col. Glenn, this is the President, <u>over</u>.' End all of your transmissions with 'over,'" it instructed, "until the last one," which should end with "out." Card in hand, Kennedy waited.

John Glenn had more pressing concerns.

Friendship 7—designed to fly itself—had begun drifting to the right, like a car with its front wheels out of alignment. The automatic control system fired a thruster to swing it back into line. Again it drifted, and again the thruster shoved it back, wasting fuel as it struggled. Glenn took the control stick—not without satisfaction. He was a pilot, by training and temperament, and pilots take control.

Yet he remained at the mercy of his machine. Glenn was on his second circuit around the Earth—traveling five miles a second, more than a hundred miles above the Indian Ocean—when Mercury Control asked him, almost casually, if he heard "any banging noises."

Not once, during nearly three years of nonstop training, during hundreds of simulations or the relentless drafting, redrafting, and rehearsal of the flight plan, had anyone uttered the phrase "banging noises." It had never appeared in NASA's vast catalog of potential catastrophes. It was the sort of phrase Glenn might have expected to hear about his family station wagon, not a spacecraft that had cost $160 million to produce and been tested as rigorously as any machine ever made.

"Negative," Glenn replied; he didn't hear any banging noises. Neither did he see any warning lights. At the Cape, though, the flight controllers did. Clusters of men in white shirts and dark ties were standing in front of instrument panels discussing, in hushed and heated voices, the light labeled "Segment 51." That signal—if correct—indicated a loose heat shield. A loose heat shield meant Glenn would lack protection against the 3,000-degree heat of reentry. As the capsule came down, he would burn up in seconds. Christopher Kraft, chief of flight operations, had a feeling the signal was wrong—a minority view. He played for time. "Don't mention it to Glenn," he told his team. "Not yet, anyway."

They didn't mention it to Kennedy, either. "Mr. President," the technician told him, "we've gotten pretty busy down here now. I don't think we've got time to talk."

"Give me a call," Kennedy replied, "if you get a chance."

CHAPTER 1

The Nearest to Heaven I Will Ever Get

When the airplane, a twin-propeller DC-3, landed at National Airport in Washington, the tripods were already in place, lining the ramp into the hangar. Newsreel cameramen, alert for any glimpse of the first corps of astronauts, tracked the plane as it taxied. The identities of the seven men were supposed to remain secret until NASA officials introduced them at a press conference that afternoon. Somehow, though, word of their flight had leaked—as the astronauts and their fellow passengers, a small contingent of minders and managers, could see from the windows of the DC-3.

The ambush threatened to upend the Eisenhower administration's plans for the afternoon, when the men would stride onto the stage like the starting lineup in a championship series. The fanfare was meant to give a jolt of excitement to a space program that had yet to capture the public imagination or to quiet its critics. NASA was nervous enough already about the pilots' readiness for the serve-and-volley of a news conference, an unfamiliar format for all but one of them; and, certainly, no one had prepared them for a crush of reporters in an airplane hangar.

The engines were still running. Before it reached the hangar, the plane turned around. It traced a slow arc on the tarmac, edged back onto the runway, and in a moment was airborne again. When it had flown far enough that the reporters, squinting upward, could no longer see it, the pilot doubled back, heading, this time, for Bolling Air Force Base, on the opposite bank of the Potomac. Some of the newsmen, acting on instinct, jumped into their vans and sped across the bridge toward Bolling. They were not quite fast enough. A NASA manager had radioed the base, and when the DC-3 touched down, the astronauts stepped into waiting cars, which carried them, anonymity intact, back across the river to the Pentagon. There, they shook hands with generals, had pictures taken, and traded their uniforms for civilian clothes—their suits and ties signifying a space program whose purposes, President Eisenhower had said, were peaceful.

THE OFFICES OF THE National Aeronautics and Space Administration were located, incongruously, in a stately if shabby mansion on Lafayette Square, a block from the White House. Built in 1820, it had been the home of Dolley Madison, the widow of James Madison, during her last years of life, and it sagged now with neglect. The floors creaked; the paint flaked; chunks of plaster fell from the ceiling. Congress had considered tearing it down. But it was here, at 2:00 p.m. on April 9, 1959, in an unadorned, unventilated assembly room, that the space agency introduced the seven men, the "astronaut volunteers," it planned to send into space.

They entered to applause. "Take your pictures as you will, gentlemen," Walter Bonney, NASA's director of public affairs, told the news photographers, who rushed and pushed and in some cases crawled their way to the stage where the seven men stood and posed, a little uneasily, with scale models of the Mercury spacecraft. The free-for-all lasted ten minutes. "These people are nuts," one astronaut, Alan Shepard, said to the others, not at all quietly. Bonney, finally, appealed for calm. The seven took their seats behind a long table covered in blue cloth and punctuated, down the line, by microphones and substantial ashtrays. The NASA administrator, T. Keith Glennan, stepped up to a podium. "These, ladies and gentlemen," he said grandly, "are the nation's Mercury astronauts." At that, the reporters applauded again.

Three of the seven were Navy men. Shepard, by his bearing, seemed to consider himself the first among equals. This was bred in the bone: the Shepards traced their roots back to the *Mayflower*. His father had fought in both world wars, and Al held degrees from the Naval Academy and the Naval War College. He had never been more than a middling student; he'd nearly been expelled from the academy. And he had never flown in combat—a sore spot. Yet Shepard had proved that he "could fly anything," a fellow test pilot once said, even if that meant flying faster or higher or dangerously lower than regulations allowed—well worth a grounding, in Shepard's view. He was here at NASA over the protests of his father, who thought that becoming an astronaut was a good way to ruin a promising military career.

Walter Schirra, like Shepard, was a graduate of the Naval Academy, though unlike Shepard, he had seen air combat—ninety missions in the Korean War. Wally wore it lightly. He laughed freely, played pranks, looked for an audience, and usually found one. His jokes, though, often had an edge. He had not only read but studied *The Theory and Practice of Gamesmanship*, a popular guide to the art of the put-down; his wit was a tool to win advantage. He had barely made the cut for Mercury: during the selection process, specialists had found a nodule on his vocal cords—the result, most likely, of cigarette smoking. Navy doctors, moving swiftly, had it removed.

At the far end of the table sat the third naval aviator, Scott Carpenter. During the Korean War, the Colorado native had patrolled the

Sino-Soviet coast; from there it was test pilot school at Patuxent and air intelligence work aboard the USS *Hornet*. Carpenter outperformed his rivals during the rigors of the selection process—broke records, in fact, in his ability to endure whatever perverse sort of stress was imposed on the pilots to see if they would crack. He had, all the same, a winsome manner—a kind of naïveté. Carpenter thought life was fair—and that the way to win a contest was to be its most gracious participant.

The Air Force pilots—there were three of them as well—were less polished and more apprehensive. Virgil Grissom, or "Gus," wore a wary smile, projecting a quiet, somber sense of purpose. None of the seven would have been happier than Grissom to wrap this up and get to work. He was the shortest of the men, but no one doubted his capacity: he had flown more than a hundred combat missions over Korea. His friend Gordon Cooper, at thirty-two the youngest of the group, was probably its rawest talent, bursting with brash self-belief, though none of that was apparent today. To Cooper the news conference was an "ordeal," as he wrote in his journal. His shaky marriage was a major source of unease, though his wife, Trudy, had agreed to get back together—a charade, they both knew—for the sake of his selection. Lastly, there was Donald Slayton, leaning back in his chair, almost scowling, slowly smoking a cigarette. "Deke" had made it to test pilot school after flying bombers in the European theater. He had little interest in spaceflight. He had applied to Mercury in part to avoid his next assignment by the Air Force: testing aircraft in icy conditions or something equally unglamorous.

The seventh astronaut was "the lonesome Marine on this outfit," as he introduced himself. But the reporters in the room already knew that; they already knew John Glenn. It had been only two years since he had set the transcontinental speed record and become, in an instant, a sensation: speaking engagements, fan mail from schoolchildren, a two-week stint on the CBS game show *Name That Tune,* tributes in Congress. "Major Glenn takes his place among the pioneers of the 20th century," one congressman had declared; Glenn was lifting mankind above "the barriers of time and space."

He had seen more combat than any other astronaut: fifty-nine missions in the Pacific during World War II, a hundred in Korea. He had also won more decorations: five Distinguished Flying Crosses, nineteen Air Medals. And now, in this other kind of rarefied air, he was the man most evidently at ease. Glenn made it seem there was no place he would rather be than right here in this room, with these pilots, answering these questions. Nearly thirty-eight, he was the oldest of the group (a point of sensitivity) but also the most conspicuously boyish: though his reddish-gold hair was thinning, his face was round and full; his freckles made it seem he was blushing. Schirra appeared sly; Carpenter looked affable. But Glenn—in a word rarely used to describe a Marine, a combat veteran—was sunny.

The first question from the press concerned the men's wives: Did they support this dangerous mission? Or, as Bonney rephrased it, "Has your good lady . . . had anything to say about this?" Yes, Carpenter volunteered, his wife was enthusiastic. "I can answer the same," said Cooper, though he couldn't. Then it was Glenn's turn. Sliding the microphone closer, he laced his fingers together. "I don't think any of us could really go on with something like this," he began, "if we didn't have pretty good backing at home, really. My wife's attitude toward this has been the same as it has been all along through my flying: that if it is what I want to do, she is behind it, and the kids are, too, a hundred percent."

Glenn, from the start, was changing the rules of engagement, and most of the others were slow—or unwilling—to adjust. "What I do," Slayton said, "is pretty much my business, profession-wise." Shepard played the premise for laughs. "I have no problems at home," he said, pursing his lips to suppress a smile but then letting it spread, wide and satisfied. Schirra, in approval, slapped him on the back. They were playing to another set of norms: fighter jock culture, flippant, irreverent. While Glenn spoke, Shepard leaned back in his chair, talking sotto voce with Slayton and others, a display of indifference or even contempt.

The questions they might have liked to answer—about the spacecraft, the training program—went to the NASA officials, while the "boys," as Bonney called them, were asked about things like their "sustaining faith." They sought safe haven in clichés. Some found themselves at a loss for words. Spaceflight, Slayton said, "is an extension of flight and we have to go somewhere and that is all that is left." As Grissom muttered about "serving the nation," Glenn—the only one with a pen and a note card—jotted something down and waited his turn.

Then he leaned forward. "In answer to this same question a few days ago from someone else, I—jokingly, of course—said that I got on this project because it'd probably be the nearest to Heaven I will ever get and I wanted to make the most of it." The reporters laughed. "This whole project with regard to space," Glenn continued, growing serious, "sort of stands with us now . . . like the Wright brothers stood at Kitty Hawk about fifty years ago, with Orville and Wilbur pitching a coin to see who was going to shove the other one off the hill down there. And I think we stand on the verge of something as big and expansive as that was."

And so it went for an hour. At least on this day, the press saw Project Mercury as a human-interest story, and the human generating the most interest was Glenn. A question about religion—a favorite subject for Glenn, who taught Sunday school—called forth a homily. "I think you will find a lot of pilots who . . . look at this thing completely from a fatalistic standpoint, of 'sometime I'm gonna die so I can do anything I want. . . .' Well, this," he said, "is not what I believe. I was brought up believing that you are placed on earth . . . with sort of a fifty-fifty proposition, and this is what I still believe. We are placed here with certain talents and capabilities. It is up to each one of us to use those talents and capabilities as best he can." A higher power, he said, "will certainly see that I am taken care of if I do my part of the bargain."

Glenn had hit, by now, nearly every note on the register. The other six were answering questions; Glenn was giving speeches—off-the-cuff (his note card aside) and brimming with humility, piety, a sense of wonder. Also humor: after his sermon, Glenn got the biggest laughs of the day when he fielded a question about the tests the astronauts liked least. "That's a real tough one," he said, "because we had some pretty good tests. . . . It is rather difficult to pick one, because if you figure how many openings there are on a human body and how far you can go into any one of 'em—" The whole room broke up, astronauts included. "You gave it away!" Schirra said. Glenn wagged his finger at the crowd: "Now, you answer which one would be the toughest for you!" He, too, was laughing now, shaking his head, raising his shoulders in a comical shrug.

The press was besotted. "The sky's no longer the limit," James Reston cheered in the *New York Times*. What made the astronauts "so exciting," he wrote, was "not that they said anything new but that they said all the old things with such fierce conviction. They talked of the heavens the way the old explorers talked of the unknown seas. . . . They spoke of 'duty' and 'faith' and 'country' like Walt Whitman's pioneers." This, however, was poetic license. All seven astronauts had sung from the same all-American hymnal, but only John Glenn knew the words by heart.

GLENN'S WORD OF CHOICE was "real." A supersonic flight was "a real kick"; it was "real fine." The word suited Glenn himself. His performance at NASA headquarters was so deft, so attuned to what America in 1959 wanted to see in its finest, fittest young men, that it could have appeared just a little too perfect. But it was real—an authentic expression of charm and self-confidence and, at the same time, a reverence for things larger than oneself.

It was real but incomplete—only part of the picture. If the press had been looking for them, they would have seen hints of Glenn's ambition and impatience, his determination to win—most of all the contest taking shape on the stage, the battle to become the first man in space. "I never saw anybody who wants as much as this man wants," a childhood friend said of Glenn. And even as a young man, before he knew exactly what he wanted, Glenn knew it meant leaving his home of New Concord, Ohio, a town that defined him but threatened to trap him.

New Concord sits amid the rolling hills of eastern Ohio, seventy miles from Columbus. Many maps of the state, even detailed ones, give no hint of the town: it is a mile across. During the 1930s, when John Glenn was a child, its population was scarcely one thousand. On the western edge of this very small town, the Glenns' house sat high on a bluff above an S-shaped stone bridge. The Glenns had moved there from Cambridge, nine miles down the road, in 1923, when John H. Glenn Jr.—called "Bud" by his parents—was two years old. Bud's father, John Sr., known by his middle name, "Herschel," had fought in France during the Great War and worked for a time on the Baltimore and Ohio Railroad, shoveling coal into the engines of locomotives. He then found work as a plumber, which was what brought him to this town. Bud's mother, Clara, had lived in New Concord before, as a student at Muskingum College, which (like the Glenns and nearly everyone in town) was affiliated with the Presbyterian church. Muskingum gave life in New Concord a different inflection from life in the factory towns, the coal-mining towns, or the farming communities nearby. The college held art exhibits and concerts; it staged public debates. "New Concord is what it is," a brochure boasted, because of its "faith in God, respect for law, and . . . love of education." This was the town's civic catechism, and Bud learned it by heart. "My early home life," he wrote in an autobiography (a school project) at eighteen, "was wholesome and above all it was Christian."

It was also close-knit. On school days, at lunchtime, Bud would often walk or roller-skate to the storefront of Glenn Plumbing, at Liberty and Main, and have lunch with both parents. The Glenns were part of the "Twice Five Club"—a group of five couples who got together once a month, kids included, for a potluck dinner or sometimes a night out in Cambridge or neighboring Zanesville, where the evening might consist of sitting in the car on Main Street and talking to friends who passed by. When the children were little they were put together in a playpen—the place where John, at the age of two or three, met a girl named Anna Margaret Castor, "Annie," who would someday become his wife.

Love of country was an article of faith. "You were either patriotic," Glenn said decades later, "or you'd practically be run out of town, I guess." His father rarely spoke about World War I, but when he did, around the dinner table at home, Bud thrilled to his stories. Herschel took great pride in his service. He had been a bugler, and he taught "taps" to his son. Every Decoration Day, Herschel would put on his uniform and march with other veterans down Main Street and to the town cemetery. There, he raised his bugle, sounded the first mournful phrase of taps, and paused. John, having practiced his part, stood alone in the woods nearby, lifted his own trumpet, and echoed his father's notes.

"Dad was my hero," Glenn later said. He picked up Herschel's interests and made them his own. After school Bud brought friends to his father's store, laid an assortment of pipes and fittings out on the floor, and set to work putting them together. When Herschel started a side business selling Chevrolets, he let Bud take engines apart and rebuild them, just to see how they worked. Father and son also shared a passion—as Glenn put it, a "mania"—for travel. He wrote in his teenage autobiography that his father had a wish to go "more places than any other man I know"—a desire that "must have been passed directly on to me." In 1934, the pair visited the Chicago World's Fair, billed as "A Century of Progress" and designed to awaken the American interest in science. The fair promised a "million-dollar thrill": a "Sky Ride" in double-decker rocket cars, made of metal and glass, that emitted colored steam as they rode in the air along steel cables.

Glenn had been fascinated by flight since he was small. His "idea of a holiday was a visit to the airport," his mother later recalled. He took an Ohioan's pride in the Wright brothers, who had come from Dayton; and one day when the radio reported that Charles Lindbergh—who had just crossed the Atlantic in the *Spirit of St. Louis*—was about to fly over New Concord, Glenn rushed outside to spot the plane. (He told his friends he'd seen it; in truth he was not so sure.) He labored over model airplanes, carving balsa-wood wings with a single-edge razor blade, then holding the plane out the car window to watch the propeller spin. At night, he slept on a pillowcase decorated with planets and stars; a koala bear, floating among them, clutched a spacecraft labeled "John" in blue embroidery.

One summer day when Glenn was eight, his father brought him along to a plumbing job in Cambridge. On the way, they passed an airstrip—a field, really—and saw an open-cockpit biplane, a Waco, sitting on the grass. Herschel stopped the car to take a look. He had seen planes like this during the war, in dogfights above the trenches, and always wondered what it would be like to fly in one. Here was a chance. The pilot, in leather helmet and goggles, was taking locals up on short trips—a couple of passes around Cambridge. Herschel turned to his son: "You want to go up, Bud?" Glenn didn't think his father could mean it. But there he was, handing a few dollars to the pilot, and a moment later the Glenns sat pressed together in the back cockpit of the plane, one strap stretching across the pair. The pilot revved the engine, guided the plane along the grass strip and then, incredibly, into the air—above the farms and the forest, above the town. Herschel pointed out landmarks, but Bud couldn't hear a word over the noise of the engine. It didn't matter. "I was elated," Glenn recalled decades later. "I was hooked."

For the next ten years, however, Glenn remained earthbound. His family muddled through the Great Depression without going hungry, without losing the house, but there was certainly no money for flying lessons. In 1939, as a freshman at Muskingum, Glenn set himself a more practical goal: to become a research chemist. "I do not hope to become a second Madame Curie or an Einstein," he conceded, "but I do want to feel that someday I can discover something or make something for the scientific knowledge of the world that will be entirely original and of my own invention." At the same time, he dreamed of becoming an airline pilot or, barring that, an aircraft mechanic. He collected glossy route maps from TWA and Pan American, listing flights he could not afford to take; he looked into a correspondence course with the Aero Industries Technical Institute in Los Angeles. Life at eighteen, Glenn mused in an essay, was a plane that was just taking off. "We have not yet established our regular cruising speed," he wrote, "but we have our power in full use."

Not yet—but soon. In January 1941, Glenn was walking by a college bulletin board when a notice caught his eye; it concerned something called the Civilian Pilot Training Program. With the air war underway in Europe, the federal government had begun an effort to train tens of thousands of pilots. Glenn was thunderstruck. Here was an opportunity to earn his pilot's license, study aerodynamics, and be at the front of the line if the country needed combat pilots. He bounded home to share the news.

It was not well received. At the dinner table that night, as Glenn effused about the program, his parents sat silent and grim-faced. Taking him on a hop over Cambridge had been one thing, but this, in Herschel's view, "was just the same as taking him out and burying him. We objected strenuously." The simple fact, Glenn's mother said, was that airplanes were "always crashing." Returning to campus, Glenn enlisted the support of Paul Martin, a physics professor who managed the program at Muskingum. He paid a visit to the Glenns and, over the course of a long talk, persuaded them that aviation was the future and that John, as one of his best students, ought to be part of it. And so John Glenn, a few weeks later, found himself on an airstrip in New Philadelphia, Ohio, learning the controls of a Taylorcraft, a light little plane with only sixty-five horsepower. Glenn had tucked away a brochure for this plane, too: in a Taylorcraft, it promised, "imaginary pleasures become realities."

GLENN SAT IN THE college chapel in a state of high agitation. His girlfriend, Annie, was playing a piece by Jean Sibelius on the organ, and brilliantly, but Glenn was hardly listening. This was Annie's senior recital; she had been offered a scholarship for graduate study at Juilliard, but her stutter was so severe that it was hard to imagine her living in New York, a place so much louder and less forgiving than home. Throughout her performance, Glenn's mind was focused on the radio news bulletin he had heard during his short drive to campus: Japan had bombed the U.S. naval base at Pearl Harbor in a surprise attack. After parking the car, he'd sat for a few minutes to learn what he could before the recital started. Difficult conversations lay ahead—first with Annie, then with his parents—but Glenn already knew he was going to enlist. A few days later, he and a friend went to Zanesville and signed up for the Army Air Corps.

And then he waited and grew restless waiting. Months passed; no orders came. Glenn had quit college when the semester ended, and now he found himself with nothing to do. He got a job plowing fields outside town. Finally, in March 1942, he went back to Zanesville—this time, to a Navy recruiting station—and volunteered again. He had come to the conclusion that "the first people who were tossed into action then were the Marines, and if I was going into the military," he recalled, "I wanted to get in and do something." He bought Annie a diamond ring, asked her to marry him—she accepted—and went off to train for war.

NEARLY TWO YEARS went by before Glenn saw the action he sought. During that time he excelled in his flight training, won the commission he wanted in the Marine Corps, and pushed his way, clumsily, into a fighter squadron. At the Naval Auxiliary Air Station in San Diego, he and a friend, Tom Miller, lobbied so hard to get out of flying transport—"boxcars," they called the multiengine DC-3s; flying one was as dull, Glenn said, as "sitting in an office"—that their commanding officer called them into his office, stood them at attention in front of his desk, and walked around them in circles, barking about insubordination. They spent days apologizing. In the end, they got the transfer—into F4U Corsairs, carrier-based, single-engine fighters with inverted gull wings. Japanese ground troops called the Corsair the "Whistling Death" for the noise it made on approach. Glenn rhapsodized about its "speed, climb, power, and firepower." There was "nothing . . . more fun," he wrote of the Corsair, "than to play around 'upstairs.' "

But the plane could be perilous. Its cockpit was set back on a long, uptilted nose, making it hard for the pilot to see below and difficult to land on an aircraft carrier. The Navy abandoned the Corsair and gave its fleet to the land-based pilots of the Marine Corps. While Vought and Goodyear, the Corsair's manufacturers, redesigned key components, Vought's parent company, United Aircraft, hired Charles Lindbergh to test and promote the new planes. In October 1943, he flew a new Corsair—painted navy blue—to the Marine Corps Air Station at El Centro, California, where Glenn was awaiting orders to ship out. Glenn and others gathered to watch as Lindbergh climbed out of the cockpit, tucking his helmet under his arm and looking every bit the part of the most celebrated pilot in the world. Glenn was overcome by a "feeling of kinship," which was easy to understand, even if it was the kind of kinship a minor-league ballplayer might feel toward Joe DiMaggio.

By the time Glenn arrived in the Marshall Islands, in July 1944, the Central Pacific Theater was moving westward in bloody, costly, grinding fashion. U.S. forces had taken the Marshalls a few months before as part of a push farther west, but thousands of Japanese troops—well armed with anti-aircraft weapons—remained on the islands and atolls, harassing the slow-moving U.S. dive-bombers and keeping alive the threat of reoccupation. Glenn commanded a division of four planes—a role he had been assigned too soon, some of the more seasoned pilots griped. But Glenn had won the confidence of his skipper.

He quickly showed why. Day after day, Glenn led his division low and fast across the islands, skimming the tops of palm trees, passing through thick clouds of tracer fire, and strafing the anti-aircraft positions. They dropped armor-piercing bombs on Japanese bunkers; blanketed towns and small islands with a horrific new firebomb, napalm; and, between missions, tested the Corsair's limits and their own. Glenn was keen to see how fast and high he could fly the Corsair before losing control, how well he could handle the plane during a dive, and how heavy a bomb load it could carry. This last question brought Lindbergh, again on United Aircraft's behalf, out to the Marshalls. Lindbergh attached a special rack to his plane, added an extra 2,000-pounder to his load, and tested the proposition on a bombing raid with Glenn's squadron. "Luggin' bombs," Glenn wrote in his diary, "gets to be sort of fun after a while. If you've ever seen a 2,000 pounder go off at close range you know what I mean." In the humid air above the atolls, a one-ton bomb sent shock waves when it exploded.

For Glenn, it was more than fun. In the South Pacific, for the first time in his life, he experienced the thrill of excelling at something dangerous and of real consequence; he felt the euphoria of becoming "a part of, and the brains of," a powerful machine that responded to his "every touch and whim." Glenn achieved, as he described it, the kind of "absolute confidence . . . that makes a fighter pilot believe that there actually is not another pilot and plane that can whip him in all the world."

Yet he knew he was not invincible. At El Centro, the left elevator had come off the tail of his Corsair in the middle of a dive at 460 miles per hour; Glenn was tossed around his cockpit so violently he could not even grab the controls. The plane leveled off just enough that he managed to land it. Others in his squadron were not so lucky. During training, one Marine had died after bailing out. At Midway, before reaching the Marshalls, Glenn had watched two planes collide on the runway, killing one of the pilots. And on Glenn's very first combat mission, his wingman, Monty Goodman—"one of the best friends I've ever had," a bunkmate, a partner in singing and clowning on the base—had disappeared in the smoke of anti-aircraft fire, leaving no trace but an oil slick in the sea. On a later mission, Glenn's own plane was hit: an explosive shell tore an eight-inch hole in the left wing. Glenn made it back to base, praying all the way.

"You need not worry," Glenn had written his pastor in December 1942. "I just know that in His keeping, everything will work out for the best." Such was the certitude of the cadet. By the spring of 1945, though, as his tour wound down, Glenn had come to the view that "stand[ing] for right and for the things of God" afforded him no special protection in battle. He had seen that luck—as much as faith and goodness, as much as skill—had a controlling hand. And Glenn's luck had held. After flying fifty-nine missions, he returned home with ten Air Medals and two Distinguished Flying Crosses. This was a start.

CHAPTER 2

Supersonic

Glenn's consuming worry about the war in Korea was that it would end before he got there. A stateside ground post, which was where Glenn was stuck in 1952—on the general staff at the Marine Corps Air Station in Quantico, Virginia—felt to him like an affront, an egregious waste of his ability. The years since World War II had been productive ones for Glenn—he had learned to fly jet fighters at 600 miles per hour, substantially faster than a Corsair; studied amphibious warfare, the art of coordinated combat by air, land, and sea; and aced (and then taught) a Navy course in all-weather flying, a training program in which flight instruments were blacked out or forced to fail. During the winter of 1946–47, he had flown patrols as a noncombatant in support of Chinese Nationalists. But now he felt that he was marking time.

Combat flying, he reflected, was "my vocation . . . , my life's work." Once a month he wrote his commanding officer, a colonel, asking to be assigned to Korea. And once a month, the letters came back, each bearing a stamp: REQUEST DENIED. After several months of this, Glenn was told to stop writing letters; the Marines, the colonel said, would send him to Korea when they were good and ready.

Glenn got his orders in October 1952—the cue for Annie to start packing their belongings into a small stack of boxes and four suitcases. Well practiced in the Marine-family move, she could manage this in half a day. Since 1943, when she and John had gotten married, Annie had moved the family and its belongings more than once a year, on average. Now the couple and their children—John David, almost eight years old, and Carolyn Ann, who was five—sang traveling songs in their convertible Buick Roadmaster as they drove back to New Concord, where Annie and the kids would live for the time being. Glenn stayed just long enough for both sets of parents—Annie's and his own—to register their unease about the war.

Most Americans felt that way. In November 1952, while Glenn was preparing to ship out, Dwight D. Eisenhower won the presidency by declaring that the Korean War—after more than two years of carnage—was a tragedy, a failure. Rejecting the notion that "America had to be bled this way," Eisenhower promised to bring the war to an honorable conclusion. During the war's first phase, in 1950, U.S. and U.N. troops had crossed the 38th parallel, the border dividing the Communist North from free South Korea, then were turned back and slaughtered by Chinese troops. The war since then had been a stalemate: deep trenches cut across the peninsula from the Yellow Sea to the Sea of Japan. Infantrymen spent long, frigid winters clutching hand warmers to keep their trigger fingers from stiffening; the summers, a correspondent reported from the front, were "an endless morass of mud under a cheerless, leaden sky."

But the war was different up in that sky. There it was all motion and speed—such incredible speed that a sky filled with jet fighters circling, tailing, and firing at one another could suddenly, in an instant, become a vacuum, empty and quiet, as the planes disappeared in all directions. In February 1953, Glenn arrived at Pohang Bay, on Korea's southeastern coast. He was assigned to a squadron that flew the U.S. Navy's F9F Panthers—straight-winged, heavy-framed, midnight-blue fighters—mainly in support of Marines on the ground. The pilots called themselves the "Willing Lovers"—after the WL tail codes on their aircraft—and wore silk scarves, sky blue with red hearts. On many missions, Glenn's wingman was a pilot from the reserves: Ted Williams, the Boston Red Sox left fielder who had hit .406 during the 1941 season before spending the next few years flying Corsairs. (On his return to baseball, Williams led the Red Sox straight to the World Series.) The two became fast friends and bunkmates, sleeping at opposite ends of a Quonset hut.

Looking back years later, Williams had this to say about Glenn: "The man is crazy." Glenn seemed almost to hurl himself at targets, flying too fast and too low through sheets of anti-aircraft fire, blasting his 20-millimeter cannons. He returned to Pohang seven times with holes in his plane. One enemy shell hit Glenn's empty napalm tank and sprayed shrapnel across the plane, leaving nearly three hundred holes, some of them in the cockpit. The squadron started calling him "Ol' Magnet Ass." On another raid, Glenn ignored strict orders against making a second attack and ended up with a hole the size of a basketball in his plane's tail. (He somehow made it back to base, then posed for pictures.) "I was rather intent on war-making," he acknowledged later. "I won't say I was trying to win the war all by myself, but I took it all very, very seriously."

During the late spring of 1953—as truce negotiations flickered back to life and the Communists agreed to exchange sick and wounded prisoners with the United States—Glenn found the fight he wanted most. Blowing up bridges and strafing anti-aircraft guns was gratifying work, but back at Quantico he had learned about an Air Force program aimed at giving Marine pilots experience in air-to-air combat—in other words, dogfights at transonic speeds against Soviet-made (and often Soviet-flown) MiG-15s. Now, as Glenn's Marine flight duty wrapped up, he got the assignment.

The Twenty-Fifth Fighter-Interceptor Squadron was based at Suwon, just south of Seoul. Its pilots flew F-86 Sabres, the Air Force jets that had been controlling the skies along the Yalu River and across North Korea, shooting down MiGs at a rate of ten to one. The single-engine Sabre was one of the most advanced planes ever sent into combat. Its swept-wing design gave it stability as it approached the speed of sound. The Sabre had set the world speed record in 1948 and had broken it every year since: in 1953 the plane reached 715 miles per hour. It could fly as high as 48,000 feet, and its computing gunsight used radar to calculate the range of a target—a novelty for Glenn, who was used to relying on his "combat eye." But on a basic level Glenn found dogfighting in Korea to be a lot like dogfighting in World War II: he trusted his instincts and flew, as the cliché had it, by the seat of his pants.

Other pilots painted pinups on the fuselage; Glenn adorned his with the names LYN, ANNIE, and DAVE. But in all other respects he fit right into a group that, as the *New York Times* portrayed a Sabre squadron, was "grimly disconsolate" unless "the hunting has been good." Glenn, in fact, was so disconsolate at his failure to encounter any MiGs by the start of the summer that his squadron's painters added under the cockpit, in big red letters: MIG MAD MARINE. Only one Marine pilot had become an ace in Korea. Glenn was set on becoming the second. He needed five kills, but by the middle of June he had yet to fire a single shot on an enemy plane. In a letter to his son, Glenn grumbled that he had only been the wingman, not the leader, on his seven missions so far. He tended, at moments like these, to sulk in letters home—referring to himself, sourly, as "second-best Glenn."

On June 16, 1953, Glenn was flying wing with his commanding

officer, Lieutenant Colonel John Giraudo, low along the Yalu River. A cloud deck made it impossible to see any MiGs, so the two went looking for enemy trucks. More than a hundred miles from base, skirting the border of China and North Korea, they spotted a small convoy. Giraudo rolled into a strafing run and Glenn followed. It was a flak trap. Anti-aircraft shells began bursting from both directions. The planes pulled up; Giraudo radioed that he had been hit. He climbed a few thousand feet, nosed over, tried to level out, swung the plane back up, and plunged—a terrifying pirouette. He had to bail out. Glenn watched from above as Giraudo and his seat shot out of his cockpit. His parachute opened, billowed out across the tree tops, and came to rest there for a moment. Then it disappeared beneath the trees. This was a good sign. Giraudo must be burying his chute, Glenn thought, to keep it hidden until the rescue helicopters arrived.

Except they didn't arrive. And Glenn, still circling, was running out of time. "I orbited in the area . . . as long as fuel permitted," he reported the next day, "to try to direct the rescue planes" to the right spot. He calculated his "bingo fuel"—a pilot's term for the bare minimum it took to get back to base—and kept on circling until he had passed that point. He needed a new plan, and urgently. With the last of his fuel, Glenn rocketed to 40,000 feet—where, as expected, his engine flamed out. He was high enough now, if his calculations were right, to glide home. Or most of the way home. If the Sabre dropped too low too quickly, Glenn would have to eject—though South Korea was seventy miles away. He called his radar controller and asked him to get another plane ready on the runway: Glenn was planning to go back for Giraudo.

He was gliding now—without power and almost, it seemed, without a sound—across the span of North Korea. Frost began to form on the windscreen, spreading across the cockpit until Glenn could see only out of a narrow gap. But the winds were in his favor. He maintained altitude long enough to reach the base at Suwon, where he attempted a "dead-stick" landing—"which means," he later explained, "you do everything right or you wind up off the end of the runway." He did it right. Without fuel, he couldn't taxi in; the Sabre rolled until it stopped of its own accord. Glenn climbed out, jumped into a waiting jeep, and several minutes later was back in the air in a new Sabre, with three others trailing, following him to the spot where Giraudo went down. The search yielded nothing—not even a radio signal. They stayed in the area until they hit their fuel limits, then flew back to Suwon.

The debriefing that night was somber. Afterward, one of the other pilots approached Glenn. "You know," he said, "you ought to say something to Father Dan. He was really in your corner today . . . he was running the rosary like a monkey climbing a flagpole." Father Dan Campbell was the squadron's chaplain; he blessed the planes, and when airmen went out on missions he sat in his jeep on the side of the runway and shouted, "Give 'em hell!" He and Glenn had spent a lot of time together over recent weeks, talking religion. Now Glenn thanked the priest for his prayers. Campbell promised they would get Giraudo back. Glenn doubted that.

"Never had anyone going down affect me like that," Glenn wrote to Annie. "Never felt so helpless in my life. . . . Fly with a guy literally fighting for his life and not be able to help." In a letter to his son, Glenn struck a more reflective note about his long glide home: "It was a funny feeling, Davey, but I actually was never afraid or scared. I knew what the plane would do and I knew what I could do, so that was all there was to it. . . . If you train yourself to do things the way you should, then when the whole operation goes haywire, your past training and experience takes over and you'll know what to do."

A few weeks later, Glenn had happier news. "I am singing a slightly different tune than my last few letters," he wrote his family on July 12.

"Today, <u>I</u> <u>finally</u> <u>got</u> <u>a</u> <u>Mig</u>, cold as can be. Of course I'm not excited at this point, <u>not</u> <u>much</u>!" No longer a wingman, Glenn was leading a formation of four Sabres near the Yalu when they encountered two MiGs. He called a "bounce"—the signal to attack—and swung wide to come up behind one of the enemy planes. "Kept 'hosing' him until I was right up his tail-pipe and had to break over him to keep from ramming him," he wrote. "What a time! Really a thrill. . . . Funny how the bullets sparkle when they hit a plane like that. . . . Just chopped it up good, until it finally flamed." The MiG hit the ground and disintegrated—spreading, Glenn thought, like a napalm drop.

And then, a week later, one day after Glenn's thirty-second birthday: "Got another one today. #2! . . . I've dreamed of deals like this." Eight Sabres, sixteen MiGs, jets screaming in every direction. Glenn picked out a MiG, but it outmaneuvered him; Glenn's wingman, Jerry Parker, swung in and shot it down. Then Parker's speed dropped dangerously—engine trouble. Glenn flew in large circles around him, protecting him, until a string of six MiGs closed in on both of them. Glenn faced the MiGs head-on, firing until they flew off. "I latched onto the last one coming past and closed on him," Glenn wrote. "Hacked him good"—smoke trailed from the fuselage as the MiG rolled over and crashed. One by one and in pairs, the others came back. Glenn fired one more burst—the last of his ammo—and turned to rejoin Parker. Miraculously, the MiGs didn't follow. "Hope this was a forecast of things to come," he added in his letter home.

"Nailed another one!" he wrote Annie on July 22—his third MiG. He marveled at the speed of combat in jet aircraft: that day's fight, from start to finish, lasted no longer than forty-five seconds. "You just run on reactions . . . just go get 'em." It was "a pretty speedy war—but fun! Guess it's not normal to enjoy killing people, but I'm abnormal then, because I haven't had such a wonderful time in years. Thoroughly enjoying this. Just need two more now to get my five for the Ace-dom. I'll get them if this war holds out."

It didn't. Several days of bad weather kept Glenn on the ground, and that was that: at 10:00 a.m. on July 27, North Korea and the United States signed a truce in Panmunjom. Glenn's first reaction was disappointment. But he took pride in his role in the first jet air war. He no longer believed that air combat was the same as it had been in World War II. It required new tactics, new techniques. So his war ended well. And Father Dan had been right: John Giraudo had indeed survived. A North Korean patrol had taken him prisoner. In October 1953, Glenn was waiting at Freedom Village in Panmunjom, where POWs were being returned, and stood there grinning as his commanding officer walked up. Giraudo was smiling, too. "You son of a bitch," he said to Glenn.

Back in 1945, at the end of World War II, Glenn had made clear—in his respectful but emphatic way—that he did not want to join, and someday run, his father's plumbing business. Now, in the same spirit, he declined an invitation from his father-in-law, Dr. Homer Castor, to take over his dental practice. "I had my sights set elsewhere," Glenn recalled.

Yet he was thirty-two and a father of two children. It no longer made sense—to him or to Annie—to bounce around from base to base, from assignment to assignment, waiting for the next war. One path would lead upward through the ranks of the Marine Corps. He supposed he could see himself, someday, as General Glenn, and it was safe to assume that others could, too. It had always been the case that people looked to him to lead. He had that air of self-possession that drew others to him, that made a man as famously prickly as Ted Williams refer to him, devotedly, as "my idol." This was also because Glenn, as

Williams said, was "the bravest SOB I ever met." As Glenn had shown, he was willing to push himself harder than almost anyone around him. (He pushed others, too, sometimes to his detriment.) And he had a cornball, unserious side, a warmth he was unembarrassed to show—traits that won him friends and, whether or not they saw themselves as such, followers.

Yet he knew he was never going to get to the top—was never going to become commandant of the Marine Corps—even if he went to all the right midcareer military schools and jumped through all the right hoops. No pilot had ever been appointed to run the Marine Corps, which, for all its proud talk of "land, sea, and air," was still mainly a ground-based operation. Glenn wanted to rise, but more than that, he wanted to fly. In late 1953, still in Korea, he applied to be a test pilot at the Naval Air Test Center, at Patuxent River, Maryland. It had taken some work to bring Annie around, but Glenn was very persuasive when his ambitions were at issue. He got his orders on Christmas Day. "We broke into the sparkling burgundy in the mess," a fellow pilot recalled. "John didn't drink much normally, but he drank the hell out of that burgundy."

He reported to Patuxent in January. For the next two years, he spent his days pushing supersonic planes to the limits of the envelope and then purposely past that. It was a disciplined kind of danger: every turn and spin was measured, calibrated, analyzed. This was the "daily business," as Glenn called it, and except for the fact that, more than once, he almost failed to get home alive, Patuxent had the feel of normal life. There were barbecues and water-skiing; Annie played organ at the nearby church. On the weekends, Glenn taught a Sunday school class that included his own children; at night, he studied calculus, preparing himself for something, though he was not yet sure what.

In November 1956, Glenn transferred to the Navy's Bureau of Aeronautics (BuAer)—Fighter Design Branch—in Washington, D.C. On paper, this was a promotion. When test pilots identified a problem with a plane, BuAer determined what to do about it. Glenn was perfectly qualified for the position—and poorly suited to it. The job was mostly paperwork, and the commute from St. Mary's County, Maryland—where his family had settled into life around Patuxent—was an hour and forty-five minutes each way. After two years of flying high-altitude tests in the F7U Cutlass and the F8U Crusader, Glenn now found himself in a Studebaker, stuck in traffic or searching for a parking spot. When his car was rear-ended one icy morning, his co-workers at BuAer found it hilarious—a decorated fighter pilot in a fender bender—and teased him about it for months.

Glenn quickly grew restless. He began waging an insistent campaign for approval of "a perfect capstone" to the testing program for the Crusader: an attempt to break the transcontinental speed record. A maximum-power, long-distance flight, he argued, would prove the worthiness of the Crusader's engine and convince Congress that the plane was a sound investment. In this way, Glenn judged, he could "kill two birds with one stone." Or maybe three birds: he also proposed that he pilot the plane. The Crusader had come to represent something significant for Glenn: a way out. Out of his desk job, out of paper pushing, out of the feeling that "I was a bureaucrat."

Project Bullet, he called the plan, and the Pentagon blessed it, including his place behind the controls. Early in the morning of July 16, 1957, at an air base in Los Angeles, Glenn put on a gold-colored flight helmet—the word NAVY and a small pair of blue wings painted just above the visor—and climbed into the cockpit. He took off through a thin layer of clouds, quickly crossed the sound barrier, and soared to an altitude of nearly ten miles. At that height, so far above any point of reference, he had little sensation of speed—less, he thought, than drivers do on the highway. Like most test flights—but even more so, given the need for three in-air refuelings—this was an exercise in precision. In Glenn's view, it was not unlike "sitting inside an IBM calculator." On the ground, it didn't seem that way. The Crusader dragged a sonic boom across much of the country: in Pennsylvania, the shock waves caused windows to shatter; in Indiana, a ceiling collapsed. Glenn's flight path took him near New Concord, where his parents and friends gathered outside to watch the plane pass. When the boom hit the ground, an elderly neighbor telephoned the house. "Mrs. Glenn," she shouted, "Johnny dropped a bomb!"

Just before 12:30 p.m., the Crusader roared past the control tower at Floyd Bennett Field in Brooklyn, made a quick circle, and landed. Glenn had done it—had set a record. He didn't expect it to last long; nor did he think that it should. This was the jet age, and it was just beginning. But for now he had given the Navy and the Marines bragging rights over the Air Force—which had set the previous record—and they were ready to make the most of it. On the tarmac, the press was permitted—encouraged—to rush the plane. Glenn opened the cockpit, removed his helmet, and waved. For the newswires, the fuselage provided a caption: MAJOR J. H. GLENN, USMC, it read in bold letters, and, on the jet intake, DANGER. Glenn hopped down, hugged Annie, Lyn, and Dave—who had been waiting there among the admirals and generals—and posed, now, in his family's embrace. Within hours the pictures blanketed the nation.

The response was disproportionate—even Glenn thought so. Maybe, he later joked, "it was a slow news day." More likely it had something to do with the match between the Glenns and the idealized Americans who now appeared nightly on television: the Nelsons, the Cleavers, the white, middle-class sitcom families. John Glenn was the same age, with the same build and the same winning, unthreatening smile, as those TV dads. And he had the same kind of family rushing to greet him after his day's work: Annie, in her white hat and gloves, got a kiss from Glenn. Dave got his crew cut ruffled and was given back the Boy Scout knife he had asked his father to bring on the flight ("Here's your supersonic knife, Dave"); Lyn, in white gloves like her mother, got a gold Siamese kitten charm ("And here's your supersonic cat, Lyn"). Glenn even talked like those dads: his flight, he said, was "a real kick." (Annie's stutter went unmentioned by the press; she looked like a sitcom mom, even if she didn't sound like one.) Exchange his tan flight coveralls for a cardigan or a suit and tie, and the entire cheerful, wholesome American tableau could have been dropped directly into one of those full-page Coca-Cola ads in *Life*.

Except that Glenn wasn't wearing a suit. And that heightened his appeal. He was conventional in all the familiar white, middle-American ways but was also, as a NASA official later put it, "supernormal": an everyman-superman. Glenn might teach Sunday school, but he had also just flown across the country considerably faster than a speeding bullet. He was in this way an answer, a rejoinder, to *The Man in the Gray Flannel Suit,* Sloan Wilson's best-selling indictment of postwar materialism and ennui. Its protagonist, a former paratrooper named Tom Rath, broods that "my profession was jumping out of airplanes with a gun, and now I want to go into public relations"—a "ridiculous enterprise" whose only virtue was that it allowed him to "buy a more expensive house and a better brand of gin." At a time when a Harvard theologian, Paul Tillich, denounced the "total patternization" of American life, Glenn, somehow, had sprung the trap. He embodied all the most traditional values while representing, at the same time, progress—literally and symbolically. He was speeding into the future, and was being called upon to describe it in terms America understood.

The invitations flooded in—interviews, speaking engagements.

Glenn was prepared. In 1952, while stationed at Quantico, he had taken a class in public speaking. The hallmarks of effective speech, the course materials explained, were "a sincere desire to get across ideas" and "close contact between the speaker and listener." These were his strengths. On September 7, 1957, addressing the National Exchange Club, he was conversational and charming, winning laughs and applause as he told stories about sonic booms. His speech built to a patriotic close: Why, he asked, does a man take such risks? He does it, Glenn said, "because it has become his way of life and he knows what is expected of him. He won't let his country down. He won't let the people back home down. He does it because he must do it as an American." It was what many Americans in the late 1950s wanted to believe about themselves but were no longer sure they did. Glenn believed—that was clear.

His note cards for the speech included a few fragments about the next great frontier: "Future—to Moon. Why? Columbus to U.S." Even a "cave man," Glenn wrote, had been driven to "look over next hill." Glenn never delivered the line, but the subject came up a few weeks later on prime-time TV. After his flight, Glenn had remained in New York for a round of news conferences, and between appearances he took Dave shopping at Macy's. A woman tried for a while to catch Glenn's eye and then approached him. She was a casting director, she explained, for *Name That Tune,* on CBS; she had recognized Glenn from his picture in the paper and now asked if he wanted to appear on the game show. Glenn—who collected big-band records and belted out old ballads at family events—was interested enough to raise the question with his superior officers. They were happy, it turned out, to keep the spotlight focused on Glenn just as long as he could sustain it.

The show's producers paired Glenn with Eddie Hodges, a ten-year-old boy from Hattiesburg, Mississippi. Starting in September, and every Tuesday night for the next five weeks, Glenn and Eddie were not only a team but a variety act: they bantered, stage-whispered as the timer clicked, delivered canned jokes like two practiced actors, and sang closing numbers (including, incongruously, the African American spiritual "Fare Ye Well"). And they kept winning, so they kept coming back—Glenn in his dress whites one week, dress blues the next, Eddie in a flannel shirt and bolo tie or his Cub Scout uniform. The show's host, George DeWitt, recounted Glenn's heroics in Korea, including the rescue of his wingman Jerry Parker, the pilot who had engine trouble while under attack from six MiGs. The camera cut to the studio audience, where Parker was seated; he gave Glenn a salute.

On October 8, Glenn and Eddie had a chance to win the grand prize: $25,000. But that night's script—which Glenn himself had a hand in writing—was not as lighthearted as the ones before; it felt, in parts, more like a news program. Four days earlier, the Soviet Union had sent the first object into orbit—a metal sphere named Sputnik, twenty-three inches in diameter—and with every pass it made over the United States, the national unease grew. Headlines sounded the alarm: CIRCLING U.S. 15 TIMES A DAY and RUSS SATELLITE SIGHTED OVER S.F. The silver glint of its booster rocket, assumed to be the satellite itself, could be seen against the darkness by the naked eye. Amateur radio operators tracked its beeps and pings. "Listen now," an NBC announcer said, "for the sound that forevermore separates the old from the new."

"Major," George DeWitt asked Glenn, "what do you think of the Russian satellite?"

"Well, to say the least, George, they're out of this world." Glenn laughed, seeming embarrassed by his corny script. "This is really quite an advancement," he went on. "It's the first time anybody has ever been able to get anything that far out in space and keep it there for any length of time, and this is probably the first step toward space travel or moon travel—something we'll probably run into maybe in Eddie's lifetime, here, at least."

"Eddie," DeWitt said, "would you like to take a trip to the moon?"

"No, sir," Eddie answered. "I like it fine right here!"

When the applause quieted, DeWitt put the same question, somewhat more in earnest, to Glenn. "Would you like to be the first man to reach the moon?"

Glenn gave that sheepish smile again. "Well, sure, I'd like to go to the moon, George. But I think like any other test pilot, I'd sure like to check out all the equipment before we started a trip like that."

"You'd like to make sure you get there?"

"Well, sure," Glenn said. "I'd like to make sure we get there, but I'd also like to make sure we get back."

The audience laughed and applauded again. But if anybody was going to travel into space—never mind to the moon—Glenn couldn't expect that he'd be the one. The *New York Times,* in a profile, wondered whether Glenn, at thirty-six, was "reaching the practical age limit for piloting complicated pieces of machinery through the air" at speeds faster than sound. The line stung, and he never forgot it. But soon he determined to prove it wrong.

CHAPTER 3

Red Moonlight

DWIGHT EISENHOWER STOOD, right hand on hip, brow furrowed, impatient. He had begun the news conference in good humor—affable, in his usual way. But only five days had passed since the launch of Sputnik, and as the satellite swept again around the Earth, the nearly 250 reporters who sat in the Indian Treaty Room—the almost comically ornate chamber in the Executive Office Building, across from the White House—showed little interest in anything else, not even the crisis in Little Rock, where the president had recently dispatched federal troops to enforce the desegregation of Central High School.

Eisenhower minimized—mocked—the Russian feat. "They have put one small ball in the air," he said, dismissing the notion that the Soviets had proved anything to anybody. "The mere fact that this thing orbits involves no new discovery to science." This had been the official line since the news had broken. The "man-made moon," as the press was calling it, had sent senior officials reaching for stock metaphors and analogies, each more contemptuous than the last. Rear Admiral Rawson Bennett, chief of naval research, called Sputnik a "hunk of iron almost anybody could launch." Sherman Adams, the president's chief of staff, said that the United States had no interest in getting "high score in an outer space basketball game."

The president, too, was making a show of his nonchalance. But in the Treaty Room, one correspondent after another rose to ask why the administration had conceded the race to the Russians; why the United States hadn't given its own satellite program more funding and assigned it greater importance; and whether the day was approaching when Soviet "space platforms" could be used as nuclear missile bases, beyond the reach of U.S. defenses. Eisenhower grew snappish. Sputnik "doesn't raise my apprehensions, not one iota," he said. And "no one ever suggested to me" that a race was on. Yes, he had expected there would be "a great psychological advantage in world politics to putting the thing up. But that didn't seem to be a reason . . . ," he said, "to grow hysterical about it."

PRESIDENT CALM IN RED MOONLIGHT, read the next day's headline in a Kansas paper. This was either an observation or an indictment. Eisenhower's failure to acknowledge, perhaps even to recognize, the apparent urgency of the situation raised doubts about his leadership. Herblock, the *Washington Post* cartoonist, drew the president in bed with his head under a pillow; a sign on the bedpost read, PLEASE DO NOT DISTURB ONE IOTA. The caricature was not far from the truth. In part, Eisenhower was downplaying the news to minimize the Russian achievement. Yet at the same time, as he told his Science Advisory Committee, he just couldn't "understand why the American people have got so worked up about this thing. It's certainly not going to drop on their heads." This was a losing argument. The national panic continued to grow; dire prophecies came in torrents. "America," declared Edward Teller, the physicist known as the father of the hydrogen bomb, "has lost a battle more important . . . than Pearl Harbor."

Even Sputnik's weight was worrisome. At 184 pounds, the satellite was more than fifty times heavier than the grapefruit-sized sphere the United States was hoping to send into space by year's end. This suggested that the Russians had not, in fact, been bluffing when they'd claimed to have built an intercontinental ballistic missile (ICBM) with enough thrust to send a nuclear payload anywhere in the world. No longer would Americans joke, as they had in recent years, that the Soviets were incapable of smuggling an atomic bomb in a suitcase because they couldn't develop a good suitcase. "We have foolishly underrated their skills," a *New York Times* editorial admitted. "We have become a little too self-satisfied, complacent, and luxury-loving."

As the news conference made clear, space exploration stirred no excitement in Eisenhower; the popular obsession with space—the comic strips, the radio and TV programs, what *Time* called the "space cult"—struck him as kid's stuff. He was not opposed to space exploration, but in his nearly five years as president he had yet to hear a convincing argument in favor of a full-throttle program, as opposed to a modest and circumscribed one. Scientists made the case that a satellite could yield useful information about the Earth's shape, its gravitational field, its weather patterns, the composition of its atmosphere. In 1955, Eisenhower had approved a plan—Project Vanguard—to develop a scientific satellite, though he'd had little interest in the results. Indeed, as it proceeded, he grew irritated by "the desire of the scientists to put more gadgets into the vehicle."

What sold him on the project was something else. Vanguard was a trial run of the idea that a nation could fly a satellite over other countries without violating their sovereign airspace. In this case, the satellite was scientific, but "the big stuff," Eisenhower told his team, was a space-based, military reconnaissance system: spy satellites. Vanguard, he hoped, could reduce the need for risky U-2 surveillance flights over

Soviet territory—guarding the United States against the one thing Eisenhower truly feared: a nuclear surprise attack.

The best way to avoid an attack, Eisenhower understood, was to promise annihilation. "Massive retaliation" was the term of art, and it required a massive buildup of new and more powerful rockets. Starting in 1955, his administration fast-tracked development of ICBMs and intermediate-range ballistic missiles (IRBMs). As spending soared, Eisenhower—whose belief in "thrift and frugality and efficiency" bordered on mania—looked elsewhere to economize. Vanguard had to take a back seat, he later told his staff, "because we had begun to pour big money into the missiles—the things that really count, at this stage." The satellite program was starved not only of funding but of personnel: physicists, electrical and aeronautical engineers, experts on radar, on weather, on liquid and solid propellants—all were pulled into the missile program. Charles Lindbergh, who served on an Air Force advisory committee, described the prevailing view in the White House and the Pentagon: "There would be time to orbit satellites after our nuclear-warhead missiles were perfected."

There was a wishfulness at work here, a kind of willful naïveté. It had never really been tenable to engage in an arms race while abstaining from a space race—not at a time when the Soviets were rushing with resolve into both. And it would be difficult to convince the public either at home or abroad that preeminence in space did not equal preeminence on Earth. Did it not seem obvious that what it took to send an aluminum sphere into orbit was the same as what it would take to send a nuclear payload halfway around the world?

To the press and the public it did, though experts knew otherwise. The R-7 Semyorka rocket that the Soviets had used to launch Sputnik was a blunt instrument—harder to move, hide, launch, or aim than its American counterparts. For these reasons it was shaping up poorly as a weapon, as a classified report by U.S. intelligence confirmed. The irony was that the rocket's scale, a mark of its backwardness, gave the Soviets a big advantage in weight-lifting capacity. The R-7 was "a hellish big booster," as one American scientist put it, and it endowed the USSR with a far greater ability to heave objects (and surely, over time, larger and heavier objects, including capsules containing humans) out of the atmosphere.

Nikita Khrushchev, first secretary of the Communist Party of the Soviet Union, had failed to anticipate the shock, the sheer astonishment, that Sputnik would create, but he happily exploited it. The Soviet Union's success in space provided effective cover for its inability, as yet, to develop a deployable ICBM. Khrushchev's chest-thumping could be heard around the world: the satellite, in his telling, was a projection of power almost without limit. The Soviets were producing ICBMs "like sausages," he insisted—adding, darkly, "We have other things up our sleeves." He spelled out the lesson for the United States: "It must be realized that the Soviet Union is no longer a peasant country—it is dangerous to take this view."

Were there many Americans left who did? Or, for that matter, many French, or West Germans, or North Africans, or South Asians? The shock of Sputnik was hardly limited to the United States. In Britain, there was widespread bewilderment that the Soviets, not the Americans, were first in space; in France, the press chided the United States for its complacency. And these were America's allies. In hostile nations, as well as ostensibly neutral ones like Egypt, the reaction was gleeful: Cairo newspapers cheered that Western bomber bases were now obsolete. Polls taken in recent years had shown a declining confidence, globally, in American leadership and strength, as well as a rising expectation that the Soviet Union was on a path to surpass America militarily and technologically. Sputnik intensified these trends.

On October 17, nearly two weeks after the launch, the U.S. Information Agency (USIA) sent the White House a confidential report titled "World Opinion and the Soviet Satellite." It was a refutation, of sorts, of Eisenhower's insistence that national prestige had little bearing on national security. Sputnik, USIA judged, had given a powerful boost to Soviet claims of superiority. The effect was pronounced among the least developed nations, the ones most eager for rapid technological advancement and the "most vulnerable," therefore, "to the attractions of the Soviet system." The concern was laced with racial condescension: "Among those least able to understand it," USIA suggested, Sputnik "will generate myth, legend and enduring superstition of a kind peculiarly difficult to eradicate or modify, which the USSR can exploit to its advantage."

But in a sense, the report acknowledged, the United States had itself to blame: for overhyping its own satellite program; for assuming, and publicly asserting, that America's lead in science and technology was insurmountable; and for responding to Sputnik with paroxysms of "public anxiety, recrimination, and intense emotional interest"—reactions that, the report scolded, "have been widely noted abroad and assiduously reported by Soviet media." Indeed they had: *Pravda* ran extensive coverage of America's anguish.

Khrushchev, exultant, pressed his advantage. He ordered a second launch in early November, to coincide with the fortieth anniversary of the Bolshevik revolution. Sputnik II, which entered orbit on November 3, did more than underscore the previous feat—it made the first Sputnik seem like a feint, a lark. The new satellite weighed half a ton—six times more than Sputnik I. And this one had a passenger: a dog named Laika. (She was destined to die in orbit.) American scientists were at a loss to explain how the Soviets had managed to lift a satellite that heavy. When a Russian scientist alluded, ominously, to a "new source of power," the *New York Times* filled its front page with panicked speculation about nuclear propulsion, chemical fuels with unknown properties, anything that might provide an explanation.

Wernher von Braun, the German rocket engineer who had helped design the Nazi V-2 before reinventing himself, after the war, as America's leading proponent of space exploration, told the Associated Press that "we are in for a few more shocks." The United States, by von Braun's estimate, was five years behind the Soviet Union. *Life* put von Braun on its cover next to a model of a moon rocket—a fantastical notion at odds with the pages that followed. The magazine featured a lengthy "case for being panicky" by George Price, who had served on the Manhattan Project. "This time we may *not* win out," Price warned. "We have seen . . . a multitude of signs pointing directly towards our defeat by Russia; yet we have pretended that these did not exist." Russia would soon have the world's top scientists and mathematicians; the United States, the "best TV comedians and baseball players." What, Price asked plaintively, did Americans want most? "A Cadillac? A color television set? Lower income taxes? Or to live in freedom?" Making their own views clear, students in a seventh-grade science class in Leavenworth, Kansas, launched a fundraising drive to "get a man-made moon into space before too long." They asked for ten-cent donations and added their own dimes.

On November 7, Eisenhower delivered an Oval Office address billed as a discussion of science in national security. The speech had been a subject of contention within the White House. Some aides urged the president to issue a Churchillian call for blood, sweat, toil, and tears; others counseled him to maintain a level tone. Eisenhower leaned toward the latter view. His speech took a dispassionate look at American capabilities—where they were strong and where they were at risk of falling short. There would be no more blithe dismissals of Sputnik as a stunt. "The Soviet launching of earth satellites is an achievement of the first importance," he now conceded. And though the satellites themselves posed no direct threat to the nation's security, they did suggest a level of technological competence that the United States could not ignore. He announced a set of actions, including the creation of a White House office to advise him on science and technology. "We need to feel a high sense of urgency," Eisenhower concluded. "But this does not mean that we should mount our charger and try to ride off in all directions at once."

THE CAVALRY, HOWEVER, had already scattered, every rider his own general. Though a White House science office was a welcome step, the president had provided no plan or timetable for catching up to the Russians. His speech did little, therefore, to quell the concerns of congressional leaders. Or to still their ambitions: it already appeared that space would be an issue in the 1960 presidential election.

Certainly the Senate majority leader, Lyndon B. Johnson, thought so. Two days after the second Sputnik went up, the newswires carried a photograph of the Texas Democrat looming over a large globe, squinting at the spot believed to be the launch site. The photo was taken at the Pentagon, where Johnson and other senators sat through seven hours of briefings. Returning to his office that evening, trailed by reporters, he said that the administration remained in "feeble denial of the facts." And the central fact, as he saw it, was that "there's a dog going around up there in the air." Catching up would require "determined men to apply money, manpower, and brain power"—making clear, if there could be any doubt, that LBJ was the most determined of men.

He had reached for the reins within hours of hearing the news about the first Sputnik. Before the evening was out, Johnson was working the phones, talking to aides, talking to senators, sketching out plans for an investigation. In mid-October, George Reedy, one of LBJ's closest advisers, wrote a long memo arguing that the issue of Sputnik, "if properly handled, would blast the Republicans out of the water, unify the Democratic party, and elect you President. . . . You should plan to plunge heavily into this one." And space, Reedy added, might change the subject from segregation, which was tearing the party apart.

So Johnson plunged in. On October 18, the day after receiving Reedy's memo, he spoke at the annual Rose Festival in Tyler, Texas. Ronald Reagan, the star of *General Electric Theater,* served as master of ceremonies. "We have got to admit frankly and without evasion," Johnson declared, "that the Soviets have beaten us at our own game—daring, scientific advances in the atomic age." He previewed the questions he planned to pursue as chairman of the Preparedness Subcommittee of the Senate Armed Services Committee: How had the United States blown its chance to beat the Soviet Union to space? Did Sputnik indicate further-reaching, more dangerous gaps in America's defenses?

The hearings opened on November 25 in an "atmosphere of another Pearl Harbor," as Johnson put it in his opening remarks (doing his part to make it so). Over the next several weeks, LBJ's daily show of doom and gloom dominated the newspapers, radio, and television. *Life* put him on its cover. A MAN OF URGENCY, the magazine proclaimed—a headline that no one, at that moment, would have applied to Eisenhower. Johnson summoned scientists like Teller and von Braun, who issued frightful forecasts of Soviet domination of the heavens; he summoned generals and admirals, who rebuked the administration for its stinginess in the national defense. ("The generals were permitted to speak freely—and they did," *Life* observed.) All told, he called seventy-three witnesses. The picture they painted was of a space program—and, by implication, a presidency—that was aimless, anemic, and plainly outmatched by the Russians.

But the most damning testimony was provided by Vanguard, the three-pound satellite the administration had promised to launch before year's end. Just before noon on December 6, as LBJ presided over the Senate Caucus Room, the rocket lifted off at Cape Canaveral. It rose four feet and then sank, collapsing onto itself and exploding in an earth-shaking cloud of flames and black smoke. The tiny satellite, scorched but intact, was somehow thrown clear of the inferno—its *beep-beep*, as it rolled, a tragicomic coda, a feeble echo of the Sputniks.

Johnson received word at his seat on the dais. He was made for this kind of moment. "How long," he lamented to the cameras and the crowd, "how long, oh, God, how long will it take us to catch up with the Russians' two satellites?" His own answer, provided a few weeks later, was tentative. In January 1958, LBJ's subcommittee issued a set of recommendations, billing them as "decisive action," though most took the form of funding increases for missile programs. In terms of space policy, Johnson offered no blueprint, no battle plan to win the "race for survival," though he did propose that the United States build a much bigger rocket and that the space program be run by a new, independent agency.

This last recommendation excited little comment in the press. It did, however, find a receptive audience in the White House—something LBJ could not have expected. Most scientists—including the members of the President's Science Advisory Committee (PSAC)—were in favor of shifting responsibility for space projects, at least non-defense space projects, out of the Pentagon. But they had gotten nowhere with Eisenhower. He had resisted, reflexively, the idea of what he called "a great Department of Space." His instincts cut against creating a new federal bureaucracy. It struck him as wasteful, expensive, and redundant—Eisenhower's unholy trinity.

Yet as a military man he was under no illusion that the Defense Department, left to its own devices, would run a cost-efficient space program. Eisenhower had done his best, with limited success, to curb the tendency of the Pentagon and its contractors to rationalize larger and larger expenditures in the name of national security. He was determined not to militarize space; reconnaissance satellites, in his view, were one thing, but a space station armed with missiles, the fantasy of some Pentagon planners, was another thing entirely. And he was growing tired of the backbiting among the military branches. One Air Force official complained to a reporter about the Army's boasts of leadership in satellite technology: "That's a lot of crap, and they know it and so do other people." On joint projects, the services even bickered over whose name—AIR FORCE or ARMY or NAVY—should be painted on the side of the rocket.

The idea of a civilian space organization was gaining momentum. By the end of February 1958, Eisenhower had approved the approach, and on April 2, he sent a message to Congress calling for a "National Aeronautics and Space Agency." DoD, he said, would retain control of all military technology, whether or not it was gravity-bound, while the new agency would run the rest: research and development, management, flight operations. To that end, NASA would absorb the National Advisory Committee for Aeronautics (NACA), which, for decades, had worked hand in glove with the military and the aviation industry. The new agency's purposes would be peaceful, Eisenhower stressed, and would meet "the compelling urge of man to explore the unknown." On April 14, he sent a draft bill to Congress, where the agency was upgraded to an "administration." Over drinks at the White House, Eisenhower and Johnson ironed out their remaining differences. Eisenhower had long distrusted the majority leader, thought him callow and phony, but the two men now had a shared investment in the bill and, no less, the success of the space program. On July 29, Eisenhower signed the Space Act into law—"for the benefit," it read, "of all mankind." Whether he believed this was an open question.

THE PRESSURE EASED, but only slightly. The advent of NASA was a belated acknowledgment that America had entered the race and aimed to compete. But the space program had yet to find its footing. Earlier that year, on January 31, the Army had succeeded in sending a small satellite, Explorer I, into orbit. But that triumph was followed, three days later, by another failure of the Navy's Vanguard: as thousands of spectators watched from the Cape, the rocket went off course; a range safety officer had to destroy it by remote control. And then, on March 5, Explorer II ended up not in space but in the Atlantic. The United States was fighting, and mostly losing, a battle of attrition against the forces of gravity, faulty electronics, and bad luck.

And the space effort was still beset by rivalries—now, over control of the ill-defined territory known as "man-in-space." The Air Force had been staking its claim to "space-equivalent" flight for nearly a decade, and showed no eagerness to relinquish its role. In 1954, an Air Force pilot had flown a Bell X-1A, a rocket-powered jet aircraft (developed jointly with the NACA), to an altitude of 90,000 feet; in 1956, Captain Iven Kincheloe Jr. had taken an X-2 to 126,000 feet—the upper reaches of the stratosphere. From that height he could see the curvature of the Earth; as the X-2 seemed to hang there, in an indigo-blue void, Kincheloe was weightless for almost a minute. By the time Congress passed the Space Act, in 1958, the research and development arm of the Air Force was pursuing two different means to send a man into orbit: the first, a rocket-launched, hypersonic glider; the second, a separable manned nose cone, or capsule, atop a ballistic missile—called "a man in a can on an ICBM" by some on the research team. The Navy, undeterred, went to work on its own plans (among them, an inflatable reentry vehicle), while the Army, playing the weakest hand among the branches, proposed a less ambitious "suborbital" flight, with the pilot in a four-by-six-foot cylinder launched by a Redstone rocket; the capsule, after several minutes in space, would simply fall back to Earth.

At the same time, the laboratories and wind tunnels of the NACA—on the eve of becoming NASA—were running tests at all hours on every conceivable element of human spaceflight: air jets and heat shields, escape systems and orbital guidance systems, nose cones and fiberglass contour "couches" in which a pilot could withstand acceleration up to 20 g's, twenty times the force of gravity. The NACA, throughout, kept a watchful eye on the Air Force, historically its partner and now its chief competitor. In March, Maxime Faget, a leading engineer at the NACA's Langley Research Center in Hampton, Virginia, met with Air Force officials and had the strong impression that they were using the manned space program "to justify . . . an extension of their realm beyond the atmospheric layer." There was nothing subtle about this. At the meeting, a colonel thanked Faget and his colleagues for their guidance while stressing, in the same breath, that the Air Force had matters well in hand. In this and other ways, officials made it known that the Air Force regarded "space for peace"—Eisenhower's phrase—as little more than rhetoric.

By August 1958, the president had to step in. The Space Act, it was clear, had not settled the issue. Eisenhower remained firm that space exploration served no national security interest. As he'd told a group of senators earlier that year, he "would rather have a good Redstone than be able to hit the moon," because the United States didn't have "any enemies on the moon!" And the United States, as he again had to point out, still lacked a reliably good Redstone, not to mention a good Atlas, or Jupiter, or Thor, or any of the other mightily named missiles that tended, all too often, to blow up on the launchpad or in flight. He wanted the military to get back to work on that and to stop wasting time and money on projects of dubious worth. Thus resolved, Eisenhower permitted the Air Force to continue development of its X-20—a manned high-altitude, hypersonic bomber that was known, without irony, as "Dyna-Soar"—while putting human spaceflight securely in NASA's domain.

Seizing the mantle, NASA's first administrator, T. Keith Glennan, committed his organization to a "Man-in-Space program on the 'capsule' technique." A bell-shaped capsule, though inelegant, was the simplest (if hardly simple) and safest (if not yet safe) means to get a man above the atmosphere and bring him back alive. It reflected Glennan's conservative approach: "to use known technologies, extending the state of the art as little as necessary." Space on a shoestring—just as Eisenhower would have it. Five days after the creation of NASA, Glennan established the Space Task Group (STG), located at Langley, and charged it with the responsibility of putting a man in orbit.

But to what end? Man-in-space was a program in search of a purpose—beyond the obvious aim of ensuring that the man in question was American and not Russian. There was now a vast governmental and industrial machinery dedicated to sending that American man into orbit, but no clear or shared understanding of why he should go. When von Braun, in 1952, had proclaimed that manned spaceflight was as inevitable "as the rising of the sun," he'd envisioned a massive space station, teeming with men and missiles. That idea (and its $4 billion price tag) had won few converts. And certainly no one thrilled to the suggestion of one Air Force general that manned rockets be used for intercontinental cargo transport, or the Army's notion of ferrying troops by way of outer space—a cosmic convoy. By 1958, all the military rationales had washed out. The focus was shifting to science. As S. Paul Johnston, director of the Institute for Aeronautical Sciences, wrote to James R. Killian, the White House science adviser, "The 'take' from the probing of outer space by rockets, satellites and interplanetary vehicles will be of more direct interest to the scientist than to the strategist. We can discount at this point most of the 'Buck Rogers' type of thinking which anticipates hordes of little men in space helmets firing disintegrators into each other from flying saucers."

Even the scientists, many of them, were tepid about man-in-space. If the principal goal, now, was to measure cosmic radiation and the Earth's magnetic field, it was not obvious what advantage a pilot presented over a Geiger counter or magnetometer. On October 28, 1958, a few weeks after NASA opened for business, one of the NACA's last acts was to provide recommendations on the shape of the program. Its directors' endorsement of manned spaceflight was grudging. They acknowledged that humans had the "knowledge and experience to apply judgment that cannot be provided by instruments." This might be useful if problems arose. Still, the role they saw for man in this machine was limited: a helpmate, a troubleshooter, a switchboard operator.

Whether this was worth the cost or the risk was hard to gauge. So was the question of whether a pilot could survive the trip. Would a man in a weightless state go mad, as some scientists feared? Would heavy g-loads at liftoff and reentry cause him to black out or put such a strain on his organs that they would simply stop functioning? Would he reel with vertigo? Would zero g cause his eyeballs to change shape, preventing him from seeing the controls? Would his autonomic nervous system fire randomly and wildly, producing, as one aeromedical physician worried, "an absolute incapacity to act"? Would cosmic rays bombard his cells and kill him slowly? None of this was known. And the United States was well behind in the race to find out. "All right," Glennan told a group of aeronautical engineers that fall. "Let's get on with it."

On December 17, as the eventful year drew to a close, two thousand leaders in aviation gathered at the Sheraton-Park Hotel in Washington, D.C. The occasion was "Wright Day"—the anniversary, in this case the fifty-fifth, of the Wright brothers' flight at Kitty Hawk. Glennan was the keynote speaker. If the crowd expected a rallying cry, it didn't get one. The NASA administrator seemed intent on deflating expectations. "The Russians know what they're doing," he said. "They know where they want to go. Let us be equally sure we know what we want to do—and then let us do it." But Glennan's focus was on what he did *not* want to do. Sounding a lot like the president who had appointed him, Glennan warned that the United States should not let itself be "panicked into spending millions in almost any sort of a frantic attempt to avoid being beaten in assumed space races with Russia." He mentioned almost in passing that the manned program would be known as Project Mercury. But again, he didn't want anybody to get the wrong idea. "At the very earliest," he noted, "success of this venture is several years away."

Glennan so undersold the announcement of Mercury that the *Washington Evening Star* relegated the news to page A-10, while A-1 carried a story about whether the residents of Fairfax County, Virginia, were permitted to shoot stray cats. (They were not.) Also on the front page was a report that the United States had resumed animal flights. The *Star* described "the short but eventful career of Gordo, the monkey in the nose cone": researchers had sealed a South American squirrel monkey in a capsule one foot long and four inches wide and sent him three hundred miles high atop a Jupiter IRBM. Gordo endured ten g's and was weightless for eight minutes, longer than any pilot had been. The monkey, however, was never recovered. The nose cone splashed down and was lost at sea. "Hard luck," a pair of military doctors told the press. But the mission, they said, was "98 percent successful."

CHAPTER 4

Zero G

THE PRESS, a bit too eagerly, called it a "space ship," yet in truth the contraption was just a metal pod, big enough to hold a man but more often carrying a chimpanzee. The pod was mounted to a fifty-foot arm, the business end of the large-scale centrifuge at the Naval Air Development Center at Johnsville, Pennsylvania. When John Glenn, in the spring of 1958, allowed himself to be strapped into the machine and spun around at such a speed that the g-forces were as great as anything he had experienced during a steep dive in a Sabre or Crusader, he did so with an awareness that the test might be of use, someday, in training men for spaceflight. Congress, that spring, had just begun to consider the administration's proposed space bill.

Glenn did one run in an upright position and another on his back; he rode facing forward ("eyeballs in," it was called, for the way the pressure felt) and then backward ("eyeballs out"). He found himself able, throughout, to keep his eyes focused on the instruments and to work the hand controller. Monitors tracked his heart rate, blood pressure, and breathing, gauging how his body responded. He felt, he said later, a bit like Laika, the Soviet dog. Still, he loved the ride, not least because it got him away from his desk at the Bureau of Aeronautics. He and

Annie had decided to relocate from Patuxent to the Washington area by summertime—a welcome move but also, he knew, an admission of defeat, a concession that at the age of thirty-six he was no longer meant to fly planes but to push paper. Glenn "could feel the time ticking away." His record-setting transcontinental flight, and the acclaim that had followed, now seemed to mark the end of something—not, as he had hoped, a new beginning.

But BuAer offered Glenn one advantage: he was positioned to hear the latest information, or at least the rumors, about the future of flight. And that included the possibility of space travel. Glenn had an eye on the development of the X-15, the experimental rocket plane that was intended to fly fifty miles high at seven times the speed of sound; and at the beginning of 1958, he heard that the NACA needed test pilots to ride a spaceflight simulator at its laboratory at Langley. Glenn jumped to volunteer. The simulator—which fed its data into a massive "electronic calculator," the IBM 704—gave Glenn an introduction to orbital flight: how to achieve it, how to maneuver along the path around the Earth, how to reenter the atmosphere without burning up. "I enjoyed that very, very much," he recalled. And he excelled at it. His performance at Langley earned him the invitation to ride the centrifuge that spring. Soon, as one of the very few pilots who had taken part in these tests, he was seen as something of an expert. Not long after his time at Johnsville, BuAer sent him to the McDonnell Aircraft plant in St. Louis, where a seventy-person team was at work on designs for a space capsule configured for fully automatic flight. Asked for his views, Glenn expressed his conviction—already strong—that "the person in that capsule was going to have to be able to control it." He also had a sense of who that person should be.

Servicemen had a term for what Glenn was doing: "sniveling." As Glenn explained in later years, "sniveling" meant "going around and getting what you want to get even if you're not slated to get it. There's nothing wrong with it—and I was superb at it." During World War II, sniveling had gotten him a slot as a fighter pilot when he had been assigned to fly transport planes. In 1957 it had put him behind the controls of a Crusader to fly a mission he himself had concocted. And now it was positioning him to compete for a role as an astronaut. For this he made no apologies. "My view," Glenn reflected, "is that to sit back and let fate play its hand out and never try to influence it at all is not the way man was meant to operate."

That fall, as the Glenns settled into a ranch-style house on North Harrison Street in Arlington, Virginia—the twenty-third move of their marriage—NASA came into being and started filling its senior posts. It also began searching for candidates for spaceflight. Glenn had a hunch that he fit the bill well. At the same time, he had two worries: first, that he was close to the cut-off age (thirty-nine, he had heard), which of course he could do nothing about, and second, that he was, by his own estimate, thirty pounds overweight. (Another reason to deplore the desk job.) The McDonnell engineers had told him about the importance of a lightweight capsule, which required a lightweight pilot. Glenn put himself on a diet and started exercising at the Pentagon gym: running and swimming, jumping on a trampoline. "When the call came," he later wrote, "I would be ready."

NASA had considered opening the application process to anyone with sufficiently "hazardous, rigorous, and stressful experience"—deep-sea divers, perhaps, or arctic explorers. Eisenhower, however, decided to limit the search to military test pilots. After the embarrassment of Vanguard, he wanted the whole business kept secret, and military officers could be expected to follow that order. And, of course, test pilots were accustomed to high-risk, high-speed, high-altitude flight. Granted, even for experienced pilots, riding a missile into space would constitute, in the words of a NASA official, “a major change in their work environment.” But the STG’s astronaut selection committee was confident that it could identify pilots with the right skills and mind-set. (And, as they saw it, the right gender and race: women, banned from combat flying, were not deemed eligible; neither were black pilots.) Over a few harried weeks in late 1958, committee members reviewed the records of hundreds of pilots and judged 110 men to be qualified.

Glenn very nearly was not among them. The Navy’s representative in the selection group, a psychologist named Robert Voas, had assumed—wrongly—that the Navy’s files included pilots from the Marine Corps. Only at the last minute did someone catch Voas’s obvious mistake. The Navy was never, of its own accord, going to put forth a Marine pilot. Calls were hastily made, files were quickly reviewed, and two candidates emerged. One was Glenn. He was so clearly outstanding that NASA’s manpower director, Allen O. Gamble, was willing to overlook the fact that Glenn had no college degree—one of the requirements. After leaving Muskingum to enlist in the war, Glenn had never gone back. Yet in the years since then, as the STG knew, he had done more course work than it would have taken him to graduate.

Thus it was probably unnecessary for Colonel Jake Dill, who had been Glenn’s commanding officer at the test pilot school at Patuxent, to show up unbidden at NASA headquarters, as he did at this time, carrying Glenn’s academic records, combat citations, and flight reports, and loudly insist that the major “was the kind of man with the kind of image that would be perfect for the program and the nation.” Glenn, thereafter, would always credit Dill with putting his name in contention—and would maintain that he himself had nothing to do with it. Whatever the case, Glenn, at the end of January 1959, received orders marked TOP SECRET and, a few days later, found himself in a Pentagon meeting room with a group of other pilots, for reasons that few but Glenn suspected.

When he got home that night, he was buzzing about what he'd been told. He was indeed being given a chance to contend for a slot—a chance to ride a missile out of the atmosphere. As he talked to Annie, he saw her mood darken. She was well aware of his interest in spaceflight; he had barely been able to contain it after his visits to Langley, Johnsville, and St. Louis. For Annie, as for any test pilot's wife, mortal danger was a constant presence; she affected, in the face of it, a stoic veneer. The previous fall, at work on a speech about Project Bullet, Glenn had drafted a response to a question he had faced many times: What did Annie think about his vocation? "Annie's stock answer to that one is that she leaves the flying up to me because she knows that's what I most like to do," Glenn wrote. But her "true feelings," he knew, were less clear-cut. "She worries some, but she rarely lets me know it. . . . During my recent three-year tour as a test pilot, twelve pilots were killed, all on active test flying. It's a little hard, I know, for the wives' feelings to become immune to such happenings." At the same time, Glenn added, these pilots had lost their lives to advance the technology that "will prevent America from being a pushover for any foreign power. Annie has expressed herself many times as being proud of the fact that we have contributed our tiny bit toward the same end, even though she sometimes isn't very favorably inclined toward the project."

Yet Mercury was certainly the most dangerous and very possibly the most reckless project that humans had ever devised. It was almost ludicrous, the whole idea. "It just sounded like something out of a book," Annie told a reporter a few months later. "But I didn't say 'no.'" Mostly she asked questions. And Glenn, mostly, offered her the same assurances that NASA had offered him: that, as Annie recalled, "they are not going to send anyone out there unless they know for sure he is coming back." Glenn's greater worry, he confessed to Annie, was that he wouldn't make the cut. "If you want it," she replied—yielding, now—"you'll make it."

THE SELECTION PROCESS was a marathon run at the pace of a sprint: Mercury was in a hurry to pick its pilots. Over the next few days Glenn went back and forth to the Pentagon. Along with thirty-one other candidates—the rest had dropped out or been sent home—Glenn was given an engineering test, an analogies test, and a mathematical reasoning test. He took a personality test, prepared by the Psychological Corporation of New York, containing true-or-false questions ("Sometimes I feel like cursing"; "Strangers keep trying to hurt me") and sentence fragments to complete ("I hope . . ."). He was interviewed by aeronautical engineers, military psychologists, and physicians. Then he returned for a one-on-one conversation with Charles Donlan, deputy director of Project Mercury. Glenn, in his uniform, entered the office with a thick manila envelope under his arm. Donlan asked him what it contained, and Glenn pulled out a stack of memos documenting his runs on the centrifuge. "He was very excited and animated in describing this thing," Donlan recalled; mainly, Glenn was eager to show that he had withstood the heavy g-forces of liftoff and reentry. In effect, he had aced a test that few, if any, other candidates had taken.

During the interview, Donlan showed Glenn some drawings of the Mercury capsule, and as the meeting ended Glenn asked whether he might be allowed to come back and review them more closely. Donlan said that was fine. And that night, as Donlan was leaving for home, long after the building had emptied out, he walked past an office and "there was John Glenn"—alone, intent, scrutinizing the drawings. "Now, those are the kind of things you look for when you evaluate a man's suitability for a job like that," Donlan recalled.

On February 13, Glenn received a letter from Donlan. "You and your record made a very favorable impression on our selection committee," it stated. "As a result, you are invited to continue on into the second and third phases of this competitive selection program." Glenn was to report on February 27 to the Lovelace Foundation, a clinic in Albuquerque, New Mexico, for a week's worth of medical tests. He would be assigned a number and should identify himself as such. Then, on March 7, he would go to Wright-Patterson Air Force Base in Dayton, Ohio, for another week of tests—psychological and "stress testing," whatever that meant.

It is possible that no human being had ever been so thoroughly examined as the thirty-two pilots—the remaining candidates—at Lovelace and Wright-Patterson. The tone was set on Glenn's arrival by a mimeographed paper with the heading "Instructions for Enema." It ended with an admonition that all enemas (there would be many) "must be very thorough." This was to clear a path for a sigmoidoscope that the pilots called the "steel eel," which pretty well captured it. The hardest part of the week, Glenn explained in a letter home, was a series of shock treatments—"almost had me crawling up the walls." He closed with a warning: "Mum's the word on all this. Three AF pilots have been dropped from the program just because their names got to the wrong people. Say nothin' to nobody 'bout nothin'!!!!!"

Lovelace, it turned out, was a warm-up act, a curtain-raiser for Wright-Patterson. "Psycho tests galore," Glenn scribbled on the back of his daily schedule at Wright-Pat, which began, on Monday at 8:00 a.m. in Building 33/1, with a session labeled "Stress & Fatigue." Glenn was Subject #1 in his group of six. (Subject #4 was a naval aviator named Scott Carpenter.) Some tests were intended to simulate, in extreme fashion, high-altitude flight: an air-pressure chamber that squeezed air out of the lungs; an injection of cold water in the ear to throw off the subject's balance, to see how quickly he recovered; a mechanical chair, outfitted with straps, that shook a man violently while it shrieked at a frequency he could feel in his bones. Other experiments seemed to be, well, experiments, with no ostensible purpose but to see how far a person could be pushed before he cracked: strobe lights were one means, also ice buckets, also a heat chamber (130 degrees). Glenn and the others, one by one, were wired up to monitors and sent into an isolation chamber—a pitch-black "anechoic" room that not only dampened but absorbed sound, denying the inner ear the cues it needs in order to maintain equilibrium; it was hard to stand without toppling over. No one told the men when they might be let out.

There was a chair in the room, and a desk, and Glenn sat down and settled in. He assumed it would be a mistake to follow the path of least resistance by putting his head down, falling asleep, riding it out. Surely the physicians and the psychologists in the white smocks wanted to see how alert he could remain. Looking for some way to keep busy, Glenn felt his way through the desk and found, in the bottom drawer, a large pad of paper. This was good fortune: he happened to have a pencil in his pocket. Glenn pressed a finger against the left edge of the pad to mark his starting point, positioned his pencil, and started to write. When the tip of the pencil slipped off the right edge, Glenn slid his finger down a notch to begin a new line. "Will attempt to keep record of the run," he wrote, "and any extra ideas that I think of on how to pass time." Ideas like "memory exercises. Think of list of things and keep adding to list to see how far down list I can go and still remember whole list. Write story, poetry, etc. Explore the area. Not really much use because I already know what is here. . . . Do exercise to help keep in shape. Lousy idea. It's too hot in here now. Probably better to stick to mental processes to keep active." His mind wandered. "Can't help but think of Helen Keller. She's this way all the time." Then he made a small discovery. "Just found a new light source. When I tore off that last page it generated a good sized spark of static electricity. Now if I could just tear off pages fast enough and had a big tablet I'd have this program knocked."

Glenn began to write some poetry, each line shooting off from the left margin at some odd angle.

Use all your inborn talents
use them each & every day.
Add to mankind's store of knowledge,
make them glad you passed this way.

There were many more stanzas. Glenn was growing sleepy now. He stood up, made a paper airplane to see how well he could do that in the dark, and estimated that he had spent three hours and twenty minutes in the chamber—almost right. At the three-hour mark, the door opened. Glenn had not only endured the isolation but enjoyed it. "It is so seldom these days," he had written on his pad, "that we get a chance to really be alone and think about the things that really count."

For now, though, there was one thing that counted, and it was close at hand. On March 17, while the next group of candidates ran the gauntlet at Wright-Pat, Glenn flew to St. Louis to represent BuAer at the first full-scale "Mockup Review Inspection" of the Mercury capsule. Over the next two days, the brain trust of the space program—the leadership of NASA, managers of the STG, generals, physicians, scores of engineers—circled around the McDonnell plant, inspecting the heat shield and the retrorockets, taking notes, huddling with designers, recommending (sometimes demanding) changes. Nearly everyone with a say in selecting the astronauts was in that room. Glenn, now, stood among them, watching a test pilot demonstrate an escape from a capsule like the one he himself might fly someday.

After the meeting, the selection committee reconvened at Langley. Of the thirty-two candidates, only one had failed his medical test. Now, after a great deal of discussion, the list was winnowed to eighteen. At this point, charts and data, all the output of all those tests, ceased to be of much use. The conversation drifted toward intangibles: mainly whether the physicians "had liked the individual," as Bob Voas recalled. Glenn, he said, had "overwhelmed everybody." On the physical and psychological tests, his scores were not always the highest, "but the physicians gave him top evaluations due largely, I believe, to his strength of personality and his dedication."

On April 6, the phone rang at Glenn's desk—that hated desk—at BuAer. It was Charles Donlan. "Major Glenn," he said, "you've been through all the tests. Are you still interested in the program?"

"Yes, I am," Glenn replied. "Very much."

"Well, congratulations. You've made it."

Glenn was so overcome that, later, he could not remember his reaction. But it was easily read on his face three days hence, on April 9, when he and six other men stepped into the assembly room at NASA headquarters not as test pilots but something new: "astronaut volunteers." Glenn was the one with the broadest grin.

THE RAPTURE OF THE PRESS, at its first sight of the astronauts, defied all reason or proportion. "My gosh!" thought Wally Schirra. "We haven't done a damn thing yet!" But in a sense they had: they had been deemed worthy of immortality. So, anyway, judged the *New York Times*. Whichever of these "outstanding human beings" was first in space, first to pit "his frail body against the terrors of the cosmos," would earn, the editors proclaimed, "immortal fame alongside Columbus and Magellan." (That is, if a Russian didn't beat him to it.) Equally awed, the *Washington Post* hoped that the astronauts "will forgive the rest of us our bad manners if we greet them with mouths agape." The *Post*'s editorial had the feel of a fan letter: "To the Astronauts!" Reporters, no less than their readers, succumbed to the need for heroes; they saw the seven as an antidote to the insecurity that had followed Sputnik. America, it was felt, needed these men—on Earth as much as in space.

"We've joined a whole new world!" Schirra said, and they had, and they were its avatars. Yet there was something familiar, reassuringly so, about them—the six unknowns as well as Glenn. Greatness (and, by extension, national greatness) was not lost: it had always been right down the block. The astronauts, in this light, represented the courage, the resilience, and, just maybe, the redemption of the typical American. "They do not have the bulging muscles, the supercilious good looks, the arrogant condescension of fictionalized spacemen," observed the *Richmond Times-Dispatch*. "They are moderately personable, of moderate height, moderate weight, and possessed of a saving sense of humor. They are all in their 30's. They are all married and fathers. They are quietly but not ostentatiously religious. . . . You put these qualities together and what do you have? You have an ordinary American. . . . And that is the amazing thing about it, is it not?"

Keith Glennan, the NASA administrator, declared himself pleased that the press had taken note that the astronauts were "seven mature men, all happily married, family men, serious, studious." Indeed it had. 7 SELECTED TO BE FIRST IN SPACE, ALL MARRIED, announced the AP—courtesy of a tip from NASA itself, even before it had revealed their names. After the April 9 news conference, where the first question concerned the support of their wives, the family life of these family men excited more commentary than their exploits as pilots did. Annie Glenn pasted these articles in her scrapbook—one with a headline asserting WIVES APPROVE, another contending that the families take SPACE FLIGHT RISKS IN STRIDE . . . THEY ARE USED TO DANGER. This was the official line, and Annie did her best to toe it. A college friend, Betty Smith, wrote in admiration: "I just shake my head with wonder that you and the other wives are so understanding and unselfish."

But Annie's good cheer was a pose; her smile was forced. Things had always been difficult when John was away for long stretches. Annie's stutter was rated at 85 percent—meaning that 85 percent of the time, her flow of speech got stuck, broke down. When she was growing up, and in all the years since, children mocked her; adults corrected her or walked away; but Glenn always acted as if nothing were wrong. Unfailingly, he helped her cope. He made phone calls for her, talked with store clerks for her. As Lyn and Dave got older, they did these things, too. But John was her link to the world. And now, at a time when most of his cohort was retiring from active service, he had volunteered for a new and dangerous mission. "It was one thing to have him sent overseas during wartime," Lyn later reflected, "but another for him to be in this country but decide to be apart."

In March, while Glenn was being examined at Wright-Patterson, Annie became so anxious that she sought solace from their minister, Reverend Frank Erwin. "We discussed everything from faith in God to faith in the government," she said later. Erwin assured her—just as Glenn had—that NASA would never send a man into space until they were sure they could bring him back alive. It helped to hear this again. Yet even as her family converged on Washington—her parents and John's were driving east to see the cherry blossoms when the car radio brought the news of John's selection—her gloom did not lift. "I know that you are very lonely these days," her sister, Jane, wrote. "You have trying & tense days ahead so that doesn't help much. These are just words but I cry for you—even tho' I'm happy for Johnny cause he's getting to do what's important to him. . . . You & the kids have the lonesomeness & heartache to withstand."

That began in short order. NASA gave the astronauts two weeks at home; they were to report for duty at the Space Task Group facility on April 27. Glenn was permitted to forgo the move rather than uproot his family a second time in less than a year. With Annie and the children nearly two hundred miles away in northern Virginia, Glenn could train in the manner he preferred: relentlessly, single-mindedly. He would bunk at Langley and go home on weekends when his schedule allowed. He traded his Studebaker for a new, mustard-yellow NSU Prinz, a West German microcar that could make it from Hampton to Arlington on little more than a dollar's worth of gas—a trinket of a car, a counterpoint, in every regard, to Shepard's Corvette and Schirra's Austin-Healey 3000. But for the most part these cars, as Glenn understood, would sit in a parking lot while the astronauts flew the simulator and while their families, on the base or farther away, led essentially separate lives.

"Monday and Tuesday: Astronauts check in and straighten out their paper work." So the memo instructed. On this mundane note began one of the most ambitious missions in human history. Each of the seven was given his own standard-issue metal desk—packed, all of them, into a single room staffed by a single secretary. Wednesday morning they gathered for the first of many lectures: a discussion of the escape system on the Atlas rocket. That afternoon they were shown the layout of the cockpit, as it now stood; the next morning brought a briefing on the life-support system; and on it went through the week, concluding with a review of operating procedures for a spacecraft that was far from operational.

Glenn took notes on loose-leaf paper, doodling the outline of the capsule and jotting down what NASA officials said about their plans—or hopes—for the future: "Space station. . . . Lunar landing." Before the week was out, he was looking for supplemental reading. "Gentlemen," he wrote to the Government Printing Office, "Would you please send me the price listing for the GPO publication 'Space Handbook: Astronautics and Its Applications.' . . . Thank you. John H. Glenn, Jr." He received a swift reply: "60 cents a copy." (No freebies for astronauts.) "He was highly motivated, to say the least," Schirra observed later—not a compliment. All the astronauts were highly, indeed supremely, motivated; Glenn's mistake, as it emerged, was his failure to hide it.

But Glenn had no greater tolerance than any of them for lectures. He was happier on the centrifuge—even though, at 16 g's, "it took just about every bit of strength and technique you could muster to retain consciousness," as he explained in a letter to James Stockdale, a friend from Patuxent. He was happier still on parabolic flights aboard a C-131 transport: at 40,000 feet, the top of the arc, its pilot began a steep dive and for ten to fifteen thrilling seconds, the astronauts floated, did flips, and walked on the ceiling until gravity kicked back in. "I have finally found the element in which I belong," he told Stockdale. Gordon Cooper was even more excited: "This zero-g is what man has been looking for for thousands of years," he wrote in his diary. "It's a complete feeling of freedom—as if shackles have been removed!!!"

The astronauts also made the rounds of the laboratories and factories that were developing the capsules, the rocket engines, the space suits. Each of the seven was assigned an area of focus, a domain in which he would consult with experts and make suggestions. Glenn's was the cockpit. These working sessions were major events—the Cold

War equivalent of a royal visit, with cheering cadres of contractors and engineers on the shop floor and photo shoots of astronauts in hard hats. But they were engaged, all of them, in a serious business. The sense of purpose, the esprit de corps, was incredible. "We were all learning together," recalled Arnold Aldrich, an engineer at Langley, and they were applying those lessons in real time.

Even so, big problems remained. The fact that the first lecture at Langley had concerned the escape system revealed more, perhaps, than the STG intended about its faith in its own hardware. The principal safety feature of the Mercury spacecraft was a sixteen-foot tower, looking much like a steeple, mounted to the top of the capsule—which itself sat on the top of a booster rocket. The tower contained a tiny "getaway rocket"—powerful enough, if triggered, to pull the capsule away from an exploding booster. Once clear of the fireball, the capsule would jettison the tower, deploy a parachute, and drift safely down to land or sea. This, anyway, was the idea. In March, a test of the escape rocket sent a capsule crashing into the surf after making two grand, gut-wrenching loops. Later tests went better. Still, these were empty capsules, fired directly from the launchpad, not aboard a missile. The idea that an astronaut could be flung safely from a giant cloud of burning fuel remained in the realm of theory. The White House science adviser, George Kistiakowsky, counted himself among the skeptics. The manned space program, he told a group of engineers, "will be man's most expensive funeral."

The press, with NASA's encouragement, focused more on the astronauts than on the machines. Photo spreads in magazines—multi-page albums of astronauts running, astronauts spinning, astronauts climbing into space suits—gave the sense that the men were ready to fly tomorrow, if required. The rockets were another matter. Within the space program, anxiety was growing. The Army's Redstone, based on the German V-2, was known as "Old Reliable," and had been put through its paces since the early 1950s. If any missile could be expected to send a man safely into space, it was the Redstone. Yet the Redstone lacked the power to lift an object into orbit. For that, the STG had placed its bets on the Atlas—a bigger, more advanced missile sponsored by the Air Force and designed, in large part, by Convair-Astronautics. They called it the "gas bag": its stainless-steel skin was thinner than paper, and unless the Atlas was pressurized, the "sucker couldn't hold itself up," as Jerry Roberts, a McDonnell engineer, recalled. It would deflate like a seventy-foot balloon.

Most missiles had their champions. Von Braun, who had designed the V-2 during World War II, took a paternal pride in the Redstone. The Atlas was unloved. Even the Air Force, which was paying for it, wanted little to do with it. (The Titan, more powerful than the Atlas, was the missile that had the generals' eye.) As everyone in the program understood, and as a NASA publication later explained, the Atlas had "problem areas." These included "structure, propulsion, guidance, and thermodynamics"—everything, in short, that made a rocket go. During the first half of 1959, six out of eight Atlas tests were failures—failures of almost every variety. Engines cut out, fuel systems sprang leaks, transistors failed, gyroscopes went awry. It was as if the missile had been designed to "give us fits," complained Christopher Kraft, NASA's first flight director. On April 14, just five days after the astronauts had stood and posed with a three-foot model of the rocket, an Atlas had to be destroyed thirty-six seconds into a test flight because its engine malfunctioned.

A month later NASA was ready to try again, this time—in a display of confidence no one fully possessed—with an audience of astronauts. On the night of May 18, at Cape Canaveral, the seven were set up on a viewing stand about half a mile from the Atlas. Glenn was struck by the beauty of it all. The night was clear and the silver rocket, on the gantry, was lit up by searchlights; pumped full of liquid oxygen cooled to 293 degrees below zero, the rocket iced over and emitted a cloud of vapor that, too, was luminescent. Had the astronauts been closer, they could have heard the metal skin stretching and heaving; it "moaned and groaned," Roberts recalled, "like it was alive." The countdown started and reached its end, the rocket lifted slowly, and the astronauts watched it rise above them. And then it blew. It exploded overhead "like an atomic bomb," Glenn later said. The astronauts ducked—a reflex—and then, seeing that the debris was falling a safe distance away, they stood and stared as the pieces rained down.

"That's our ride?" one of them said. "Well, I'm glad they got that out of the way," Shepard replied, and the others chimed in with cracks of their own. The gallows humor was well practiced; this was how test pilots coped. But from where they now stood, it took a great deal of faith, or maybe foolishness, to imagine themselves as pilots and not as pawns. There was no getting around the fact that their fate lay in the hands of engineers who were not yet done finding ways to fail. Shepard dropped his bravado. "I hope they fix that thing," he said to Glenn, "before we sit on it."

Later, behind closed doors, the seven had a frank talk—made a cold assessment—about the odds they faced. It was inevitable, they said, that one of them would die before Project Mercury was complete. The only question was which one.

CHAPTER 5

A Seven-Sided Coin

IT WAS SOMETHING SHORT OF a hero's welcome, but cordial enough. On May 28, 1959, ten days after watching the missile explode above their heads at Cape Canaveral, the astronauts flew to Washington to spend the day on Capitol Hill. A meeting at the White House had been promised and then canceled. The president, that day, had the grim duty of attending the funeral of John Foster Dulles, his secretary of state, who had died of cancer; Eisenhower then had to play host to the foreign ministers of Britain, France, and the Soviet Union. But the astronauts' impression, which was hard to contest, was that Eisenhower had little interest in them anyway. For all the fanfare of their announcement, he had not even asked to have his photo taken with them. That he left to his vice president, Richard M. Nixon, who seemed perfectly eager to pose with the astronauts on the Capitol steps. Nixon was less eager, it turned out, to talk with them after the photographers left.

The seven made the rounds. They sat briefly in the gallery, like tourists, and watched the Senate in session; they paid a visit to Lyndon Johnson, who (as his manner made clear) had done more than anyone to shape the space program. The astronauts stood around the "Taj Mahal"—as LBJ's palatial offices were known—looking "awestricken," as Wally Schirra put it, then made their way to the New House Office Building for a closed-door meeting with the Committee on Science and Astronautics. When James G. Fulton, a Pennsylvania Republican, asked them whether Mercury was so important to the national defense that they were willing to risk their lives for it, "Yes, sir" was the resolute response—a verbal salute, delivered in unison all down the line. Invited, as they each were, to elaborate, Glenn conceded that he was new—"fresh cotton, more or less"—to the whole idea of manned spaceflight but declared himself "very much impressed so far" that NASA was proceeding with such caution.

A question about the men's motivation elicited, mostly, the same artless answers they'd offered at the press conference in April (all test pilots want to fly "considerably higher and considerably faster," offered Schirra). But Glenn, after uncorking an old reference to the Wright brothers and their pioneering flight, did his best to play to this particular crowd. "Volunteering for this sort of thing puts us in the vanguard of our chosen profession," he said. "So I look at this as being the same thing as possibly you gentlemen looked at when you were aspiring young lawyers, you looked at Congress and thought you could possibly never even hope to attain such heights. But here you sit, gentlemen." He had laid it on, perhaps, too thick. "We are not all lawyers, thank God," one congressman replied.

By the time the astronauts left the hearing room, America had a pair of new "space heroes"—so ordained by *Life,* the arbiter of such things—but none of the Mercury Seven were among them. That morning, near the island of Antigua, Navy frogmen had recovered the eight-foot-long nose cone of a Jupiter missile, whose occupants—Able, a female rhesus monkey, and Baker, a female squirrel monkey—had just survived a fifteen-minute trip into space. After the nose cone bearing the monkey Gordo had been lost at sea the previous December, and after remonstrations by humane societies, all animal flights had to be personally approved by the president. Able and Baker had been given four-to-one odds of a safe return—good enough. With that, the monkeys were strapped into separate compartments, wired up with instruments to measure their vital signs, and fired at a speed of ten thousand miles per hour to an altitude of 360 miles above the Earth. Accompanying them, in the nose cone, was what *Life* deemed "a biological Noah's Ark": specimens of yeast and corn, human blood cells, fruit fly larvae, and the eggs and sperm of several sea urchins (to be fertilized in space), all to assess the effects of weightlessness, radiation, and reentry.

"Able Baker perfect," the recovery ship radioed Cape Canaveral. At the control center, technicians cheered like football fans. The flight dispelled, for the moment, an assemblage of demons, from exploding Atlas missiles to space-traveling Soviet dogs—two more of which had gone up in 1958 and, unlike Laika, had come back alive. Able and Baker were flown to Washington under military escort and brought to NASA headquarters, where they were introduced to the press corps in the same room, at the same table, where the astronauts had lined up less than two months before. As the monkeys made their entrance, reporters applauded, just as they had for the Mercury Seven. Photographers elbowed each other and climbed over chairs to get a better shot—another replay of April, except in this case their subjects sat impassively, eating peanuts. (Able also bit her trainer's handling stick.) "It would be dangerous to extrapolate from monkeys to men," cautioned Colonel Robert Holmes of the Army Research and Development Command, but "one can reasonably hope" that humans would fare as well as the monkeys had: their heart rate and respiration had remained normal during their nine minutes of zero g.

This, for the astronauts, was the most important finding from the trip—possibly worth enduring all the "monkey-shit talk," as Slayton called it. "Real" test pilots, as the stalwarts at Patuxent and Edwards Air Force Base styled themselves—the kind that still actually flew aircraft instead of riding in wingless pods—had been merciless, relentless, in reminding the astronauts that "a monkey's gonna make the first flight." And now a pair of monkeys had done just that. Able and Baker, as the test pilots told it, had done exactly what the astronauts were expected to do: sit quietly, flip switches, and let the ground controllers do the real work. Even the public relations side of the job, it appeared, could be managed by monkeys: the news conference received rapturous coverage, and the cover of *Life* pictured Able and Baker in front of the American flag, looking pensively into the middle distance. The test pilots' needling reflected, to be sure, intense jealousy. Not since Lindbergh had any pilot been as celebrated as these seven—for a flight they hadn't taken yet, a flight that sure looked, to Chuck Yeager and Scott Crossfield and other test pilots, like a fool's (or monkey's) errand. It was galling to the extreme.

But the monkey-shit talk had the ring of truth—not least to the astronauts themselves. "It made them madder than hell," Jerry Roberts of McDonnell recalled. They had, after all, been promised a significant role in operating the spacecraft. The promise had been made, and made repeatedly, that morning at the Pentagon back in February to the entire pool of candidates. It had, it seemed, been made in earnest. And in one way it had: it reflected the hopes of certain officials. But it was also misleading. The role of man in the manned space program was an unsettled question. It was hotly debated within NASA, and the biomedical labs, and the engineering facilities. It continued to be debated even after the seven had been selected. It did not take them long to suspect that they'd been sold a bill of goods. And once that feeling settled in, there was no end of tension between the men and their managers.

A passive role for the pilot was in a sense predestined: it was built right into the capsule's design. In 1958, NASA had rejected the more thrilling notions of Air Force planners—hypersonic gliders that looked like a manta ray—in part because a capsule was more aerodynamically stable. It could right itself during reentry, making it less likely to burn up. But the engineers' aim, as it evolved, was a capsule that could not only right itself but fly itself. It was probably not a coincidence that the driving force behind the design was a team drawn from the Pilotless Aircraft Research Division (PARD) of the NACA. "Sort of ironic," said Bob Voas of the Space Task Group. PARD was full of "very fine engineers," but they had "no background, really, in dealing with pilots." The NACA did have test pilots, but they had no special status, no exalted role as pilots did at Edwards and Patuxent. To Joachim Kuettner, a Project Mercury engineer at the Army Ballistic Missile Agency, the astronaut was a "problem child": a "living payload which, unfortunately, is very sensitive and positively inexpendable. As such, he is eyed with misgivings by missile engineers." At most, the astronaut was a "redundant component" in an automated system. In the fall of 1959, McDonnell's Edward R. Jones, an expert in flight safety, told a meeting of the American Rocket Society (ARS) that "if everything goes well and the operator desires, the mission may proceed through launch, boost, orbit, reentry, and rescue without the astronaut turning a hand." Even, they reasoned, if the astronaut was dead.

A question remained: "What," Voas asked, "was the man going to be allowed to do?" For some researchers, though, the issue was how to stop the man from doing anything at all. "All we need to louse things up completely is a skilled space pilot with his hands itching for the controls," said John R. Pierce, the research chief of Bell Telephone Laboratories, which was designing the Mercury Control Center at the Cape. One sure way to keep the pilot's hands from itching: shoot him full of sedatives. As Jones revealed to the ARS, "serious discussions have advocated that man should be anesthetized or tranquilized or rendered passive in some other manner so that he will not interfere with the operation of the vehicle." This had not, needless to say, been part of the sales pitch at the Pentagon.

Even awake and alert, the astronaut would largely be at the mercy of mission control. The volume of complex information being transmitted—telemetered—from the spacecraft to the control center would be too great for any person to absorb. Physicians would track the astronaut's pulse and breathing; engineers, armed with wiring diagrams, would monitor the electrical system, the stabilization control system, and the levels of oxygen, carbon dioxide, and moisture in the capsule; weather stations would track storm systems wherever the capsule might come down. And flight controllers, on the basis of all that, would make decisions. The pilot would be told what he needed to know—in the view of people who were not the pilot. Increasingly it seemed there was no other way.

This emerging order of things was alien and unwelcome to a group of pilots who had only ever been firmly in control of their machines. The principle of "pilot in command" was fundamental to flight. It was instilled in pilot training, embodied in rules and regulations, bred in the bone. Mastering the mechanics of flight was seen as the easy part; the focus of any good training program was decision-making, risk management—judgment, in other words, not just skills. Even in controlled airspace, when air traffic control came over the headset, ultimate authority rested with the pilot to take action or not, as the pilot saw fit. Mercury threatened to turn that on its head. Control was not merely a matter of ego—it was a visceral need. Voas remembered an experiment in which a pilot and a co-pilot were hooked up to monitors to measure their stress levels during flight. Whenever the co-pilot took the controls, the pilot's palms started to sweat; he was anxious only when someone else was in charge.

During the summer of 1959, Wernher von Braun began warning his colleagues that the astronauts were chafing, already, at these constraints. Von Braun, who was overseeing rocket development at an Army facility in Huntsville, Alabama, wrote the director of NASA's Jet Propulsion Laboratory that the astronauts had come to "deeply resent the suggestion that they are human guinea pigs." A few months later, at a party in Dallas, von Braun got into an argument with the flight director, Chris Kraft, over the mission control concept—a fight that grew so heated that Maria von Braun had to pull her husband away. Kraft thought von Braun's position had more to do with "Teutonic arrogance" than anything else, but there was plenty of bullheadedness to go around.

That included the astronauts. They were not by nature acquiescent. As the first months of training and the first flush of excitement passed, they began to test their collective power. In late 1959, they went to NASA's design team and insisted that a window be added to the Mercury capsule. They refused to accept that the best NASA could do, given concerns about the capsule's weight, was a periscope with a fish-eye lens or a tiny porthole positioned so thoughtlessly that an astronaut had to crane his neck to see out of it. Without a proper window, the astronaut would, in effect, be flying blind—incapable of orienting the spacecraft visually if the automatic control system failed. He would also be unable, in Glenn's words, to "describe a sight that had been an ageless dream of humankind," the Earth from the heavens. They got their window, despite the cost and complication of the redesign.

Technical issues could be dealt with this way: a quiet application of pressure. Others required a more public approach. When Glenn told reporters that the astronauts weren't going "to just sit there . . . aboard this thing" but "will be working the controls," he wasn't stating a fact; he was making a demand. The press was a willing accomplice. There was never any question whose side it was on. On July 8, the *Washington Evening Star* put a startling claim on its front page: SPACEMEN'S ABILITIES WITHERING IN RED TAPE. The story described a bureaucratic passing of the buck between NASA and the Pentagon, which, on top of a relentless training schedule, was preventing the astronauts from flying even the minimum hours required to maintain proficiency and receive flight pay. When they complained to NASA, they were given occasional use of a subsonic jet trainer. The *Star*'s scoop was not the product of sleuthing—it was a deliberate leak by Cooper followed by a full-on, public relations ambush by the entire cohort, who spoke to reporters on condition of anonymity. "We were brought into this project because of our proficiency in high-performance aircraft," one of them said. "Now we can't get any time in high-performance planes. Isn't that a hell of a note?"

The story had its intended effect. On July 9, George Miller of the House Committee on Science and Astronautics announced that he would "send some of our men down there to Langley to find out about this." The astronauts, he said, "can't be allowed to get rusty." Before another day had passed, a committee staff member was poking around the STG, asking questions, taking notes. The seven, informal and unconstrained, took the opportunity to air other grievances: they lacked sufficient "think time" and free time. They had no say in their own schedules. They were spending forty hours a month on commercial airlines, hopping around the country to meet with engineers here, aviation executives there, eating at restaurants on the road on their own thin dime. They spilled all this out in apparent good cheer. Their morale, they insisted, was as good as ever; their faith in the program was high.

Meanwhile, NASA managers stewed in their offices, perplexed and annoyed by this turn of events. These alumni of the Pilotless Aircraft Research Division had failed to anticipate the astronauts' power in the public sphere—or how willing they would be to use it.

The astronauts felt they had cleared the air. Their bill of particulars was dressed up with narrative detail and submitted to the committee—without a countervailing word from NASA—as a confidential report. The staffer who prepared it noted that he had been deeply impressed by the astronauts' "professional attitude." In truth, their morale was shaky and would remain so. Over the next six months, they would spend nearly a third of their time on travel, though most of this was for training purposes. More frustrating still, they were no closer to assuming control of the capsule. In the margin of a memo, in a tiny scrawl, Glenn would later write the words, "Astro vs. mgt." It was an abiding conflict, a chronic tension.

IT WAS ALL FOR ONE at times like that. One unit, one "conglomerate," as Slayton put it, one unpronounceable abbreviation—CCGGSSS—composed of their last names. "We were a close-knit group," Glenn said later. In a way, they had to be: they were undergoing the same training for the same larger purpose. "We got along very well," Gordon Cooper judged. "We were kind of like a bunch of brothers." Or like their own, new branch of the military—a branch with no ranks or pay grades, no officers, no noncommissioned officers, no enlisted men, no chain of command beyond the fact that they all reported to Robert Gilruth, the balding, pipe-smoking, forty-six-year-old engineer who had spent nearly half his life supervising flight research and was now director of the Space Task Group.

There might be no distinctions among them now, but there would be, before long. One of the seven would become the first American, possibly the first person, in space. The national parlor game—who will win the competition?—began right away. 7 VIE TO BE FIRST MAN IN SPACE, declared the *Washington Post.* Most of the seven talked a good game about teamwork, but all were frank about their ambition to come out on top. During their closed-door meeting with the House committee, Schirra put his own spin on Glenn's reference to Orville and Wilbur Wright, who had flipped a coin to see who would fly first. "The Wright brothers had only a two-sided coin to worry about," Schirra said. "We have a seven-sided coin." He put it more bluntly in *Life:* "I want to be first. The impact of being first is something that will be only yours for the rest of your life and on into history." Like Schirra, Glenn made no apologies. "Anyone who doesn't want to be first," he said, "doesn't belong in this program."

And Glenn was the man to beat. To those who had known him and watched him for years, he was indomitable, inevitable, the "chosen one," as folks back home called him. Time and again, hadn't he shown it? "I've been telling everybody," a Marine buddy wrote him, "'I'll bet anything you want to bet that John Glenn goes first.'" "With your phenomenal gift of salesmanship," a friend at the Naval War College predicted, "I know who'll fly that thing." Of course, the other astronauts had their own cheerleaders, their own true believers; every one of the seven, as *Life* rhapsodized, had reached "the summit of inquiry and hope, where there is only room for a handful of people." But Glenn, it seemed, stood higher still. The others sensed it. Every one of them "singled Glenn out as a rival," reflected Kris Stoever, one of Scott Carpenter's daughters. "Everyone did. It was John's force of character. And he was famous coming in."

Glenn's celebrity colored—and therefore threatened to skew—the contest. It was well known that reporters were rooting for Glenn. "In the advance betting," Loudon Wainwright of *Life* recalled, Glenn "was the favorite, picked as most likely to win the space sweeps." It all just made sense. It seemed preordained. "If central casting had chosen our first astronaut to orbit the earth," said Paul Haney, a NASA public information officer, "they couldn't have done better than John Glenn." It was also the case that Glenn, compared with the others, was a pleasure to deal with—a willing interviewee, a source of good quotes. Shepard they found regal, cold, on edge. Schirra, wisecracking behind the scenes, was usually stiff in public, as were Grissom and Slayton. Carpenter and Cooper were more appealing, but their public personas were less fully formed. With the press—just as he had with the physicians at Lovelace and Wright-Pat—Glenn got points for personality.

But not with most of his peers. Glenn's original sin, it seems, was his outsized performance at the April 9 news conference, when he committed the offense of being more at ease than the others. "We all had these dumb answers," recalled Schirra—all, he said, except for Glenn. The resentment hardened into ill will. Even decades later, the other astronauts came back, again and again, to that moment. While they had squirmed in their seats, Slayton complained late in life, Glenn "ate this stuff up." Every question was cause for another "damn speech about God and family and destiny." To Cooper—who had found the day an ordeal, and who feared that he had come across as a rube—Glenn was "a hungry mountain lion that had just spotted dinner."

Glenn, they thought, had been pouring it on. Word among the astronauts was that during his fighter pilot days, Glenn had been just like them—foul-mouthed, irreverent—but fame had changed him. *Name That Tune*, in this telling, marked the debut of a new John Glenn: the Boy Scout, the saintly Sunday school teacher, the aw-shucks, homespun hero. Of course, Glenn had never been a coarse, hard-bitten Marine. Even in wartime, as members of his squadrons would (and did) attest, he had been every bit this upright, this devout, this unselfconsciously corny. But to his new rivals, Glenn came off as too good to possibly be true. Worse yet, it seemed to be working. He was beating them at a game they had little interest in or aptitude for playing but that might determine who flew first. No one at NASA had said as much, but the agency's emphasis on public relations suggested that performance on camera, no less than performance in the simulator, mattered a great deal.

In an April 24 memo titled "Public Relations," John A. "Shorty" Powers, NASA's chief spokesman, promised the astronauts that NASA would "protect you" from a voracious press corps and would keep the demands for interviews to a minimum. On both counts it did a terrible job. Reporters staked out the men's homes, trailed them into restaurants and public bathrooms, followed their kids to school and asked the teachers whether the sons and daughters were as smart as their fathers. To stem the tide—and to earn some supplemental income—the astronauts signed an exclusive contract with *Life,* but it did little to shelter them, even in the supposed sanctum of their office. Inevitably one of the seven, in some corner of the room, was being buttonholed by a reporter. As Loudon Wainwright put it, the Mercury astronauts were a "journalistic commodity," and NASA marketed the men as such. It was starting to feel, one of them said, like the whole "space program was being run as a public relations exercise."

It was Glenn, in fact, who made that complaint. It was also Glenn who said that the deluge was "fierce and distracting," that it was cutting into their time to train, to think, to rest. He joined the six others in staging another of their minor rebellions—demanding that NASA limit their public exposure to one day a week. Again they prevailed. The conglomerate reinforced its clout. Still, the more press-averse astronauts refused to believe that Glenn was on their side. The spotlight sought him and suited him, and they held to their view that he was dragging them, all of them, into its glare.

What he was trying to do, ironically enough, was to share center stage. In May, when Lawrence Spivak, the producer and moderator of NBC's *Meet the Press,* invited Glenn—and Glenn alone—to appear on the show, the astronaut sent a surprisingly candid reply: "Would you possibly be interested in having the group of seven on the program together? I realize this is a little larger group than you usually interview. We are, however, making every effort to maintain the seven Astronauts as a team in virtually all our activities. . . . This is solely an effort to eliminate as much friction within the group as possible, which I am sure you will agree is highly desirable in a group that must work together as closely as we are on Project Mercury."

Whether NBC or NASA made the call, the seven never appeared together on the show, and Glenn did not go solo. (Nor would he for another four years.) Yet the friction, clearly, could not be dispelled. "Glenn's willingness to make the astronauts national symbols and public figures never sits well" with the rest of the astronauts, observed Walt Williams, director of operations for Project Mercury. (It never sat well with Williams, either.) The notion that Glenn had the power to "make" them symbols—that America's obsession with the astronauts was somehow his design—was absurd, but it had taken hold.

Being a public figure was not all bad, of course. None of the astronauts minded a little adulation. With the exception of Glenn, they had never experienced it before, which made it even sweeter. As test pilots, they had achieved incredible things without anyone outside the military or the aviation world taking notice, let alone treating them like movie stars. Seeing themselves in newsreels did not leave them pining for their old obscurity. But there was a downside: the difficulty of maintaining a private life and, moreover, doing with it whatever they pleased. "Extracurricular activities" was Shepard's term. "Socializing" was another, but one of the astronauts' wives put a finer point on it. There was "a lot," she said, "of screwing around."

"We were always looking for ways to let off steam," Slayton recalled. Opportunities were many. As the months passed, the seven spent more time at the Cape and less at Langley—a welcome change, in that there were actually things to do at the Cape after spending ten or twelve hours squeezed into a pressure suit. The astronauts would jump into the Corvettes some of them had bought (for a dollar a year from a local Chevrolet dealer) and speed down to Cocoa Beach, to the strip along Route A1A, where they would make the rounds of the motel pools and tiki lounges. "A harlot of a town," one journalist called Cocoa Beach. Depending on your point of view, this was an epithet or a term of endearment. For the astronauts, every night was shore leave. They were their own, middle-American Rat Pack—would-be Sinatras in Ban-Lon shirts ("jet-fast wash and wearable"). Women were heard to say, "Four down, three to go," and at least some were keeping accurate count. Shepard was rumored to have an understanding with his wife, Louise; if so, it was an understanding that Al was going to do what Al was going to do. There was nothing subtle about it: as Rene Carpenter, Scott's wife, recalled, "Al was about sex and business."

And Glenn? He didn't flirt, not even with the glamorous Rene—a restraint that few men, in that louche time and place, managed in her presence. If Glenn stood apart as a bit of a square, happier singing barbershop songs at a cookout than drinking at a poolside bar, he had never been a wet blanket. He had lived on and off military bases, stateside and overseas, for well over a decade and had always been well liked, even if he wasn't seen as one of the boys. At El Centro in 1943, awaiting orders to ship out, he and his equally straitlaced roommates were known as the "purity kids." Still, he had always been expansive and fun, a storyteller, an extrovert. Others sought his company and fed off his energy.

Less so on Cocoa Beach. Glenn hit the strip with the rest of the astronauts, but increasingly they saw him as a prude and a scold. Glenn's position was that their private business was no longer private: they weren't unknowns anymore, throwing back beers on a dust flat somewhere; they were the focus, now, of the nation's hopes. A scandal could sink the Mercury program—or, at least, their place in it. In July, when the House committee checked up on their morale, it found them "fully aware of their responsibilities to the project and the American public, particularly with regard to the heroic role they are beginning to assume with the young people of the country. They have imposed upon themselves strict rules of conduct and behavior." With this in mind, Glenn began warning—hectoring, really—the other six. "I was afraid that any bad publicity might have an impact on whether we really had a manned program," he recalled years later. "I talked about this at some of our meetings. I said I thought that everybody should be more careful about their outside activities. . . . Everybody said, 'Well, you're probably right,' but there was still more activity than I thought there should be. We talked about this a number of times."

Glenn's concerns were brushed aside. The others had a hard time imagining that what happened at the Carnival Club or the Starlite was a risk to anything except, in theory, their own marriages. Shepard might or might not have had an understanding with his wife, but they all seemed to have one with the press. Reporters looked the other way—or joined in. One night, Shepard and a few other astronauts corralled Walter Cronkite, the CBS News anchor, and some other journalists and went looking for nesting sea turtles—one of the milder forms of recreation on Cocoa Beach. As they reached a remote section of the beach, the sounds they heard from the palmetto scrub were not, it turned out, leatherback turtles but a man and a woman: specifically, Shorty Powers and his secretary in the back of a convertible. As Powers flailed his arms at the group, yelling something about being "entrapped," a pair of photographers started snapping pictures—just to provoke him further. For Cronkite it was "one of the grander evenings" he had spent at the Cape. He and Shepard were laughing so hard they gave up on the turtles and headed, instead, for the bar.

THE LINES OF DIVISION were forming. "We have a split," Walt Williams, the head of Project Mercury, observed, "with Scott Carpenter becoming a sort of follower of John Glenn and the other [four] aligned more or less with Al Shepard."

"Follower" put it too strongly, perhaps. Carpenter's easygoing, just-happy-to-be-here demeanor suited him well with all of the others. But he did become Glenn's sidekick. To Rene Carpenter, John and Scott were "Gloryhound and Tonto"; as Carpenter's daughter Kris Stoever later described it, "John was Scott's big brother." The two had crossed paths at Patuxent in 1954—Glenn was in the class one year above Carpenter's—and fell into an easy rapport during the astronaut-selection process when they were assigned to the same group at Wright-Patterson. Later on, their wives, too, got on well, further drawing the families together. In Stoever's assessment, her father was loyal to Glenn, admiring of Glenn, but also, she thought, "half-jealous. Because John was John." Glenn saw Carpenter as "outgoing and thoughtful"—not unlike himself. They were the only two astronauts who took much interest in the arts or showed any real enthusiasm for the scientific side of the mission. They had been willing—even eager—to do business with the doctors at Wright-Pat, to analyze ink blots and submit urine samples, as if being tested was a grand old time. This, to the others, was a kind of betrayal; the doctors were the enemy and best kept at bay.

Shepard's relationships with his fellow astronauts were complicated. They were always eager to cruise around the strip or play handball with him; they loved his impressions of José Jiménez, the hapless would-be astronaut played by the comedian Bill Dana. Some of them looked up to him. But as one of the astronauts—anonymously—told *Life,* Shepard was "always holding something back." He himself was aware of it. "I was friendly with several of them," he said later. But "I was now competing with these guys, so there was always a sense of caution"—particularly regarding technical subjects, the kind of knowledge that could give a man the edge. When it came right down to it, Shepard was less interested in being the leader than being the winner. And he saw Glenn as his principal rival.

It was clear that they were the standouts: the strongest personalities, the most willing to speak their minds, the least willing to defer. It was there for all to see. In the time since Glenn had outshone everyone at the first press conference, the laconic Shepard had become loquacious. Now "he talked his head off at the press conferences," remembered Paul Haney of the NASA Public Information Office. "You literally couldn't shut him up. . . . It was laughable." Glenn and Shepard had a lot in common. Yet they mystified each other. "Al was . . . an enigma," Glenn said later. Shepard, for his part, couldn't understand how Glenn could be so cornball and so cutthroat at the same time. In the guise of bonhomie, they belittled each other. Shepard was constantly poking Glenn about his "junk heap" of a Prinz and saying he ought to get himself a man's car—like Shepard's Corvette. Glenn managed to grin and bear it for a while, but the ribbing rankled him, maybe even rattled him. It was obvious that the car was a proxy for the vehicle they really wanted to ride—the capsule—and whether Glenn had what it took.

For now, Glenn would simply respond in kind. One morning Shepard entered the astronauts' office to find a message on the blackboard: DEFINITION OF A SPORTS CAR, it read. A HEDGE AGAINST THE MALE MENOPAUSE.

ON SEPTEMBER 12 the astronauts began a West Coast swing—one of those exhausting odysseys from contractor to contractor, with stops in between at city hall and a local TV station. After a week of this, they traveled to San Diego for briefings at Convair-Astronautics, which was building the troubled Atlas rocket. When they landed at Lindbergh Field, they were a man short: Slayton was in the hospital recovering from a viral infection. Convair put the group up at the Kona Kai, a cluster of cottages on Shelter Island, in San Diego Bay. The thatched-roof cottages were enshrouded by palm trees; umbrellas ringed a C-shaped swimming pool. Glenn bought a postcard of the hotel and sent it home to Annie. She put it in her scrapbook.

When the astronauts arrived in San Diego, Convair was still bickering with NASA and the Air Force over who was to blame for the Atlas and what could be done about it. By the perverse logic of the Mercury tour, the astronauts were there to buck up the engineers, not the reverse. Grissom made the first attempt. Addressing six hundred supervisors on the floor of the plant, he delivered a three-word speech—"Do good work"—and sat down. It was both an admonition and a plea.

In the off-hours they went water-skiing and skin-diving. At night there were cocktail parties, catered dinners with executives, jazz bands at the Del Mar Powerhouse. Local papers covered it all, and rapturously. The astronauts, pronounced the *Union,* were "seven of the most important men of the century." The *Evening Tribune* felt, in their presence, "a glow of patriotic pride." The group attracted so much attention that Shorty Powers grew uneasy. The whole business was getting harder to control. At a party one evening, he threw up his arms and refused to allow any more pictures or interviews. "I've already been more lenient with the press here than any other place," he snapped.

After midnight one night, Glenn heard a knock on the door of his cottage. It was Shepard, and he had been drinking. "I think I got myself in trouble," he said. He had been at a bar across the border, in Tijuana, where he'd met a woman. At some point they had been alone, or thought they were, and that was when the flashbulbs popped.

Shepard was worried that the photo would find its way to his wife. He remained, all the same, pretty pleased with himself; he took, as ever, a professional pride in his virility. With a fresh tale to tell, he roamed the Kona Kai in search of an audience—a more appreciative audience than Glenn had been. He banged on Scott Carpenter's door. Carpenter was bunking with his friend Don Schanche, a correspondent for *Life* who had been sent along to cover the trip. As Shepard came in, Schanche later recalled, he "confessed to a case of nerves"—then proceeded, with relish and in detail, to recount his "hilarious but satisfyingly carnal liaison." It was "a pretty funny story," in Schanche's view, and unlikely, he thought, to result in any harm.

Carpenter felt differently. As Shepard regaled him, Carpenter seethed. His own zeal for clean living—in the face of "temptations that were coming fast and furious," as Rene put it—was a tenuous thing. Shepard's slick self-assurance, "that ease he had with women," both awed and repelled him. He envied Al, but the rectitude, the moral clarity of John Glenn was the standard he strove for. Mercury's poles pulled in opposite directions—but not with equal force. Carpenter lashed into Shepard: he had dishonored his uniform; he had dishonored his marriage. Shepard, in response, barely registered the attack. To Schanche he appeared unmoved, bemused, "utterly unchastened."

There was nothing to do now but wait. At 2:00 the following night, Glenn was awakened again, this time by a ringing phone. Powers was on the line. His voice was tense and thin. "Well, it's happened," he said.

"What do you mean?" asked Glenn, but he knew.

"What you've been talking about has happened," Powers said. An editor at a major West Coast paper had called to say he had compromising photos of Shepard. Not only that: a reporter, having joined the photographer in trailing Shepard, had written a detailed account of the night's debauch. The scoop, if it hit the newsstands in the morning, would provide an answer of sorts to that week's issue of *Life*—which featured, on its cover, seven prim and smiling women and a caption in bold: ASTRONAUTS' WIVES: THEIR INNER THOUGHTS, WORRIES.

Glenn asked Powers to come by his room. They resolved to kill the story if they could. Glenn now made a series of calls: to the reporter, the photographer, the newspaper publisher. He reached all of them. He gave them one of those John Glenn speeches—the kind he gave to the press corps or the House committee. This one, though, had an edge. The United States, Glenn said, was falling behind the "godless Communists"; the press, he insisted, had to "let us get back in the race," had to help America catch up. "I pulled out all the stops," he recalled.

When his alarm clock went off a few hours later, Glenn headed straight for the newsstand at the Kona Kai. The morning paper said nothing about Shepard. The publisher had suppressed the story.

Glenn's relief gave way to anger. "I was mad," he wrote later. Mad not just at Shepard but at all of them who had repeatedly, habitually, put the mission one headline away from disgrace and dissolution. He demanded a meeting right away. As the group gathered in a hotel suite, Glenn's lips were tight and his face was flushed. He lit into his fellow astronauts—lectured them as if they were his charges, mere schoolboys. They had "dodged a bullet," he said, because he had been there to clean up Shepard's mess. He had said it before: there was a limit. The press would indulge only so much carousing before a single story shut the whole program down. They had all worked too hard, Glenn said, for themselves and their country, "to see it jeopardized by anyone who couldn't keep his pants zipped."

The others sat sullenly and took it for a while. Finally Shepard stood up. "Why is this even coming up?" he said. "Doesn't everyone have the right to do what they want to do?" It wasn't a question; it was a pronouncement, an impenitent creed. Despite his narrow escape, despite the fact that the night of the incident he had been so shaken he had sought Glenn's help, Shepard was taking nothing back. More galling for Glenn, everyone other than Carpenter sided with Shepard—Schirra and Cooper strongly so. Grissom acknowledged Glenn's point but was plainly unwilling to live by his code. "My views were in the minority, but I didn't care," Glenn concluded decades later. "I had made my point." Indeed he had. And it would cost him more than recklessness had cost Al Shepard.

CHAPTER 6

Second in Space

WHILE THE ASTRONAUTS WERE ON their West Coast swing, Nikita Khrushchev boarded a plane in Moscow for a tour of his own: a thirteen-day, coast-to-coast journey across the United States. The Soviet premier would visit factories, farms, schools, and, incongruously, Hollywood before a round of talks with Eisenhower at Camp David, the presidential retreat in Maryland. As he prepared to depart, Khrushchev announced that he was bringing a gift for the president: a five-sided medallion inscribed with the Soviet hammer and sickle and the words USSR, SEPTEMBER 1959. It was a replica, he said, of a medallion that, one day earlier, a Soviet spacecraft had delivered to the moon.

The spacecraft, Lunik II, had been traveling seventy-five hundred miles an hour when it hit the moon somewhere between the Sea of Tranquility and the Sea of Serenity. (A soft landing was—for now—well beyond Soviet capabilities.) It was the first man-made object to touch the lunar surface. Whether or not the medallion survived the crash, it had served its purpose. In Manhattan, an audience at the Hayden Planetarium gasped when an astronomer told them the news. A Roman Catholic bishop in New York said it had "the psychological impact of Hitler's march into Poland." Khrushchev's insistence on calling the medal a "pennant" did not go unnoticed: a State Department spokesman warned that if the USSR had any notion of claiming the moon as Soviet territory, it was going to have to do a lot more than "stick a Red flag" in the lunar soil.

On the morning of September 15, Eisenhower greeted Khrushchev at Andrews Air Force Base in Maryland, welcoming him to the United States in a spirit of "common purpose." Khrushchev gave lip service to the same. Then he twisted the knife. Soviet hearts, Khrushchev said, were filled with joy that their country had "blazed a road from the earth to the moon." Yet he had "no doubt," he assured Eisenhower, who was standing beside him, that someday America would plant its own pennant on the moon. And when that day came, he said, in his guileful way, "the Soviet pennant, as an old resident of the moon, will welcome your pennant and they will live there together in peace and friendship, as we both should live together on the earth."

Eisenhower got in a few digs of his own. Two days later, while Khrushchev was riding a special train to New York, Eisenhower held his weekly news conference at the White House. Khrushchev's "little memento" was interesting, Eisenhower said, adding—to appreciative laughter from the press corps—that the real medallion, when it hit the moon, "probably vaporized." The administration's effort to shrug off—or laugh off—the Soviet feat recalled its reaction to Sputnik. Vice President Nixon, in New York for a speech to the American Dental Association, told reporters that Lunik II was no reason to get "hysterical." "In science," he said, "sometimes we're ahead and sometimes they're ahead." And overall, he insisted, "we are way ahead." In a rebuttal of sorts, an Atlas rocket blew up during an engine test on September 24—one of three rockets that failed during Khrushchev's visit.

The next blow fell swiftly—and with exquisite timing. October 4 was the second anniversary of Sputnik. Had all gone according to plan, a U.S. satellite would have eased into orbit around the moon just in time to welcome, with fanfare, what the *Washington Evening Star* was calling "Year III of the spectacular space age." The exploding Atlas ended that hope; NASA had no backup rockets. Thus Year III began just as Year I had: with the Soviets exultant and Americans abject. On October 4, Lunik III—a four-foot-long cylinder that the Soviets described, a bit grandly, as an "automatic interplanetary station"—was sent on its way to photograph the moon's far side, which had never before been studied or seen. As it swept around the moon, the spacecraft stopped its rotation, held steady for forty minutes while its two cameras snapped, then flew back in the direction of Earth, transmitting the images to receivers on the ground. The photographs were blurry, the display of proficiency breathtaking.

Americans were left, once again, to gaze at the heavens with fear and frustration. Officials were left, once again, to issue statements of congratulations—weary statements that, the *New York Times* pointed out, were "reminiscent of those issued on previous such occasions." Experts debated whether the United States was a year or a decade behind the Soviet Union. Editorials assailed the administration for its manifest failures of imagination and blasted Congress for cutting NASA's budget so deeply that the agency had to reduce the number of rocket tests (and, therefore, the number of rockets). America's allies, meanwhile, wondered anew whether the nation was simply outmatched. Kenneth Garland of the British Interplanetary Society spoke for many: he was, he said, running out of superlatives to describe Soviet achievements.

The picture was not entirely bleak. Over the past two years, the United States had sent twice as many satellites and probes into space as the Soviets had. It had discovered belts of radiation around the Earth; it had transmitted the first television images of the Earth from space; and on October 4, it had successfully launched a booster rocket carrying a mock-up of the Mercury capsule. Lunik III, said Keith Glennan, NASA's administrator, had left America "no further behind than we were yesterday. But that's far enough," he was quick to concede. And "we cannot run second very long and still talk realistically about leadership."

It was a candid assessment. And it was undercut, with due haste, by Nixon. The vice president was not merely carrying water for Eisenhower. He was running to succeed Eisenhower, and as the election year approached, Nixon understood that the space program was a liability. He responded by becoming its cheerleader—unyielding in his optimism, sanguine about every setback. On October 6, at a fish fry in Indiana, Nixon said the space program was "moving along at a reasonably good pace." There was no need, he insisted, for a "massive crash program." He was the picture of confidence. This had less to do with the space race, surely, than with another piece of news that made the papers that week: favorable reporting on Nixon's foreign trips was bearing fruit politically. A new poll revealed that after lagging for months, he had surged ahead of his strongest Democratic opponent for the presidency: Senator John F. Kennedy of Massachusetts.

IN 1940, at the beginning of World War II, John Kennedy—at the age of twenty-three—published a book called *Why England Slept.* In an earlier, unpolished form, it had been his senior thesis at Harvard. The title was a play on *While England Slept,* Churchill's 1938 prophecy of impending war. The young Kennedy took an unsparing view of Britain's unpreparedness in the face of Nazi rearmament. His book began with the story of the British retreat from Flanders, Belgium, in May 1940: "The one remark on every soldier's lips was, 'If we only had had more planes.'"

In 1959, as the space race began in earnest, as the nuclear arms race intensified, the question Kennedy posed had a familiar ring: *Why didn't America have more missiles?* On November 13, the candidate battled his way across southeastern Wisconsin through a heavy snowstorm, stopping to make speeches along the way. He drove the icy roads from Portage to Watertown and then to Milwaukee, arriving late for an address to the state Democratic Party Convention. He reminded the crowd that in 1952, when Eisenhower was first elected, Americans had been "the unchallenged leaders of the world in every sphere—militarily, economically, and all the rest. And now it is 1959. The Russians beat us into outer space. . . . They beat us to the moon." These achievements, Kennedy noted, were not ends in themselves; they were aimed at proving the superiority of the Communist system. While Americans became "complacent, self-contented, easygoing," the Soviets had pressed, with single-minded intensity, "to take away from us our prestige and even our influence." But it was not too late. "The Russian pennant on the moon has shown us our task," he said. "And I think we can live up to it."

The irony was that Kennedy had no greater interest in space exploration than Eisenhower did. In the late 1950s, as a member of Harvard's Board of Overseers, Kennedy served on the Astronomy Department Visiting Committee, but he does not seem to have engaged with the subject matter; he never spoke with wonderment—at least that anyone recalled—about the moon and stars. What had engaged him since his teenage years was not science but matters of war and peace, diplomacy, great-power politics. Neither had his service in the Navy exposed him, in any meaningful way, to state-of-the-art technology or modern weaponry. Quite the contrary: in 1943, JFK was sent to the South Pacific to command *PT-109*, an eighty-foot, wooden-hulled patrol boat armed, or perhaps burdened, with Mark 8 torpedoes, decried by sailors as worse than useless. In the Senate—where Kennedy had served since January

1953, following three terms in the House—his principal interests were labor reform and foreign affairs. Even after Sputnik, whenever he spoke about Russian satellites he was most likely referring to Poland, East Germany, Czechoslovakia.

Kennedy shared the president's instinct that space shots were a waste of money. Still, he was quicker than Eisenhower to understand their impact on world opinion, and quicker than any of his potential rivals for the Democratic nomination—any but Lyndon Johnson—to see the utility of space as a campaign issue. "He thought of space primarily in symbolic terms," recalled Ted Sorensen, Kennedy's close aide and speechwriter. "He had comparatively little interest in the substantive gains to be made from this kind of scientific inquiry. . . . Our lagging space effort was symbolic, he thought, of everything wrong with the Eisenhower Administration: the lack of effort and initiative, the lack of imagination, vitality, and vision." The space gap, which was evident, and the missile gap—which was widely (and erroneously) reported—were painted as symptoms of the same malaise. Sounding a generational call to arms, Kennedy promised "leadership for the '60s"; his speeches heralded a "New Frontier." "Our democracy," he proclaimed on January 1, 1960, "must demonstrate that it has the capacity to defend itself in a world of intercontinental ballistic missiles; and that it has the energy and the sense of adventure—as well as the technical skill—to play a role of leadership in the exploration of space."

HE HAD FRAMED the challenge; now he scrambled to understand its dimensions. In 1959 and well into 1960, Kennedy sent out a flurry of inquiries to his informal brain trust of academics, defense analysts, physicists, and other experts. In November 1959, he asked Paul Nitze, the noted authority on national security, whether the United

States should continue trying to catch up to the Soviets in non-military space exploration—whether it was possible or even all that important to match them shot for shot, as opposed to establishing a military presence in space. "I doubt whether we can catch up in any short time period," Nitze replied. Still, he thought "the risks of not making the attempt are greater than the foreseeable risks of making it." The dangers of inaction—and of failure—remained a concern throughout the campaign. In September 1960, Kennedy wrote Archibald Cox, a Harvard Law School professor and trusted adviser, to seek his "judgment of the significance of our being in a secondary position in space in the sixties. Example: the military effort. Will the Soviet Union have a reconnaissance satellite before we do, and what will it mean? What military advantage does this give to the Soviet Union?"

The short answer—actually, in 1960, it might have been the full answer—was that it was difficult to say. The implications of superiority in space were "impossible . . . to foresee," Nitze told Kennedy. This did not stop the campaign staff from speculating, in a briefing paper for JFK, that the Russians would send a man into orbit before the end of the year and might, before long, "be able to orbit space platforms and use them as military bases and even use the moon as a military base." There was no evidence for any of this beyond the feeling, shared by most Americans, that the Russians could do pretty well as they pleased in space. All that could be said with certainty—as Donald H. Menzel, director of the Harvard Observatory, wrote Kennedy that fall—was that the situation was urgent. "The most dangerous thing that we, as Americans, can do," Menzel warned, "is to underestimate the enemy."

For now, though, Kennedy's chief concern was to win the election. By July 1960, when the Democrats arrived in Los Angeles for the party's convention, Kennedy had knocked his main rival, Senator Hubert Humphrey of Minnesota, out of the race, then suppressed, with steely efficiency, eleventh-hour challenges by Adlai Stevenson and Lyndon Johnson. In a blur of events that no one, not even the protagonists, could see or ever reconstruct with clarity, Kennedy emerged with Johnson as his running mate. Johnson's strength in the South, not his burst of leadership on space policy, drove Kennedy's decision. Still, his presence on the ticket made it more likely that the Democrats would campaign on the issue. As the fall campaign began, Kennedy's staff prepared for a broad, sustained assault on the space program. Ralph E. Lapp, a physicist who had served on the Manhattan Project, prodded him to go after Project Mercury. By laying the groundwork now, Kennedy could lock up the election if the Soviets put a man in orbit before November 8. An aide readied talking points for Kennedy to use in that event: "Mercury was energized too late to have a real chance of winning the orbital race. . . . The United States made the mistake of engaging in the space race in an area where it had the least chance of success."

Was Kennedy prepared to go this far—to question not only the execution of the man-in-space program but its premise as well? His views were unclear—even, perhaps, to himself. At a time when, as Kennedy said in his acceptance speech in Los Angeles, "a revolution of automation finds machines replacing men in the mines and mills of America," a time when racial discrimination was rampant, when "a dry rot, beginning in Washington, is seeping into every corner of America," when one-third of the world's people were victims of repression and poverty and hunger, there was a limit to how much he felt he should say about space. He continued to use space as a cudgel. "Can you tell me," he shouted at an outdoor stadium in Wichita, "how any citizen can vote for a political party and leadership which permits us to be second in space?" But this, he judged, would suffice. When Walt Rostow, an outside adviser, proposed that Kennedy deliver a "space speech" on "the [New] Frontier theme," he shrugged off the suggestion.

Space, as Sorensen said, was a symbol—not just for Kennedy but also for the American people. "Second in space" spoke volumes about the nation's failure, after eight years of Republican rule, to seize the opportunities of the new age or to meet the challenge posed by Soviet science, arms, and resolve. "Second in space," Kennedy said in Wichita, ought to be reason enough to vote against Nixon. The question of how to make America first in space could wait, he concluded, until the ballots had been counted and a winner declared.

As a young man, before his tenure as a university president and his service on the Atomic Energy Commission, Keith Glennan had been a studio manager at Paramount Pictures. He had overseen set construction, lighting, and sound for directors like Cecil B. DeMille. If Glennan, back then, had been asked to cast the role of administrator, leader of a new national effort at the frontiers of science—a levelheaded man, hale and composed—he could not have been faulted for selecting himself. But in 1960, after nearly two years at the helm of NASA, Glennan sensed he was losing his audience. On April 10, he went on NBC's *Meet the Press* and tried, in the face of growing public concern, to reassure the country about the state of its space program. "I don't think we would be in mortal danger if the Russians gained control of outer space," Glennan said. "I'd put it the other way. I think the nation which is the leader on earth is going to control outer space."

No matter how Glennan put it—and surely there was a better way—few Americans appeared to be buying it. He seemed to spend half his time trying to convince people that whatever they thought, whatever they had heard, the opposite was true. Critics, he complained, thought that if the Russians can hit the moon, "they can hit a target here on Earth—in the Washington area—with pinpoint accuracy." In February, he debunked that claim over lunch with leading journalists, explaining, as patiently as possible, that the competencies it took to launch a satellite were not the same as what it took to target a nuclear missile. Then, just as the lunch was ending, a reporter's question made clear that Glennan had left at least part of the group confused. "I went through it once again," he noted, wearily. A speech he gave a few months later in California, at the Bohemian Grove—that wooded conclave of the well-heeled—failed, again, to fully persuade. After his talk he was cornered by David Sarnoff, the chairman of the Radio Corporation of America and founder of NBC, who insisted—with vehemence—that Russia was about to launch a hundred nuclear-armed satellites and blackmail the United States into submission.

Mercury, meanwhile, proceeded at its own pace—brisk, determined, and well behind schedule. By midyear, it was clear that ongoing debates about the role of the astronauts, continuing troubles with rockets, and other problems had put NASA's goal of a manned ballistic—suborbital—flight in 1960 out of reach. Don Flickinger, a brigadier general in the Air Force and an expert in aviation medicine, informed the astronauts that the Soviet Union had probably already launched ballistic shots of its own and was preparing to orbit a two-man capsule before year's end. But this was surmise: as John Glenn wrote in his notebook during the briefing, "factual data limited. Most info by inference." Given that, there was no call—and, thanks to cuts by Congress, no resources—for a crash program. NASA maintained its emphasis on incremental gains—and on safety first. In early May, NASA conducted a successful "beach-abort test" of a McDonnell spacecraft; its escape system, landing system, and parachutes all performed perfectly. The next month brought another success: a test of the environmental control system, which regulated the levels of oxygen, carbon dioxide, and water vapor in the Mercury capsule. "To evaluate human tolerance," a NASA memo drily explained, the astronauts were put in pressure suits and subjected to "post-landing conditions"—specifically, the condition of being sealed in a titanium can for twelve hours amid the heat and humidity of coastal Florida. The environmental control system worked; even better, "no serious physiological effects" were reported.

The same could not be said a few weeks later, when the astronauts underwent desert survival training in Nevada—one of "the worst experiences of our training phase," Wally Schirra recalled. On July 11, while Democrats gathered in Los Angeles for their convention, the astronauts arrived at the Air Training Command Survival School to spend five days preparing for what NASA assured them was a remote possibility of crash-landing in the Sahara or the Australian outback; a Mercury capsule, along its orbital path, would pass over those areas. During their first few days in Nevada, the seven learned how to fashion tents and clothing out of nylon parachutes and listened to lectures about the signs of an approaching sandstorm ("orange-looking cloud," Glenn wrote in his notebook), the personality of camels ("cantankerous"), and rules for handling encounters with "natives": "Take it easy. Don't be overfriendly. . . . Do as they do." Also, as Glenn duly recorded, an astronaut should avoid giving offense: "Don't draw on sand with feet. Don't show bottom of feet—disrespectful—lowest part of body. . . . Accept hospitality—eyeballs of goat. Belch good. Pass wind—no."

Air Force helicopters flew the astronauts to a remote stretch of desert about sixty miles east of Reno. Carrying only their parachutes and survival kits—the same small package they would find, someday, in the Mercury capsule—the men fanned out to separate spots a quarter to a half mile apart. Before arriving, Glenn had come up with an experiment of his own. It had occurred to him that he had never been truly dehydrated. "I just wanted to see what it was like a little bit," he said later. He also, no doubt, wanted to see—and wanted Gilruth and Williams to see—how much further he could push himself than any other astronaut. He decided to go without water or food for his first twenty-four hours in the 120-degree heat. Shepard thought this was idiotic; Glenn, in his view, took training too seriously anyway, and this was just a stupid risk. But Glenn had cleared the idea with William Douglas, Mercury's flight surgeon, who accompanied the astronauts into the desert; Douglas promised to keep an eye on him.

Glenn found a spot near some sagebrush and dug himself a place to sit; the sand just below the surface was cooler. The parachute, draped over the sagebrush, provided a small bit of shade. It didn't take long for Glenn to realize that by going without water, he was flirting with real physical harm. "Getting rough after exertion," he wrote in a tiny spiral-bound pad. As the hours passed, he could barely raise his arms to shade his eyes from the sun. "It was hard to believe that the human body could deteriorate so rapidly," he wrote in a memo after the program had ended. "This fact . . . [was] forcibly impressed upon me." When twenty-four hours had passed, Douglas intervened. He told Glenn to drink as much water as he could possibly consume.

The training went more smoothly after that. Glenn recorded stray observations in his notepad ("B.O.," read one entry), drank pint after pint of water, did some "star-watching," and noted how strange it was to "lack news" while the Democratic convention was in full swing. On Friday, July 15, Glenn was still in the desert when the head of the astronauts' training program, an Air Force colonel named Keith Lindell, dropped by and filled him in. Glenn noted this, too, in his journal. "Lindell told about Lyndon Johnson being V.P. candidate w/ Kennedy," he wrote.

"We have been 'stewing' for the last two or three days over the proper kind of response to make to Lyndon Johnson," Keith Glennan wrote in his diary on October 27, less than two weeks before Election Day. LBJ, ostensibly in his capacity as chairman of the Senate Aeronautical and Space Sciences Committee, had sent Glennan a letter requesting a report on all key aspects of Project Mercury, including details on any problems and the actions taken to address them. He also asked for a complete schedule of any launches—manned, unmanned, or animal—that NASA planned to make before the end of the year. The idea that Johnson needed to know any of this on a day when he was giving a speech to five thousand farmers at the National Corn Picking Contest in Chillicothe, Missouri, struck Glennan as implausible. "Surely he cannot want this information for anything other than political purposes," Glennan observed, "perhaps to reassure himself that the administration is not going to pull a fast one on him and make a significant launch just before the election." If a Soviet man in orbit could tip the election to Kennedy, an American in space could give it to Nixon.

After further discussion, "we have decided to play this in a low key," Glennan wrote. NASA agreed to give Johnson its launch schedule in the hope that it "will relieve his mind as to any possibility that the administration is planning some startling space shot as a clincher." As for Johnson's other requests, Glennan intended to drag his feet for two or three weeks and then provide the report—after the election. Of course, Johnson already had enough information in hand—any cognizant American would—to deliver an indictment of the administration's record on space. This had been his aim all along. On October 30, the Space Committee issued a white paper that could have been, and possibly was, drawn from Kennedy campaign position papers. The administration, it charged, had been guilty of "drift, delay, and dilution" in the face of the Soviet threat. Most of the progress the U.S. program had made since Sputnik had come only "through Congressional prodding. . . . The Republican presidential candidate," the committee noted pointedly, "has not at any time assisted in this prodding."

In the campaign's closing weeks, national security remained the dominant topic. Kennedy was relentless on the issue of the missile gap—or, rather, the non-issue; the CIA director himself, in an effort to get Kennedy to let up, had shown him classified information indicating that the missile gap was in America's favor. He hammered at it anyway, and continued to equate it with the space gap. As he said at a rally in Canton, Ohio, the Soviets understood what Nixon did not: that "people around the world equate the mission to the moon, the mission to outer space, with productive and scientific superiority." Thus "the impression began to move around the world that the Soviet Union was on the march, that it had definite goals, that it knew how to accomplish them, that it was moving and that we were standing still."

Nixon's strained insistence that "we are first in the world in space," his boast that the United States had run up a "space score" of twenty-eight successful shots compared to eight for the Russians, only reinforced Kennedy's argument that the administration was adrift in a dangerous world. Kennedy mocked what Nixon had said in 1959, in his "kitchen debate" with Khrushchev at an exhibition of American goods in Moscow: "There are some instances where you may be ahead of us," Nixon had acknowledged, "for example, in . . . the thrust of your rockets for the investigation of outer space. There may be some instances, for example, color television, where we're ahead of you." This, to Kennedy, said it all. He quoted it all the way to Election Day, November 8.

That morning, while Americans went to the polls, NASA readied another test of the Mercury capsule's abort system—a crucial step toward its goal of a manned orbital flight in late 1961. At Wallops Island, Virginia, the booster rocket ignited and lifted off just as planned. At the sixteen-second mark, the escape tower was supposed to separate from the rocket, pulling the capsule to safety. It failed to detach. From an altitude of 53,000 feet, the entire contraption—booster, spacecraft, and escape tower—pitched over and plunged into the Atlantic.

Somewhere in the wreckage, seventy feet beneath the sea, was, perhaps, the answer to what had gone wrong. Recovery crews combed the area, collecting fragments, looking for clues. But for now, the old, nagging questions had acquired a new urgency. As the next morning's papers proclaimed, John F. Kennedy had been elected president of the United States.

CHAPTER 7

Suspended Animation

"WE ARE IN A STATE of 'suspended animation,'" Keith Glennan wrote in his diary shortly after the election. No one, from the administrator on down, knew what to expect or who might have a clue. Robert Seamans Jr., NASA's associate administrator, was "desperate for intelligence." "What," he wondered, "is Kennedy's attitude to our work? Will he support or accelerate our programs? Will he cut them in part or altogether?" All anyone knew about JFK's views on space was what he had said on the campaign trail: space exploration was important to America's prestige. The program was lagging. He would not stand for second place.

Amid this ambiguity, NASA readied MR-1 (Mercury-Redstone 1) for its first full test. The space capsule–booster combination would someday carry the first American into space. Orbital flights would have to wait for the Atlas, a missile that was still not "man-rated," NASA's term of art. On the morning of November 21, as Kennedy met with his transition team at his family's estate in Palm Beach, MR-1 stood on the pad at Cape Canaveral, 150 miles north. All seven astronauts were on hand for the launch. Glenn, in a hard hat, climbed into the capsule to inspect it, posed for wirephotos, then joined the others in the blockhouse, the concrete bunker at the base of the launchpad.

When the countdown reached zero, the rocket ignited. It rose imperceptibly, then dropped back into its launch cradle, hissing, teetering, as its engine cut off. In a plume of smoke, the capsule's escape tower shot several thousand feet into the air, while the capsule itself—still attached to the booster—deployed its parachutes, which fluttered upward and then flopped uselessly against the side of the rocket. At this point the scientists, as *Life* observed, began "losing their minds." The telemetry circuit had blown; they had no data to explain what had gone wrong. There was no time for that anyway: the Redstone, fully fueled and pressurized, was still on the launchpad. If the wind caught the chutes, the Redstone could topple over and explode. In the blockhouse, someone started shouting about finding a rifle and shooting holes into the rocket's tanks; others suggested putting a man on a cherry picker to cut the parachutes down. In the end, mission control decided to just let it be: in time, the rocket's batteries ran out and the pressure in the booster released.

To Gene Kranz of the flight operations team, it was "a fiasco . . . the most embarrassing episode yet." Shorty Powers christened it the "four-inch flight" and herded the astronauts away from the press. They would not, he said, be available for comment. Later that day, Gilruth was asked if the failed launch would delay the program even further. "It certainly doesn't help," he conceded. Another NASA official said the United States had probably blown its last chance to send a man into space ahead of the USSR. Less than two weeks later, the Soviets faced a failure of their own: a spacecraft carrying two dogs and other animals ("another space zoo," Glennan scoffed) lost control and burned up on reentry. But a similar test, back in August, had been a success, and no one could possibly conclude that the Soviet program was as star-crossed as its American rival. Underscoring the point, on December 15, an Atlas rocket that was supposed to send a satellite into orbit around the moon blew up barely a minute after liftoff. Out of eight attempts to send a payload close to the moon, this was the seventh U.S. failure. "Obviously," a *Time* reporter noted, "the Mercury program is in very serious trouble."

It was an inauspicious start to the transition. There was little to do but get back to work. "We trained harder than ever," Glenn recalled. "We were afraid that any lapse in our efforts could provide the president-elect an excuse to drop or delay the manned space program in favor of machines." Kennedy would soon have his say. But until then, NASA still had Eisenhower to worry about. Before leaving office, he had decisions to make about his budget for fiscal year 1962—among them, how much to invest in space exploration beyond Project Mercury. He was unmoved by NASA's ambition to send men into orbit around the moon—even less by the agency's rationale of "prestige" or, worse, "adventure." Within NASA, the discussion had focused on the mechanics of getting men to the moon, not the reasons for going there.

On December 16 the President's Science Advisory Committee delivered its report. The group knew its audience. Man-in-space, they told Eisenhower, was a "costly adventure" that would yield less scientific value than a satellite could. As for Project Mercury, PSAC dismissed it as "a somewhat marginal effort," lacking a clear purpose and hobbled by insufficient boosters that were probably unsafe to fly. The cost of sending men to the moon—$26 to $35 billion more than the country had already spent—was, the group concluded, a pretty steep price to pay to satisfy "emotional compulsions."

Three days later—in what must have seemed, to PSAC, an act of defiance—NASA managed to get its next Mercury-Redstone, MR-1A, more than four inches off the pad. It made it all the way into space.

Every system worked perfectly. The capsule, beneath its cheerful red-and-white parachute, splashed down near Grand Bahama Island and was ferried back to the Cape by the carrier *Valley Forge*. "I called the president and gave him the good news," Glennan recorded. "He seemed quite excited about it."

Not excited enough. The next day, December 20, Glennan went to the White House to make presentations to the outgoing cabinet, the National Security Council, and Eisenhower about the state of the program and the future of manned spaceflight. At the NSC meeting, George Kistiakowsky, the White House science adviser, cited PSAC's cost estimates. Eisenhower's response was short and sharp: he "couldn't care less whether a man ever reached the moon," and he "wasn't going to hock the family jewels" to try it. Glennan, with satisfaction, declared that the group had managed to resist "the clamor for 'spectacular accomplishments' that had no basic scientific interest." He made his point perhaps too emphatically. Shortly after the meeting, Eisenhower slashed NASA's request for post-Mercury research by 96 percent, from $23 million to $1 million. It would be his last word on space: in effect, a veto of Apollo.

"THERE IS KIND OF A HUSH over the whole scene," Glennan had noted in late November, and the silence persisted into January. As the inauguration approached, NASA's leaders had still not heard a word from Kennedy's transition team about his intentions; their calls went unanswered. Glennan was asked neither to stay nor to leave. A Republican in the budget-conscious Eisenhower mold, he let it be known that he planned to resign. Even this failed to raise a response. Glennan began the practice of ending his diary entries—every day—with the line "No word from the Kennedy administration!"

Kennedy was not, as Glennan imagined, indifferent; he was overwhelmed. James Reston, in the *Times,* inventoried the most pressing problems the new president was facing: concerns about the scale of federal spending and the adequacy of the education system; growing tensions within the NATO alliance; an outflow of money and jobs to Western Europe, where labor was cheaper; and, not least, Soviet aggression. Space was not on Reston's list. Neither was the civil war in Laos or the Communist uprising in South Vietnam. Or civil rights. Of course, others would have a say in what defined the Kennedy years—others including the black college students whose sit-ins at segregated lunch counters in Greensboro, North Carolina, were sparking nonviolent protests across the South. That fall, the President's Commission on National Goals, appointed earlier that year by Eisenhower, issued a widely read report identifying discrimination against women, religious minorities, and, most of all, black Americans as "morally wrong, economically wasteful, and . . . dangerous." Yet at the same time, the commission recognized that most concerns would take a second seat to the existential threat America faced from the Soviet Union.

Which brought the conversation back to missiles. This, indeed, was where virtually every such conversation began and ended on the eve of Kennedy's inauguration. MISSILES SPEAK LOUDER THAN WORDS was the *Times*' assessment, in bold type, of Soviet foreign policy. After the election, Kennedy asked his choice for secretary of defense, Robert McNamara, to conduct a comprehensive review of U.S. defenses. McNamara confirmed what Kennedy already knew: the missile gap favored the United States. But he did find shocking deficiencies in "usable power"—combat-ready troops, modern fighter-bombers, air-to-ground missiles, and other conventional weapons. It was also plain that the Eisenhower doctrine of massive nuclear retaliation, for all its vaunted power as a deterrent, had hardly rendered the nation invulnerable.

Beyond that, every Soviet "first" in space, and every sputtering or exploding or errant U.S. missile, made it harder for anyone to believe that America was ahead technologically—even if the new Zenith television set featured a "Space Command remote control TV tuner." ("Settle back in your easy chair!") For three years now, since Sputnik, international opinion polls had shown deepening doubts about America's strength. In late 1960, a classified USIA report—which caused a stir after being leaked—described Western Europe's declining confidence in U.S. leadership, due in no small part to Soviet superiority in space. A similar effect was seen among the so-called uncommitted nations—nations said to be weighing which great patron, the United States or the Soviet Union, offered better protection and a better shot at more rapid advancement. All this, for America, was a real and rising threat, whatever the count of ICBMs.

No one understood this better than Khrushchev. The Soviet space program was theater—an audacious bluff. It hadn't begun that way. Nor did its engineers, drawing on decades of Russian passion for and expertise in rocket science, see it that way. But Khrushchev had been quick to perceive that Sputnik provided cover for the crudeness of Russian missile technology. What the Soviets had was a Potemkin program. They made fewer launches than the Americans because they lacked the infrastructure. They relied heavily on one rocket, the R-7 Semyorka, rather than developing new technologies. They achieved a string of firsts, but at the expense of long-range planning. And they suffered catastrophic failures: in October 1960, at the Tyuratam test site in Kazakhstan, a rocket exploded on the launchpad, incinerating nearly a hundred people. All that was left of the Strategic Rocket Forces commander was a shoulder strap and a melted set of keys. The news was suppressed, the illusion maintained.

"Launching Sputniks into space," Khrushchev conceded late in life, in his memoirs, "didn't solve the problem of how to defend the country." But it did defer, or perhaps elide, other problems. In Europe, it weakened NATO's resolve in the face of Communist adventurism; in the Politburo, it quieted rivals. In some ways, the most important audience for a Soviet space shot was the Soviet people: every achievement implied that the nation's technology was so advanced that its benefits would soon spread across the economy—to agriculture, to manufacturing. As Khrushchev traveled across the Soviet Union in 1960, he saw shortages of meat, milk, butter, eggs; he saw the second failed harvest in as many seasons. Soviet agricultural productivity was a third of that in the United States. But a farmer on the steppes could look to the heavens and see the approach, ever nearer, of the workers' paradise.

"SPACE PROBLEMS." That was the heading on the memo, and Lyndon Johnson was acquainted with the subject. Better acquainted, in fact, than any member of the incoming administration, including the president-elect. "Of all the major problems facing Kennedy when he came into office," reflected Hugh Sidey of *Time* magazine a few years later, JFK "probably knew and understood least about space." As the space program slid down Kennedy's list of priorities, Johnson moved to seize it. In December 1960, Ken BeLieu, staff director of the Senate Committee on Aeronautical and Space Sciences, sent its outgoing chairman a series of alerts. "MERCURY (the Man-in-Space one) . . . is slipping badly," he told LBJ, and so were efforts to launch communications and weather satellites. NASA, he wrote, lacked leadership and competence; and the White House had allowed the National Aeronautics and Space Council—created in 1958 at LBJ's insistence—to atrophy.

All these problems were now Kennedy's problems. If "improvements do not come within the next 18 months," BeLieu observed, "the situation will provide an opening for political exploitation by the opposition." Also, he warned, the Air Force—backed by the aerospace industry and its allies in Congress—was making a grab for the entire space program. The existence of NASA was at risk—as was the principle of civilian control and the peaceful use of space. BeLieu was cheered by Kennedy's decision to place Johnson at the helm of a revived Space Council and to conduct a top-to-bottom review.

That effort was already underway. In early December, Kennedy had asked an MIT professor of electrical engineering, Jerome Wiesner, to chair an ad hoc committee to assess the state of the program. For the past several years, Wiesner had been the one Kennedy called to ask questions on topics from medical research to nuclear fallout. On Wiesner's wall was a photograph inscribed by JFK: "To Jerry, who makes the complex simple." They were almost the same age—Wiesner, forty-five in 1960, was just two years older than Kennedy—and had a similar cast of mind, curious but practical. Wiesner could talk too much for Kennedy's taste; he would stride back and forth across a meeting room as if it were a lecture hall, gripping his pipe in his palm. But like Kennedy, Wiesner was always driving at something: action, solutions.

Wiesner was not an expert on space, though he did know a great deal about military technology. As part of the Manhattan Project, he had devised electronic equipment for the nuclear weapons tests at Bikini Atoll—including an underwater detonation that sent up such a vast plume of radioactive sea spray that the follow-on test was called off. Glenn Seaborg, the chemist who discovered plutonium, later called it "the world's first nuclear disaster." It surely played a role in Wiesner's emergence as a proponent of a test ban. In 1959, Wiesner made enemies at the Pentagon by proposing a moratorium on missile flight tests. He liked the idea for another reason: it might, in his words, "get the

United States out of the space race, which otherwise will continue to be a serious source of embarrassment and frustration." When James M. Gavin, an Army general, argued rather grandly that "the nation that first achieves the control of outer space will control the destiny of the human race," Wiesner had a tart response: "Jim, I don't understand what the hell you're talking about."

Nothing Kennedy had said on the campaign trail in 1960 indicated he wanted to withdraw from the space race, and he had selected as vice president a man who, since the launch of Sputnik, had been urging an all-out effort to win it. But now, at least initially, Kennedy was putting space policy in the hands of a skeptic. On December 9, Ted Sorensen wrote to thank Wiesner, on JFK's behalf, for agreeing to chair the Ad Hoc Committee on Space. "We are interested only in the most urgent and significant steps on which you think the next Administration must decide in its first 100 days," Sorensen added, "particularly those which Congress is most likely to approve, in terms of political and public appeal, and their effect on the budget" and the economy. Still, Kennedy wanted him to think big. "This is the best time," Sorensen advised, "to get done those more difficult and controversial items which must be done."

A month later, on January 11, 1961, Wiesner presented his findings to Kennedy, Johnson, and the chairs of the House and Senate space committees. His report, cobbled together without input from NASA or the Pentagon, was unsparing. The space program—from deep-space probes to military surveillance systems to man-in-space—was deficient across the board. The Russians, Wiesner said, would almost certainly put a man in orbit before the United States did, and an expedited effort would not close the gap. The ad hoc committee did see man's exploration of space as inevitable—"this will be done"—but the concession was grudging. Wiesner's report was dubious about the value of sending men into space and America's ability to accomplish this safely.

Why, then, even bother to try? National prestige was one possible reason, though the committee was unpersuaded. (In an earlier draft, Wiesner had crossed out a line asserting that "great prestige will be gained by the nation which sends the first explorer into space.") Neither did the committee members see a convincing scientific rationale. The real reason, they suggested, had more to do with the "motives that have compelled [man] to travel to the poles and to climb the highest mountains of the earth." Still, facts were facts: "Because of our lag in the development of large boosters, it is very unlikely that we shall be first in placing a man into orbit around the earth." Rushing to match that feat invited even greater risks, the report warned: the death of an astronaut on the launchpad, for example, or, "even more serious," stranding a man in orbit.

Here its tone shifted from skeptical to scolding. The United States, the committee believed, had only itself to blame: by celebrating the astronauts—by putting them in the spotlight—the Eisenhower administration had "strengthened the popular belief that man in space is the most important aim of our non-military space effort." It had built up a project where the country was likely to fail and overlooked areas where it was already succeeding: meteorological satellites, for example. With respect to man-in-space, the committee made two recommendations: First, "stop advertising Mercury as our major objective in space activities" and try, instead, to "diminish [its] significance." Second, conduct a thorough appraisal and accept its conclusions, whatever they might be. "If our present man-in-space program appears unsound," they advised, "we must be prepared to modify it drastically or even to cancel it." But time was short. By leaving Mercury "unchanged for more than a very few months," the new president would "effectively endorse this program and take the blame for its possible failures."

"Highly informative" was Kennedy's only response. Johnson, too, kept his views to himself. He had voiced many of these concerns about the program himself. But Wiesner's critique, he thought, went too far. The professor's dismissive take on manned spaceflight was at odds with Johnson's belief that the race against the Soviets was a "race for survival." The terms had been set for a serious debate. And these two men, Johnson and Wiesner, would lead it: Kennedy had already told Johnson he could run a revived Space Council, and after hearing Wiesner's bleak presentation, he named the professor his science adviser.

"What a slap in the face," Glennan wrote in his diary on January 13, seething after two days of terrible headlines about the Wiesner report. Newspapers ran excerpts; an unclassified version, retaining the full force of the original, had been released to the press. "We thought it quite unfair to NASA and quite personal," recalled Bob Seamans, NASA's associate administrator—referring, especially, to its insistence that the agency's leadership needed major improvements. Space Task Group officials protested that Project Mercury had been unfairly maligned. Gilruth issued a terse statement that "manned orbital flight in 1961 is still in the cards."

It seemed far from certain. A suborbital flight, however, was nearer at hand, and it had come time to choose its occupant. Before the holidays, NASA—in keeping with its affinity for panels, committees, and working groups—had created a five-man board to assess the astronauts' readiness to fly. Expert witnesses would make presentations on the men's command of mission procedures and capsule systems, their proficiency on the simulators, and, among other things, their "maturity . . . , ability to work with others," and "ability to represent Project Mercury to [the] public." The board would also consider, as Walt Williams, Mercury's head of operations, put it, "how cooperative [each astronaut] can be expected to be from our viewpoint." In the end, the group would make three recommendations to Gilruth, its chairman: who should fly first, who should fly second, and which astronaut should serve as the backup for both.

One other group—the astronauts themselves—had a voice in the decision. Each of the seven was given a one-page form to fill out.

> You are asked to submit a list of three Astronauts, excluding yourself, who in your judgment are best qualified for the first two manned Redstone flights. We assume you would rate yourself in this group, therefore, please omit your name from the list.
>
> #1. ______________________________
>
> #2. ______________________________
>
> #3. ______________________________
>
> Comments, if any:
>
> Signed __________

The sheet did not say, and the men were not told, that only Gilruth would see their answers. Only Gilruth, moreover, would know what weight those answers were given.

"It took a moment to sink in that he was asking for a peer vote," Glenn recalled years later. "I was enormously disappointed and very upset. . . . After my comments . . . about everybody's needing to keep their pants zipped, I could imagine where I'd stand in a peer rating."

There was nothing to be done but to vote and hope. By prearrangement, he rated Carpenter first on his list. Carpenter did the same for him.

Weeks passed without an announcement. Though the first flight had yet to be scheduled, the window was narrowing for the first astronaut to become fully familiar with the capsule's systems and procedures. Gilruth wanted to see a chimpanzee make—and survive—a Mercury-Redstone flight before he would be ready to launch a man. That was slated for the end of January, leaving ample time for what Williams called a "long tiresome debate" on the question of whether NASA's selection (when it had one) should be announced to the public—or even to the man himself—before the last possible moment. Some argued that the buildup, the pressure, might cause the astronaut to crack. Williams thought this was nonsense. So, of course, did the men in question. "If we don't have the confidence to keep from getting clutched at that time, we have no business going at all," Glenn had told *Life* in 1959. "As for keeping this a big secret from us and having us all suited up and then saying to one man 'you go' and stuffing him in and putting the lid on that thing and away he goes, well, we're all big boys now."

In mid-January, Gilruth finally summoned his board, listened as they made their presentations, then sat back and let them debate. This was his management style: placid, inscrutable, seemingly passive. "He never once told me what I should do," recalled Guy Thibodaux, an engineer who had worked with Gilruth since the 1940s, but Thibodaux would end up, invariably, doing what Gilruth had wanted him to. "I don't know how he did that." In this case, as in most, Gilruth's views were completely opaque. Williams, more of an open book, said outright that the choice had narrowed to Shepard and Glenn. Shepard had done little, in nearly two years of training, to win friends in the program. But he had won admirers. Williams thought him "one of the best test pilots I have ever known." Shepard had performed brilliantly in the simulator, had taken an active interest in the engineering and the flight plan, and had raised his concerns, when he had them, with a military briskness. All business. Shepard had never pretended otherwise.

Much the same could be said of Glenn. Richard S. Johnston, an engineer who helped manage the environmental control system and the space suits, rated Glenn the best of the seven (and Shepard second). Glenn's focus on training was almost obsessive; he spent more time in the simulator and on physical conditioning than any other astronaut. With his family in Arlington, he lived like an ascetic. "At this time," Williams judged, "he is certainly more prepared for a 'Go' than almost anyone else in the program." He also, of course, "has enormous popular appeal and is certainly the American public's choice."

But this last quality was not, in Williams's view, an asset. It was an irritant. He was not the only NASA manager who saw it that way. "John Glenn's goal," Chris Kraft later wrote, "was to cozy up to our top management and thus improve his chances." When it came right down to it, for Kraft and Williams, there was something about Glenn they just didn't like. Kraft didn't have a vote on the selection, but Williams marked his own ballot for Shepard. He added that Grissom, not Glenn, ought to get the second flight, because Grissom, unlike Glenn, had a degree in engineering. Gilruth, as ever, kept his own counsel.

January 19 was the day before John Kennedy's inauguration. Snow paralyzed the capital: only a few inches had fallen, but more were expected, and stalled cars were lining Pennsylvania Avenue, the next morning's parade route. By late afternoon, National Airport had shut down; flights were diverted or sent back. Preinaugural parties were being canceled; the Kennedys sent their regrets to two. Meanwhile, at Langley, shortly after 5:00 p.m., Gilruth told the seven astronauts to report to the office they shared, the one with the seven metal desks. It was important, he added. He didn't say why, but it was well understood.

The office was empty when they arrived. They stood around uncomfortably, waiting for Gilruth. Shepard broke the silence. "Deke," he said, "what do you think?"

"I think I wish to hell he'd hurry up," Slayton answered. Grissom added: "If we wait much longer, I may have to give a speech!" The others broke out in welcome laughter.

Gilruth entered the room and shut the door. Like Grissom, he was not a man to make speeches. He opened with a few words about how capable they all were, how any one of them would make a great pilot for the first flight, and explained that he had made his decision, though he did not intend to announce it publicly. He said he had put aside plans to withhold the information from the astronauts until the morning of the flight; in the weeks beforehand, the pilot and his backup would need to have priority on the procedures trainer.

"So," Gilruth continued. "Shepard gets the first flight, Grissom gets the second flight, and Glenn is the backup for both." The group was silent. "Any questions?" More silence. With that, Gilruth made his escape. "Thank you very much," he said. "Good luck!"

"I DIDN'T THINK that was necessarily the end of it," Glenn observed later.

This became clear the morning after the meeting. It was January 20: Inauguration Day. The snow had stopped and the sky was bright; an icy gale blew down the coast. Glenn got into his Prinz to make the four-hour drive home. He had a passenger: Loudon Wainwright, the *Life* reporter who had become his friend. As the tiny car started north, it was obvious that something was wrong with Glenn. Typically talkative, he was stone silent today, unsmiling, "just pissed off," Wainwright recalled. Attempts at conversation went nowhere. Wainwright fiddled with the radio dial, trying to find Kennedy's inaugural address, but the sound cut in and out; they caught only fragments. Every so often Glenn slammed the steering wheel, hard, with the flat of his hand. Was he annoyed by the static? Moved by what little he could hear of the speech? Wainwright couldn't read him at all.

What Glenn was thinking was that he'd been stabbed in the back. "I kept thinking that Bob Gilruth wouldn't have asked for the peer vote unless he'd meant to use it," he wrote in his memoir, "and it must have been a big part of his decision." None of the others thought so. Slayton was no happier about the result than Glenn—"I was shocked, hurt, and downright humiliated," he said later—but suspected, probably rightly, that Gilruth just wanted to know whether the astronauts agreed with his judgment. In Williams's own account of the selection process, written a few years later, he did not even mention the peer vote. But to Glenn, it was the only way to explain Gilruth's choice. He felt sure that he had outperformed Shepard. And *Grissom*! Gus had never looked like a contender—not to Glenn, not to the press. It had played out just as Glenn had feared: "months of training were being reduced to a popularity contest." The five, all but Carpenter, had avenged Kona Kai. They had denied Glenn immortality and awarded it to their standard-bearer.

Unless Glenn could turn things around. After dropping Wainwright at National Airport, he drove home, sat down at his desk, and wrote Gilruth a letter. "It was a strong letter," Glenn recalled, "because I felt strongly." He explained exactly why his peers might have ranked him low. His only offense, he wrote, was to try to protect the program, and it was unfair, he argued, to be penalized for that.

Seeing the starkness of what he'd just written, he debated whether to send it. He talked it over with Annie and the kids. And when he got back to Langley, he handed the letter to Gilruth. There was no discussion of its contents, and there would never be a discussion of its contents.

HAD GLENN BEEN ABLE TO HEAR the inaugural address on his disconsolate drive home, he might still have missed Kennedy's mention of space. It was a passing reference—one more item in a grab bag of ambitious (and ambiguous) goals: "Together," the new president declared, "let us explore the stars, conquer the deserts, eradicate disease, tap the ocean depths, and encourage the arts and commerce."

Keith Glennan listened to the speech in his station wagon. Having resigned his post at NASA and said his good-byes, he drove home to Cleveland. The speech was excellent, he had to concede. Still, Kennedy's neglect of NASA had taken a toll. The agency remained in suspended animation. When Glennan got home, he added a few final words to his diary. "And still," he wrote, "no word from the Kennedy administration!"

CHAPTER 8

The Problems of Men on Earth

"Americans . . . ," John Kennedy said in October 1959, "like to picture hostile dictators as unstable and irrational men, the almost comic captives of their moods and manias." This was a few days after Nikita Khrushchev had finished his tour of the United States. Kennedy, along with the rest of the Senate Foreign Relations Committee, had met with Khrushchev and the next week told an audience in Rochester that he found the Soviet leader "no fool." On the contrary, Kennedy said, "he is shrewd, he is tough, he is vigorous." And he was not a man of "limited vision."

It was unlikely that Kennedy made an impression on Khrushchev during that brief encounter; he was one of seventeen committee members. A year later, in 1960, Khrushchev still had no fixed notion of JFK. He thought he had come to understand Nixon, but Kennedy remained "an unknown quantity," Khrushchev confessed to Llewellyn Thompson, the U.S. ambassador to the Soviet Union, just before the election. In the weeks that followed, scrambling to make sense of the president-elect, Mikhail Menshikov, the Soviet ambassador to the United States, advised Khrushchev that JFK was weak and pliable, "an inexperienced upstart." This meant the Russians might be able to manipulate

Kennedy—though the hard-liners in the Pentagon and the CIA would surely look to do the same.

Whatever the case, Khrushchev needed Kennedy's help in Berlin. "He had nightmares about it," Khrushchev's son Sergei remembered. "The German problem gave him no peace." Berlin was a thorn pressed deep—impossible, it seemed, to extract. Located in the heart of Communist East Germany, the city was encircled by hundreds of thousands of Soviet troops and divided, according to the terms reached at Potsdam in August 1945, between a free, western part and a Communist east. The right of free passage between the two parts of the city meant that, as one East Berliner recalled, "you could go from socialism to capitalism in two minutes." It was like passing through a looking glass. The contrast—in living conditions, in dress, in pace, in mood—provided as pointed an indictment of the Communist system as could be found anywhere in the world.

By 1961, 2.7 million East Germans had gone from socialism to capitalism, never to return. An exodus of professionals, intellectuals, and businesspeople, among others, was draining the German Democratic Republic (GDR) of its educated elites, while rich West Germans were buying up cheap East German goods, exacerbating shortages in the GDR. The East German economy seemed close to collapse. In 1958, Khrushchev had threatened to give the GDR control of all access routes, but when Eisenhower held firm, the Soviets backed down—incensing the leader of East Germany, Walter Ulbricht, who was losing patience with his sponsors.

The election of Kennedy—for all the stridency of his campaign rhetoric—seemed to offer an opening. At a New Year's celebration in the Great Kremlin Palace, Khrushchev made a toast: "We hope that the new U.S. president will be like a fresh wind blowing away the stale air between the USA and the USSR." Like an awkward suitor, he offered flattery and made solicitous gestures—more student exchanges, less jamming of Voice of America broadcasts—seeking to draw Kennedy into an early summit meeting, focused on Berlin.

Kennedy, too, wanted to meet soon for talks—though not about Berlin. He and his advisers had hope, however slight, for progress toward arms control and a nuclear test ban treaty—more hope, certainly, than they saw in Berlin. So Kennedy made concessions of his own, for example, directing the U.S. Postal Service to stop censoring Soviet publications that passed through the mail. But the goodwill gestures faded. In foreign affairs, Kennedy thought himself a realist: cold-eyed, unblinking, quick to see past posturing and through the abstractions of ideology. He was not, however, adept at reading signals, particularly when those signals were mixed. In an address on January 6, Khrushchev called nuclear war an "incalculable disaster" but expressed enthusiasm for "wars of liberation"—promising, by those means, to "nibble" America "to exhaustion all over the globe." In a cable from Moscow, Ambassador Thompson warned JFK that Khrushchev had posed his challenge in stronger terms than before. Kennedy took this personally: as a test of his resolve. In truth, the speech was largely rote, as Eisenhower understood; but it startled Kennedy, who instructed his staff to "read, mark, learn and inwardly digest" it—a religious exhortation, drawn from the Episcopal Book of Common Prayer. He read portions of the speech to his National Security Council. He was steeling himself. "I have to show him that we can be just as tough as he is," Kennedy told Kenneth O'Donnell, one of his closest aides.

Distrust and incomprehension, so long the central elements in U.S.-Soviet relations, ran in both directions. The Kremlin did not know what to make of Kennedy's inaugural address—so forceful in tone, so vague in intent. "Now the trumpet summons us again," he declared, "not as a call to bear arms, though arms we need; not as a call to battle, though embattled we are; but a call to bear the burden of a long twilight struggle . . . against the common enemies of man." Ten days later, in his first address to Congress, Kennedy was more direct. "World domination," he said, was the Soviet goal. "Our task is to convince them that aggression and subversion will not be profitable routes to pursue these ends." In the most sobering passage, Kennedy warned that "each day we draw nearer the hour of maximum danger."

The point was reinforced, in two days' time, by the United States' first test flight—a successful one—of a Minuteman ICBM. "Brother," someone cheered from the launchpad, "there goes the missile gap!" The White House announced a crash program to deploy the missile a full year ahead of schedule, along with the nuclear-tipped, submarine-launched Polaris. Now it was the Kremlin's turn for alarm. Conciliation never had a chance. Kennedy and Khrushchev slid, as if compelled by laws of motion, toward confrontation, acute and existential.

A PATHÉ NEWSREEL, taking liberties, announced, in capital letters: U.S. SPEEDS MAN-IN-SPACE PROGRAM. The new president, it said, was giving "fresh attention" to Project Mercury. Kennedy had done nothing of the kind. He had made, so far, only a single statement on space, and that was to distance himself from the ad hoc committee, the source of so much vexation at NASA. "I don't think anyone is suggesting that their views are necessarily in every case the right views," he said at his first presidential news conference, on January 25. Hugh Dryden, NASA's acting administrator, was not consoled. Like Glennan before him, Dryden had been unable to get Kennedy's advisers to return his calls. NASA had scheduled a chimp flight on the Mercury-Redstone for January 31—a test of the rocket that would later launch the first American into space—and Dryden feared that Kennedy would learn about it only after the fact, by reading the newspaper. Finally, Jerome Wiesner consented to see Dryden, who told him the news.

If fresh attention was being paid to Mercury, it came from Lyndon Johnson, now chairman of the National Aeronautics and Space Council, who was asserting control of the space program and making sure reporters knew it. The White House staff had questions about the scope of his authority; Kennedy's adviser Richard Neustadt saw the chairman's role as advisory, certainly not a "Secretary of Space." But Johnson, in nearly a quarter century in public office, had never paid much mind to jurisdictional lines.

Still, he had been unable to fulfill one of his most important responsibilities: to select a new NASA administrator. Kennedy had charged him with the task. By mid-January, Johnson had considered nineteen men for the job; every one had been rejected or had turned it down. It was all too clear that NASA's administrator would sit atop three fault lines: the first, between LBJ, who wanted a political operator, and Wiesner, who wanted a technical expert; the second, between PSAC and NASA over the fate of manned spaceflight; and the third, between NASA and the military over control of the program. Also, the prospect of working with—or under—LBJ had not been a selling point. The delay in appointing an administrator, duly noted in the press, was becoming a source of embarrassment for Kennedy. He told Johnson and Wiesner that he might have to find someone himself. That did it. LBJ placed a call to Senator Robert Kerr of Oklahoma, who had succeeded him as chairman of the Space Committee, and Kerr suggested one of his own adherents: James Webb.

Webb and Kerr were mutually indebted. In the early 1950s, Kerr had hired Webb to turn around a subsidiary of his family's company, Kerr-McGee, one of the country's largest independent oil producers. Webb had done that successfully. Webb had, in fact, done most things successfully—in business, in government, and at the junction between them, where, Webb understood, much of consequence got done. During his fifty-four years Webb had propelled himself from Tally Ho, North Carolina, into a dizzying succession of roles: law clerk, Marine pilot, congressional staffer, executive of an aeronautical equipment manufacturer. During the Truman administration, Webb served as White House budget director and then undersecretary of state. Throughout, he was an enthusiast, a promoter, of technology, though what truly excited him was *management*—the theory of it, the practice of it—particularly management of "large-scale endeavors."

Square in frame, Webb was formidable, voluble, intense in focus. "By God," recalled Paul Haney of the NASA public affairs office, "you hadn't lived and felt the enthusiasm and the hot breath of somebody from North Carolina till you met Mr. Webb. He could talk. He'd burn the wall. . . . But he sure as hell understood how Washington worked and where the money came from and how it got spent." NASA needed this sort of administrator. Kennedy was eager to give him the job. "I've been turned down too much," Johnson said to Wiesner. "You call him."

The White House operator reached Webb at a luncheon honoring Kerr. Someone handed Webb a note saying the science adviser was on the line. "The president would like you to take on this job," Wiesner said, getting right to the point. "Would you like to do it?" Porcelain clinked in the background, filling Webb's silence. "Not particularly," he finally replied. From his view, NASA looked like a dead-end job. He did, however, agree to go to the White House to talk about it, and on his arrival he could see that this was less a job interview than an anointment. He met with Kennedy on January 30 and was nominated that day. Kerr's Space Committee approved the nomination without asking Webb a single question about his plans for the program.

The next day began with the flight of the chimpanzee. Ham, a thirty-seven-pound male, had been selected from a colony at Holloman Air Force Base in New Mexico. NASA had decreed early on that its primates—which had "led sheltered lives" and might be "influenced by excitement"—would not be offered up to the press, but the policy proved impossible to maintain. After Able and Baker, with NASA's consent, had made the cover of *Life* in 1959, interest in the primates had only grown as they'd moved up the anthropomorphic ladder: an "astrochimp," as reporters insisted on calling them, was better copy, any day of the week, than a squirrel monkey. The training of the chimps, conducted by serious-looking men wearing white lab coats and wielding stethoscopes and tongue depressors, was played for comic effect. "They get intellectual training!" a newsreel narrator proclaimed over footage of chimps flipping switches. "The reward is a tasty pellet!"

Just before noon on January 31, Mercury-Redstone 2, or MR-2, launched Ham 157 miles above the Earth. Seventeen minutes later the capsule, Ham safely inside, splashed down in the Atlantic. He had done his part almost perfectly, pulling levers every twenty seconds and making clear that he could function while weightless. Gilruth proclaimed the test a success "from an aeromedical point of view." Yet minor malfunctions had delayed the launch by four hours while Ham sat in the capsule, strapped to his custom-fitted contour "couch"; at liftoff, a throttle control jammed open wide, giving the booster extra fuel and sending it higher, faster, and 132 miles farther than the planned splashdown site. When the capsule hit the water, the impact drove the heat shield into the titanium bulkhead, punching two holes into it and knocking the capsule on its side. Which is how it remained for nearly three hours, bobbing and leaking while the waves ripped its inflatable landing bag to shreds. By the time the Navy helicopter reached the scene, the capsule was half-submerged. Ham was alive, though, and when the crew finally liberated him from his restraints, they offered him an apple. The chimp extended a hand, stepped out of his couch, and burped a reproach.

An exploding rocket provides a certain clarity: more tests are in order. But the flight of Ham—a qualified success, partial failure, or near disaster, depending on how you looked at it—offered only fodder for argument. Gilruth and the STG saw that crucial systems had functioned well, and that even when a number of things went wrong, a living being could be launched into space and still be living when it got back. The astronauts were heartened; Carpenter told *Life* he would welcome a chance to ride the next one. But von Braun and his team, responsible for the Redstone, voiced concerns about its overacceleration—the same thing they'd seen in December. Two weeks after MR-2, Gilruth, Walt Williams, Maxime Faget, and others met with von Braun and "the old German group," Gilruth recorded in his diary. The Germans felt strongly that another chimp should be sent up, and "we should continue the unmanned flights until we had an absolutely clean flight, that is, a flight in which no unexpected happenings occurred." This was less about safety, Gilruth observed, than about cover. "Should a failure occur in a manned flight," he wrote, the engineers wanted to "be able to say that we had a completely successful and uneventful flight in the books to fall back on."

The foot-dragging and further discussion "drove Al crazy," Glenn later recalled. Shepard's flight was slated now for March, six months later than first projected. The growing series of checks and fixes and redesigns threatened to push it later still.

IF SHEPARD WAS ANTSY, Glenn was despondent. But his funk had nothing to do with the launch schedule. Williams wrote that "he is bitterly disappointed about being passed over and . . . is not making much of a secret about it." Loudon Wainwright, the *Life* correspondent who had taken that awful ride with a sullen Glenn on Inauguration Day, thought Glenn was not just disappointed but full of rage—a pretty "extreme response to rejection," in the reporter's view. Even at home, Glenn seemed beyond reach. "It was pretty sad around the house when Dad wasn't picked," Dave Glenn, fifteen years old and guileless, told a reporter that fall. "He was very withdrawn," Glenn's daughter, Lyn, later recalled. "I remember all of us talking about it." When his close friend and neighbor Tom Miller invited the family to go water-skiing—a favorite weekend activity—Glenn mumbled excuses. Annie and the children left him at home. Miller confronted him one day in the front yard. "You're making everybody miserable because your damn pride is so high," he yelled at Glenn. "Now, get off it, dammit!"

It was no use assuring Glenn—as friends did—that being the backup for two suborbital shots put him at the front of the line for the one that really counted, the orbital flight. "First is first," Glenn snapped. In a sour letter home from Korea, in 1953, he had referred to himself as "second-best Glenn" after being relegated to the second seat on a mission; when he was then sent to Hong Kong for R&R, he carped that he had been handed a "consolation prize." And now, when the first prize was immortality, Glenn was second-best twice over, backup to Shepard and then again to Grissom. "There was a lot of backbiting," Slayton recalled. "It was clear to the rest of us that he was lobbying for a reversal of Gilruth's decision."

It was clear, in fact, to everyone who knew about the decision. Williams observed that "John really isn't accepting the fact that he has not been chosen. He is still genuinely convinced that <u>he</u> is the best man for the job, and he almost seems to feel we have put the program in jeopardy by passing him over. Instead of accepting the decision, he continues debating it heatedly with Gilruth and others." Whether Glenn was indeed "debating it heatedly" or, more likely, raising it as dispassionately as he could manage, he was still refusing to take no for an answer. The injustice of the peer vote was his principal grievance. But he continued to argue that character mattered—namely, who had it and (by implication) who didn't, who could best represent (and not risk embarrassing) the United States before the eyes of the world. Getting nowhere with Gilruth, Glenn raised his concerns with Webb, Gilruth's new boss. Though sympathetic, Webb was facing enough difficulties in his first days in office that he was not about to pick a fight with the director of the Space Task Group. The decision stood.

As the weeks passed, the wound remained raw, inflamed by every story in the papers—and there were many—predicting, insisting, that Glenn would fly first. NASA's policy that the selection be kept secret meant there was no end to this sort of speculation. "*Everybody* assumed it was John Glenn," Slayton wrote in his memoir decades later, his schadenfreude undiminished. "I'm sure it annoyed the hell out of him." Even worse, NASA demanded to have it both ways: Shepard got to be the first man, but Glenn was still the front man, the face of Mercury, the one they put forward to look into the cameras and talk expansively about "the real goals of our program, what we're trying to do in Mercury." When Wiesner and his committee arrived at the Cape as part of a five-day fact-finding tour of NASA facilities, Glenn was the one Gilruth sent to greet them in Hangar S; Glenn was the one who was suited up and was asked to give the visitors a rundown of what, exactly, would occur the day of a launch; Glenn was the one, the next day in the control center, who was asked to speak about the role of the astronauts and vouch for their readiness. The others were given supporting roles. For Glenn, the irony—and the pretense he maintained with a smile—was nearly as galling as the decision itself.

"VENUS," OBSERVED THE *New York Times* on February 12, "is shrouded in mystery." Such great mystery, it appeared, that the paper felt compelled to clarify that Venus was the planet nearest to Earth. Venus was in the headlines because the Soviet Union had just sent a seven-ton probe in its direction. The probe had been launched from within a larger satellite, Sputnik VIII, that was orbiting the Earth at a speed of eighteen-thousand miles an hour—a matryoshka-doll approach that showed the growing sophistication of the Soviet program. The purpose of the mission, according to Soviet officials, was to scan the cloud-covered planet for evidence of "flora and fauna, if they exist," but the *Times* discerned that the probe had two additional targets: world opinion and John F. Kennedy. "With this latest shot," the editors wrote, "Premier Khrushchev is in effect flexing his muscles before President Kennedy. We do not believe Mr. Kennedy will be either frightened or awed by this latest Soviet feat; he long ago came to accept Soviet superiority in rocket power as one of the unhappy—though hopefully temporary—facts of modern life."

Kennedy was told the news at Glen Ora, the horse farm he and Jackie were leasing in Middleburg, Virginia. It was a Sunday; he made no public statement. By Monday he was already half a step behind the pace of events. He received a sharply worded letter from Overton Brooks of the House Space Committee. "NASA and the civilian space program badly need a shot in the arm," Brooks wrote. "We cannot impress upon the Executive too strongly the need for urgency." At a hearing on February 15, his committee showed its impatience by grilling Hugh Dryden—now NASA's deputy administrator—about the U.S. rocket program. "We must accept the hard fact," Dryden advised, "that it will be four or five years before we have the booster capacity to perform these difficult missions. . . . We're just going to have to sweat it out."

That evening Kennedy held a press conference, his third in three weeks, broadcast live from the new auditorium at the State Department. The idea of turning presidential press conferences into prime-time television was new and not without risk. Journalists, who might have been thought to welcome the idea, worried in print that JFK was cheapening the currency of his office. Kennedy trusted in his own quick wit, his mastery of the facts, and his directness—in contrast to Eisenhower, who had talked himself into tangles. But tonight Kennedy found himself in Eisenhower's shoes: on the defensive, confronted by news of another Soviet achievement in space, and facing the question of when—or whether—the United States would catch up. Where his predecessor had conveyed indifference, Kennedy offered a severe sort of candor: "It is a matter of great concern. . . . The Soviet Union has been ahead and it is going to be a major task to surpass them."

But Kennedy had not defined this task, and he had already lost the initiative. Only three weeks into the new administration, a fatalism was taking hold in the White House. "We must expect continued embarrassments of the present type for some time," Wiesner warned the president a few days after the launch of Sputnik VIII. Yet Wiesner—who had done more than anyone to identify the problems plaguing the program—was only beginning his effort to dig deeper into every aspect of Project Mercury and to make recommendations.

Johnson, meanwhile, was preoccupied with personnel decisions, and the rest of the White House churned in neutral, waiting for a signal, a plan, even a hint. "We are still pretty much in the dark," a staff member complained in a memo listing unresolved issues: the respective roles of NASA, Johnson's Space Council, Wiesner's PSAC, and the Pentagon; NASA's management troubles and its budget request; the status of Saturn I, the long-delayed, heavy-lift booster; the fate of Mercury and Apollo; and the question of whether the United States intended to go to the moon.

In the vacuum left by Kennedy, military planners stirred. In January, Overton Brooks had publicly warned of a stealth campaign to "downgrade the civilian space agency" and to reassign the man-in-space program—and, for that matter, most of the rest of NASA's portfolio—to the military. The effort was not especially stealthy. The ambiguity of Kennedy's views stirred the hopes of any general or admiral who saw NASA's mission—the peaceful exploration of space—as a dangerous indulgence; this new president, they believed, might be persuaded to see things as they did. To that end, the Air Force enlisted Trevor Gardner—its former assistant secretary, who had advised Kennedy on missile policy—to argue the case for collapsing the distinction between civilian and military activities in space and for loosening the restraints on the latter.

The battle for control recommenced. The first skirmish concerned the Atlas missile. On February 18, just four days after Webb had been sworn in, NASA officials asked him to approve a test of Mercury-Atlas—the launch vehicle that, if it ever proved reliable, would carry an American into orbit. In the two months since an Atlas had last exploded, the engineers had made adjustments, and they were ready to try again. Webb gave his quick assent—a vote of confidence in his new team. The decision drew a vehement protest from the Air Force: another failure, the generals argued, would undercut the missile's credibility as a deterrent. The Atlas had been designed to carry a nuclear payload, not a man. In that capacity—as an ICBM—it had performed well in test firings. The problems emerged when a space capsule was stuck on top. Yet to most observers, this was a distinction without a difference. The Atlas appeared a colossal dud. The Air Force went over Webb's head to the White House, where it had an ally in Wiesner. But JFK let Webb's decision stand—this, too, a vote of confidence, one with high stakes.

On the morning of February 21, after a brief delay, Mercury-Atlas

2 (MA-2) launched from the Cape. About a minute after liftoff, when the booster reached max Q—the point where the force of the atmosphere on the missile was at its peak—and kept going, the sighs of relief in the control center were audible. And then, exultation. The test was almost perfect—"nominal in nearly every respect," as the flight engineers put it. At the press conference that followed, Bob Gilruth actually smiled. Asked whether a man would have survived the flight, he replied yes, he would. Then he said he had an announcement to make.

Shepard's flight aboard Mercury-Redstone 3 was imminent enough that Gilruth felt compelled to say something about his selection. Still, he was not prepared to be fully honest about it. Naming Shepard publicly would expose him to a whirlwind of attention that, NASA managers feared, could disrupt his training and peace of mind. Gilruth, therefore, decided to identify three men—Glenn, Grissom, and Shepard—as finalists, suggesting that the competition was still open. Gilruth and his team had been laboring over their statement for days, a measure of their apprehension. When NASA's Public Information Office pressed him to make an event out of it, a coronation, Gilruth demanded a "routine message with no fanfare." He tacked the announcement onto the end of the Mercury-Atlas press conference, as if it were an afterthought.

"In the Redstone program," Gilruth declared, "we have selected Astronauts Glenn, Grissom, and Shepard to begin concentrated training immediately with the spacecraft programmed for the initial flight and with the personnel who will be involved in the launch, tracking, and recovery operations. The specific pilot who will make each flight will be named just before the flight," he added. The gambit worked. Reporters rushed to file stories on the three candidates—christened by *Life* the "Chosen Three." "Which will go," the magazine wondered, "the cool one, the self-denying one, or the quiet one?" For Gilruth, the morning had been a success from start to finish. "The rest of the day," he recorded in his diary, "was spent in celebration."

Not so for Glenn. He put on the bravest face he could manage. "It's an understatement to say that I'm happy" about the Atlas test, he said. But when the subject turned to Gilruth's statement, Glenn bristled. A reporter asked him what Annie's reaction had been to the news. "I would rather not get into . . . such things as that," Glenn said crisply. The other four astronauts, reduced to also-rans, were even more aggrieved—"devastated," Schirra later wrote. Gilruth's decision had been hard enough to accept. Two of them were so upset that they had threatened to quit the program. But now—worse—their second-tier status had been made public; their personal humiliation was national news. As the three so-called finalists disappeared into a bubble of specialized preflight training, Slayton recalled, "Wally, Scott, Gordo, and I even had to have a press conference . . . to reassure everybody that we weren't depressed or something."

WEBB UNDERSTOOD THAT he, no less than the Mercury-Atlas, had passed a test that day. In a show of strength, he asked the White House that $308 million be added to the NASA budget—30 percent more than Eisenhower's proposal. It was time, Webb said, that the program be "substantially accelerated."

David Bell, Kennedy's budget director, had served under Webb in the Truman administration. Webb might, on that account, have expected better than the peremptory treatment Bell gave his request: he offered $50 million. The discussion was closed, Bell said, until the Space Council had finished its review of the program that fall. Besides, he added, Kennedy was busy with genuinely urgent issues—especially Laos, where the United States was pressing for a negotiated settlement between Soviet-backed guerrillas and the Laotian government. "You may not feel he has the time," Dryden said with prophetic force, "but . . . he is going to have to consider it. Events will force this."

Webb took his case to Kennedy directly. On March 22, he and other NASA officials met with the president in the Cabinet Room. Kennedy sat in his usual place, his back to the Rose Garden, which could be seen through the windows. Around the table, in high-backed leather chairs, were LBJ, Wiesner, and others, including McGeorge Bundy, the national security adviser. Webb had stayed up half the night preparing his presentation. Shrewdly, if a little obviously, he referred to "pioneering on a new frontier" and tailored his pitch to Kennedy's concerns, starting with boosters. He noted that the Atlas produced less than half the thrust of Russia's booster. Saturn I would be more powerful than the Semyorka, but as Webb noted pointedly, Bell had rejected NASA's request to step up its development.

The president listened closely, tapping his front teeth with a pencil. When Webb was through, Kennedy led the group in discussion—asking questions, drawing out conflicting opinions. Then he raised an issue that Webb had not: the first manned flight was only weeks away if the launch schedule held. The United States, Kennedy said, had much to gain from a successful flight—and much to lose if it failed. In that light, he asked whether the event should be closed to the press. NASA's view, as a spokesman had stated it, was that the openness and candor of the U.S. program revealed "the difference between a civilization that is sure and proud of its strength and a dictatorship whose insecurity must be protected by secrecy." Except that Kennedy was unsure of America's strength in this arena.

After the meeting, Bell quickly sent Kennedy a memo casting doubt, even scorn, on Webb's arguments. The Soviets, Bell believed, were almost certain to beat America to every important milestone in space. And what did it matter? Since Sputnik, Bell argued, space exploration had offered diminishing returns in terms of prestige; those returns would diminish further "as the novelty wears off." The idea of investing so much of the federal budget in manned spaceflight struck Bell as wasteful. By redirecting that money to other priorities, he concluded, the administration could show "that we are more concerned with the problems of men on earth . . . than we are with placing men on the moon."

The next day, March 23, Kennedy called a meeting to settle the matter. Bell, LBJ, and Wiesner were present, as was Edward Welsh, just named executive secretary of the Space Council; no NASA officials had been invited. Wiesner suggested a compromise: fund Saturn and other booster programs now, but defer decisions on Apollo until Johnson's council had finished its review. This suited Kennedy. He approved an increase of $126 million. For NASA, this was less a victory than a reprieve. Having weighed the agency's request, Kennedy revealed his doubts about the program's goals—and its ability to reach them. Given a chance to repudiate the Eisenhower budget, he had, in effect, ratified it: $126 million was only 10 percent more than Eisenhower allotted. It was just enough that Webb could claim, as he did a few days later, that the administration had resolved to close the gap in booster capacity. Yet the larger question—of what payload these boosters were meant to carry—had again been deferred.

NASA's ambitions had not been curbed. Officials continued to look beyond Mercury to exploring the moon. Of course, this depended on a large, long-term budget increase—a commitment for the duration. But by the end of March 1961, President Kennedy had made it plain: no such thing could be safely assumed.

CHAPTER 9

The First Man

JOHN KENNEDY ENTERED the ballpark almost without notice: no grand announcement, no "Hail to the Chief." April 10, 1961, was opening day for the Washington Senators, and Congress had adjourned at noon so members could attend the game; much of the cabinet and the White House senior staff, along with Vice President Johnson, trailed Kennedy into the stands. If the crowd was muted, the weather, no doubt, had something to do with it. Though the rain had stopped, the air remained damp and cold, and the outfield was sodden. At the designated time, 1:25 p.m., Kennedy removed his topcoat and threw the first pitch: "a beaut," one sportswriter called it, but an errant one; the ball cleared the heads of the players who stood along the first base line and bounced off two gloves before being caught, barehanded, by Jim Rivera of the Chicago White Sox, the opposing team. The president took the liberty of throwing a second pitch. This one went straight to Hal Woodeshick, Washington's left-handed pitcher.

At the end of the second inning, JFK's associate press secretary, Andrew Hatcher, made his way to the president's row. Kennedy straightened a bit in his seat as Hatcher leaned over to whisper in his ear. United Press International, Hatcher said, was about to report that a Russian had just become the first man in space and had landed safely after orbiting the Earth. Kennedy squinted and said nothing. U.S. intelligence had been telling the White House for weeks that a manned mission was imminent. Rumors to the same effect had been circling in Moscow, London, and elsewhere. Toward the end of the game, Hatcher handed Kennedy a typewritten update: "Elaborate Russian plans to make this anticipated announcement have been abandoned for today. However, all Soviet TV and radio networks have been put on a 24 hour alert. CIA at 3:45 p.m. could not confirm or deny the report," and UPI, for now, would hold its story.

Other news outlets showed less restraint. The next morning, April 11, above a big photo of Kennedy hurling the baseball (and Johnson hooting approvingly), the *Washington Post* ran a banner headline: REDS POSE SPACE MAN MYSTERY. The paper described "a day of feverish activity" in Moscow—excitement fueled, in part, by a Russian taxi driver, who had passed along some gossip to one of his passengers, a Reuters correspondent. The story built momentum as it traveled: the cosmonaut had orbited one time, he had orbited five times. A Soviet camera crew set up lights and cables outside Moscow's Central Telegraph Office and kept a vigil, waiting for the historic news to be confirmed. The hours passed without word.

Back in Washington, at 10:00 that morning, Overton Brooks opened a hearing to consider NASA's budget and its plans for unmanned lunar exploration. Congressmen sat—for a time, patiently—through a presentation about the gamma-ray spectrometers and single-axis spectrometers that the space agency might, someday, land on the moon's surface to assess its "magnetic permeability, electrical resistivity," and the like. Before long, Brooks broke in: "Have we got anything this morning . . . on the Russian man in space?" Hugh Dryden, NASA's second-in-command, was quick to allay concerns. "As near as we can tell," he said, "it is purely rumor based on conversations with Russian taxicab drivers who had conversations with somebody else."

"It made a good story," a congressman said.

"Yes," Dryden replied, and nudged the discussion back toward NASA's own efforts, particularly its assessment of the biological effects of spaceflight. He provided a magnified photo of the brain of a mouse—an unfortunate mouse that had been sent, by balloon, 130,000 feet into the stratosphere and brought back down to be dissected and examined for signs of cosmic radiation. The photograph, a scientist told the committee, was being reviewed by the Armed Forces Institute of Pathology. This did not placate the congressmen. A mouse, it appeared, was the wrong subject to introduce on a day when a man, a Soviet man, was said—rightly or not—to have orbited the Earth. George Low, NASA's chief of manned spaceflight, would not estimate when the United States might be ready to send a man into space; more tests, he said, were needed first. When Low used the word "success" to describe the flight of Ham, which had splashed down 132 miles off target, George Miller, a California Democrat, had reached his limit. "I think you are trying to fool us," he snapped.

By late afternoon the CIA had determined that the Soviet flight was, in fact, hours away. Pierre Salinger, the White House press secretary, called in a few press aides from the Pentagon and the State Department to help draft a pair of presidential statements: one to use if the launch succeeded, the other if it ended in tragedy. Tragedy required a deft touch. It also—distasteful as it might be to admit it—presented opportunities. Such was the counsel of Edward R. Murrow, the esteemed news broadcaster who now ran the U.S. Information Agency. Murrow advised Mac Bundy that in the event of a cosmonaut's death, the United States should express sorrow "with all the sincerity we can muster," while, at the same time, covertly encouraging "commentators in other countries to deplore the low regard for human life which prompted the Soviets to attempt a manned shot 'prematurely.'"

Salinger put his feet up on his desk and lit a cigar. The drafts were not coming easily. Wiesner wanted to downplay the Soviet achievement; others thought that churlish. Excusing himself from the discussion, Wiesner went to the Oval Office at 6:20 p.m. to tell Kennedy that the flight would almost certainly take place that night. A few minutes later Salinger walked in, draft statements in hand. The president read them in silence. He said they would do, adding that the United States should certainly congratulate the Russians. He moved on to other matters.

At 8:00 p.m. Kennedy returned to the White House residence for the night. Major General Chester Clifton Jr., the military aide who delivered Kennedy's daily intelligence briefings, asked whether he wanted to be awakened if the launch took place. "Give me the news in the morning," Kennedy said. He would reckon with it then.

A WEEK EARLIER, on April 5, the Soviet director of cosmonaut training, Lieutenant General Nikolai Kamanin, had written the names of two pilots in his diary: Yuri Gagarin and Gherman Titov. Soon one would be chosen to make the first flight into space; the other would serve as his backup. "So, who will it be . . . ?" Kamanin wondered. "It is difficult to decide who to send to certain death."

Not all the Soviet planners and "designers," as they were known, shared his pessimism. But they understood that the booster that one of the cosmonauts would ride had a terrible record: out of sixteen tests of the launch vehicle—the conjoined rocket and spacecraft—half had failed. The failures were covered up, maintaining the veneer of Soviet infallibility; the program operated under such strict secrecy that the identity of its chief designer, Sergei Korolev, was unknown to the public. But the engineers felt the pressure. The previous month, a cosmonaut had been consumed by fire in an oxygen-rich isolation chamber; he died of shock from his burns. News of that, too, was suppressed (for the next quarter century); but the specter hung over the program as it prepared, with single-minded intensity, its bid to make history.

The choice was Gagarin. He had a simple, unaffected charm—the right sort of man, the designers believed, to represent the Soviet Union before the world. In 1961 he was only twenty-seven, born seven years after Gordon Cooper, the youngest astronaut. His performance as a pilot would not, had he been an American, have put him in the ranks of the Mercury Seven. But he had a quiet resilience—the product, perhaps, of his childhood, when the Nazis occupied his farming village and put him to work as a laborer. The Russians shelled the area with some frequency to drive the Germans out; "we lived," he recalled, "in fire and smoke." The crash-landing of a Soviet fighter plane nearby gave Gagarin—who brought food to the pilot and was rewarded with a look inside the cockpit—a goal, a direction in life. By April 1959, when NASA named its astronauts, Gagarin was a lieutenant in the Soviet Air Force, flying reconnaissance missions over Murmansk, in the Arctic Circle. Later that year, he was recruited for the role of cosmonaut.

At 5:30 a.m., Moscow time, on April 12, 1961, he was awakened in his cottage near the launchpad in Tyuratam, Kazakhstan, site of the Soviet space complex. He was greeted by a woman who had once lived in the house; she brought a gift of wildflowers and said that her son, also a pilot, had died in the war. Gagarin ate breakfast, passed a medical examination, and was helped into his pale blue space suit and orange coveralls. At the last minute, out of a concern that he might land in a foreign country, the letters "CCCP" were hastily painted on his helmet in gold. Once he had been strapped into the spacecraft—*Vostok,* it was named, meaning "east"—he whistled and sang softly to himself before launch. Korolev, at the flight control center, took a tranquilizer; it did not stop him from shaking. Nearly two thousand miles away, at a dacha on the Black Sea, Nikita Khrushchev was pacing, impatient for any news.

"Off we go!" Gagarin cheered as the rocket lifted off. It was 9:07 a.m. in Moscow—1:07 a.m. in Washington. An hour and forty-eight minutes later, after a single orbit around the Earth, Gagarin ejected from his spacecraft, as planned, and landed by parachute, more or less softly, in a field near the village of Smelovka, five hundred miles southeast of Moscow. Climbing a small hill, he saw a woman approaching, trailed by her granddaughter. At the sight of the cosmonaut in his orange suit, the girl turned and ran. "I'm a friend," he shouted, waving his hands. "I'm Soviet!"

And that, in a sense, was his message to the world. Radio Moscow, announcing the news, put it just as plainly: "The first man to penetrate outer space is a Soviet man." This was not, for the Soviets, a triumph of humanity, or even of technology; it was the victory of one system over another. The achievement, the Politburo announced, "embodied the genius of the Soviet people and the powerful force of socialism." Khrushchev drove the point home: "Let the whole world see what our country is capable of. Let the imperialist countries try to catch up."

The United States had tracked the flight almost since its launch. An electronic intelligence station in Alaska had intercepted its television transmissions, revealing a human being inside it. At 1:35 a.m., Wiesner called Salinger and told him that the Soviet shot had gone up. They agreed to say nothing publicly until the Soviets had confirmed the news. Salinger went back to bed, but not for long. A reporter called at 2:00 a.m. to tell him Moscow radio had announced that the country had a man in space. Salinger declined to make a statement. He had no peace after that; his phone rang on and off through what remained of the night. The White House, for now, had no comment.

Phones were ringing at Langley as well. At 3:00 a.m., Shorty Powers, NASA's spokesman, was jarred awake and reached for the receiver at his bedside. A reporter from UPI wanted to know if Powers was aware that the Russians had put a man in space. Powers wasn't. Well, the reporter said, the Russians had done it, so he needed comments from Powers, Gilruth, and each of the seven astronauts—within the next twenty minutes. "It's three o'clock in the morning, you jerk," Powers shouted. "If you're wanting something from us, the answer is we're all asleep." He slammed down the phone. UPI put this on the wires. For the next several hours, while Salinger maintained his silence, Powers's outburst would serve as the official reaction of the U.S. government.

President Kennedy was already awake when his valet, George Thomas, knocked on his bedroom door just before 8:00 a.m. Thomas told him that Salinger was hoping to speak with him by phone. "Do we have any details?" Kennedy asked Salinger. Just Gagarin's name and the time of his flight, the press secretary replied. He read aloud the statement Kennedy had tentatively approved the night before:

> The achievement by the USSR of orbiting a man and returning him safely to ground is an outstanding technical accomplishment. We congratulate the Soviet scientists and engineers who made this feat possible. The exploration of the solar system is an ambition that we and all mankind share with the Soviet Union and this is an important step. . . . Our own Mercury man-in-space program is directed toward that same end.

Kennedy authorized Salinger to issue the statement, along with a short message congratulating Khrushchev and expressing a "sincere desire that in the continuing quest for knowledge of outer space our nations can work together to obtain the greatest benefit to mankind." Murrow's memo on expressing regret in the event of a disaster had been rendered moot. Mac Bundy added a postscript before relegating it to a file drawer: DEAD LETTER—LIVE SPACEMAN.

When Kennedy arrived at the Oval Office, just before 9:30 a.m., he faced two more reminders—if any were needed—that the United States was unlikely to catch up to the Russians anytime soon. First, he spent ten minutes with James Webb and Senator Kerr. The meeting had been put on the schedule before the Soviets had sent their man into space, and now it proceeded as if that hadn't occurred. Webb, who had gone on television that morning to project an air of unconcern, showed the president a model of the Mercury spacecraft. It failed to impress. Kennedy said it looked like something Webb had picked up at a toy store on his way to the White House.

On the president's desk was his second reminder: the report from his Ad Hoc Mercury Panel. Like the rest of the space program, the report was late—held up while the panel's members had debated and reworked their conclusions. "Project Mercury," the report began, "has reached a stage where manned suborbital flight is being planned within months and manned orbital flight within a year." On a day that a Russian had entered the "untrespassed sanctity of space," in the words of a World War II–era poem, this now read like an instrument of surrender.

The panel was impressed, they declared, by the manned program's progress and had come to accept the idea that man had a role to play in space exploration. Yet as endorsements went, this was less than absolute. The "uncertainties," the biomedical experts argued, "are awesome," and the mortal dangers were many: brain seizures, blackouts, respiratory collapse, heart attacks, shock—a catalog of horrors, provided here for Kennedy's review. A forty-four-page addendum, twice as long as the report itself, sounded further alarms, casting doubt on the idea that any man was fit for orbital flight. "It is the main objective of test pilot selection and training to separate 'tigers' from 'rabbits,'" the panel wrote. "Yet, it is recognized that every 'tiger' has his limits above which he converts into a 'rabbit.'"

The wires that morning carried a rejoinder: *Izvestia*'s description of the returning cosmonaut as "stocky and happily smiling," his eyes "shining as though still reflecting spatial starlight," his arms hugging a friend "so vigorously that they seemed to be wrestling." The account was, most likely, embellished. But the fact remained: Gagarin was alive and manifestly no rabbit.

"The Russian accomplishment was a great one," Glenn told the AP, reading a prepared statement over the phone. "I am naturally disappointed that we did not make the first flight to open this new era." The adjective—*disappointed*—was accurate enough, but it did not begin to capture the feeling at Langley, at the Cape, anywhere NASA employees or contractors were giving their all to the enterprise of putting an American in space. The troubles with the rockets, the calls for more animal tests, the debates between von Braun's team and the Space Task Group—all this, as everyone at NASA understood, had slowed the pace of the program, yet the news that the Soviets had actually done it, and done it first, delivered a concussive shock.

The astronauts, in fact, were furious. That night, a few of them had dinner with William Shelton of *Time*. After a few drinks, they got to talking. (It was understood that Shelton would never break their trust by quoting them. "They would get in trouble and it would sour our delicate relations with them," Shelton wrote his editor.) Slayton railed for a while about "those people who fell in love with the black boxes," the engineers and space officials who wasted time arguing for unmanned, computerized shots. "All of us," Cooper complained, "were longing for somebody in authority to say to us, 'O.K., boys, let's go. Let's get this program in high gear. Let's get there first. We're going to beat 'em. But nobody ever said that—or acted as if they thought that way." Eisenhower and Kennedy, added Cooper, denied that the United States was in a race. "What I want to know is why *shouldn't* we be in a race? . . . Morale," he sighed, "just dropped clear to the bottom." It could not have surprised them when Eisenhower, a few days later, shrugged to reporters and said, "It is not necessary to be first in everything."

Shepard in particular could not contain his rage. Not only had he lost his shot at immortality, but the Russians had managed an orbital flight, something the United States would not be able to accomplish until the Atlas had been man-rated. If Gagarin was a space-age Lindbergh, the first to cross the Atlantic, Shepard would be turning back at Greenland. The indignation was widely shared. Most Americans believed that to be second in space—as Kennedy had argued in 1960—was to be second in science and technology, second in military strength, second in the esteem of the world, second in the existential struggle between freedom and totalitarian rule. Although the country had been conditioned to expect this news, it was nevertheless cause for anger, self-castigation, "a loser's deep frustration," as *Life* described it. The space race was being lost for "lack of imagination, lack of initiative," said Edward Teller.

At 10:00 a.m., Overton Brooks began a second day of hearings. "Since we adjourned yesterday," he said, "much has occurred in the way of developments in space." Edward Welsh of the Space Council was asked to comment on the Soviet achievement. "I think we would be wrong to minimize it," he replied. "I think we would also be wrong to overstate or exaggerate its importance." This neat bit of evasion only incited the committee. "I am darn well tired of coming in second," thundered James Fulton, a Republican from Pennsylvania. The United States could have beaten Russia to space, Fulton added, if NASA engineers hadn't been in the habit of "knocking off at 5 o'clock." Victor Anfuso, a Brooklyn Democrat, put it more strongly: "Unless we do something very imaginative, Mr. Welsh, America is lost. America," Anfuso added, "is too young to die."

All the old fears received a fresh airing: manned space platforms; bombs in orbit; space-based nuclear blackmail. Emerging from closed-door meetings with military leaders, Senator Richard Russell of Georgia raised anew the specter of a Soviet missile base on the moon. Experts debated whether a missile fired from space would be any more accurate than a missile fired from a silo on Earth, but few doubted that Russia would test the proposition. At the Pentagon that morning, Air Force officials issued a call to step up deployment of MIDAS, the Missile Defense Alarm System of satellites designed to detect a Soviet launch. But MIDAS, to date, had been an expensive failure, and hours later came news of another: the Air Force had launched a ninety-pound capsule into space to collect radioactive particles, evidence of Soviet nuclear tests—then lost the device, as it had so many others, in the Atlantic.

BERLIN, NOT SPACE, dominated Kennedy's day. Just before 10:00 a.m., he welcomed Konrad Adenauer, the chancellor of West Germany, to the Oval Office to begin a series of meetings about pressing issues, including the Soviet threat to seize West Berlin. It could not have escaped either man's notice that in East Germany, Walter Ulbricht, the head of state, was crowing about Gagarin's flight. It was a day of exultation across the Communist world—in Beijing, in Havana, especially in Moscow. Thousands paraded through Red Square, chanting Gagarin's name—the largest spontaneous celebration since the end of World War

II. Far outside the city, factory workers and farmers staged spontaneous rallies. *Izvestia* ran a headline in red: GREAT VICTORY, OUR COUNTRY, OUR SCIENCE, OUR TECHNIQUE, OUR MEN.

If the White House was looking for consolation, it found little in Western Europe, where editorial opinion was heavy with scorn. In London, the *Evening Standard* called the U.S. effort inefficient, wasteful, and self-defeating—comments echoed across the continent. Observers anticipated a harder line from the Soviets in the test ban talks in Geneva, in cease-fire negotiations in Laos, in the mounting conflict in Vietnam. The Russian message to "the weaker members of the free world," the *New York Times* believed, was blunt: "join the bandwagon of Moscow—or at least get off the Western bandwagon—before it is too late."

JUST BEFORE 4:00 P.M. the president arrived at the State Department for a press conference. This had been a forum for the Kennedy charm, but today he shared in the general gloom. The first question concerned Cuba, not the cosmonaut. On April 7, the *New York Times* had published a front-page story describing America's covert training of anti-Castro Cuban commandos at camps in Florida, Louisiana, and Guatemala; CBS had reported that an invasion was imminent. "Has a decision been reached," a correspondent began, "on how far this country will be willing to go in helping an anti-Castro uprising or invasion of Cuba?" Kennedy responded that "there will not be, under any conditions, an intervention in Cuba by the United States armed forces"—a less than categorical denial.

Then the inevitable question: "Could you give us your views, sir, about the Soviet achievement of putting a man in orbit and what it would mean to our space program?" Kennedy stuck, at first, to the language of his press release: the flight was an "impressive scientific accomplishment"; he had sent congratulations to Khrushchev and "the man who was involved." Kennedy added, matter-of-factly, that it had been expected all along that the Soviets would be first. As for the United States, "we are carrying out our program and we expect to"—he caught himself—"*hope to* make progress in this area this year ourselves."

This failed to put the matter to rest. Could the United States catch up? reporters demanded. Did Kennedy share the frustration of the member of Congress who had said he was tired of second place? "The Soviet Union," Kennedy replied, "gained an important advantage by securing these large boosters . . . , and that advantage is going to be with them for some time. However tired anybody may be, and no one is more tired than I am, it is a fact that it is going to take some time and I think we have to recognize it. . . . The news will be worse before it is better."

Kennedy professed confidence without doing much to project it. His answers were as crisp as ever; his command of the issues was firm; but as he fielded questions about a tangle of problems, from unemployment at home to the continued fighting in Laos, he seemed beset and already, less than three months into his term, a bit weary. Whether these problems would yield to reason and determination or the United States itself would yield to the tides of history was an open question. "The Communists seem to be putting us on the defensive on a number of fronts—now, again, in space," a correspondent asked the president. "Do you think that there is a danger that their system is going to prove more durable than ours?"

"Well," Kennedy replied,

> I think that we are in a period of long drawn-out tests to see which system is, I think, the more durable. . . . A dictatorship enjoys

> advantages in this kind of competition over a short period by its ability to mobilize its resources for a specific purpose. We have made some exceptional scientific advances in the last decade, and some of them—they are not as spectacular as the man-in-space, or as the first Sputnik, but they are important. I have said that I thought that if we could ever competitively, at a cheap rate, get fresh water from salt water, that it would be in the long-range interests of humanity which would really dwarf any other scientific accomplishments. . . .
>
> I do not regard the first man in space as a sign of the weakening of the free world. But I do regard the total mobilization of men and things for the service of the Communist bloc over the last years as a source of great danger to us. And I would say we are going to have to live with that danger and hazard through much of the rest of this century. . . . But in the long run I think our system suits the qualities and aspirations of people that desire to be their own master. . . . Our job is to maintain our strength until our great qualities can be brought more effectively to bear.

The press conference was panned. Kennedy's passivity—his apparent willingness to accept that the United States was behind in the space race—struck many as dispiriting. That same day, when a leading Russian scientist declared that the race to the moon was "half won," who could doubt him? The feeling was underscored, in sinister fashion, by an ABC News program that night. "This," announced the host, Jules Bergman, "is what the Russians have in mind for us—a film preview of the future from Soviet science." Its animation was crude, and its soundtrack, an unearthly wail, could have been borrowed from *Destination Moon* or any other sci-fi B-movie. "The Soviet scientists who designed this project describe it as fully possible," Bergman said, as a cartoon rocket lowered itself onto the moon and its fuselage unfolded—gently, as flowers do—to reveal a lunar tank.

"Ten days ago," Bergman continued, "I asked Lieutenant Colonel John Glenn—who may well be our first astronaut into space—how he would feel if the Russians beat us into space." The film cut to footage of Glenn in his space suit, standing in front of the Mercury capsule, his helmet under his arm. "Well," Glenn told Bergman, "there's only one answer to that: it doesn't change our program one bit. This is like saying because Henry Ford started a new car first, no one else should be in the automobile business today; General Motors should have dropped out before they started. Well, this is ridiculous, of course; it's probably a ridiculous example. But the fact that the Russians happen to get a shot off or may not get a shot off a little bit before we do does not alter the objectives of our program a bit."

"We're not going to change our plans," Bergman said.

Glenn smiled, cocked his head, thought for a second. "No."

THIS WAS THE OFFICIAL POSITION. Webb had stated it himself, and now, the next morning, had to say it again, this time to the House Space Committee. The United States, he insisted, was "sparing no effort." The trouble was that he didn't believe it. Webb had come to Capitol Hill to laud a program he knew was falling short and to defend a budget request he considered too meager. Webb was so hamstrung that he couldn't even say "space race"; in keeping with Eisenhower's example, Kennedy avoided the phrase, as if fuzzier ones, such as "broad competition," could take the edge off the contest.

Whatever one called it, the United States was losing; and Congress, as Webb reported to the White House, was "in a mood to try to find someone responsible." To Webb, the hearings now had the feel of a full-blown investigation. James Fulton badgered and scolded and contradicted his witness so persistently that Webb, finally, lost his cool and snapped, "Do you want me to delegate to you my responsibility for ordering the first shot?" That lapse aside, Webb never lost sight of the fact that Congress, not the White House, had the last word on his budget—and seemed inclined to increase it. "I want to see the country mobilized to a wartime basis," said Victor Anfuso. "Because we are at war." Webb put it differently. "What we are doing here in the United States," he concluded, "is having a good hard look at how a great nation faces a great problem."

While Washington churned, Moscow, under sunny skies, was a scene of jubilation: crowds waved placards bearing portraits of Gagarin. A twenty-two-foot model rocket stood in the middle of Red Square; across from Lenin's Mausoleum a massive crimson banner proclaimed, FORWARD TO THE VICTORY OF COMMUNISM! In 1956, Khrushchev had denounced cults of personality, but today, with his blessing, Gagarin's seemed firmly established. The Soviets had already named a mineral after him (gagarinite) and a twelve-thousand-foot peak (Mount Gagarin).

And now the man in question, behind the controls of a fighter jet, appeared above Moscow, escorted by a contingent of planes; and soon he was marching, alone, down a long red carpet. Khrushchev, bearing flowers as a suitor might, wiped away a tear, theatrically. He beckoned the cosmonaut to join the lineup of leaders atop Lenin's Tomb. Later there were fireworks, followed by a gala reception for Gagarin, at which Khrushchev boasted to the assembled diplomats and apparatchiks that a country regarded by the West as "barbaric," a country whose people had not long ago gone "barefoot and without clothes," had conquered the heavens. "That's what you've done, Yurka!" he shouted, using the diminutive for "Yuri." "Let everyone who's sharpening their claws against us know, let them know that Yurka was in space, that he saw and knows everything."

In Congress, in the press, impatience was growing. Only two days after the flight of *Vostok,* the posture adopted up and down the chain of command—that nothing could or should be done to accelerate Mercury—was already proving untenable. Sensing the shift, Kennedy told his advisers to gather that evening in the Cabinet Room. He also asked Hugh Sidey of *Time* to join. Sidey was already scheduled to interview JFK about space; the invitation to the meeting was now sprung on him at the last minute, as if unpremeditated, a reflection of trust in him and not, in fact, what it was: a contrivance, an effort by Kennedy to be caught in the act of grappling with (and therefore not neglecting) the issue.

Webb and Dryden arrived early to huddle with a group of advisers. Ed Welsh tried to rouse the men from their malaise. Speaking, he stressed, for the vice president, who was overseas, Welsh admonished them to stop thinking that the United States couldn't catch up. It could, he argued—if the nation made an all-out effort, if the president lifted the restrictions on working overtime at NASA, if he demanded more money from Congress. Sorensen, picking up a point Wiesner had made in a memo, asked if America might conceivably beat the Soviets to the moon. Webb and Dryden said yes: the timetable was long enough, and the undertaking complex enough, that the Soviet advantage could be overcome.

The sun had set before Sidey arrived; the West Wing grew quiet as phones stopped ringing and secretaries went home for the night. Sorensen led Sidey past the door to the Oval Office and into the Cabinet Room, where the group sat solemnly around the long table. As Kennedy entered, the others stood to greet him. He eased into a chair, put his foot against the edge of the table, and tilted his seat back as the others leaned forward. "Now," he said to Sidey, "what questions do you have about our space program?" But answers were lacking. Everyone was equivocal. Kennedy himself seemed severely on edge—rocking back and forth, running his fingers through his hair, fiddling with a piece of the sole of his shoe. "We may never catch up," he grumbled under his breath. He looked to all appearances like a man in a bind.

"What can we do?" he demanded. "Can we go around the moon before them? Can we put a man on the moon before them?" Dryden responded that a moon landing was possible, but even at $40 billion—the estimated price tag—there were no guarantees of success. "The cost," Kennedy said. "That's what gets me." Wiesner, who considered the project a colossal waste of energy and money, slumped so low in his chair that his head barely cleared the surface of the table. Pulling on his pipe, he muttered darkly, "Now is not the time to make mistakes," but he didn't specify which ones he meant.

Every option looked like a mistake, every direction a dead end. Kennedy scanned the faces of the men in the room, then brought the meeting to a close. "When we know more, I can decide if it's worth it or not," he said. "If somebody can just tell me how to catch up. Let's find somebody—anybody. I don't care if it's the janitor over there, if he knows how." As he stood up, the others jumped to their feet. He looked around once more. "There's nothing more important," he said quietly, and walked out.

THIS LAST COMMENT WAS a pretense. Kennedy had never seen space as the most important thing; and events kept pressing to ensure that he wouldn't. On April 14, a brigade of fourteen hundred Cuban exiles boarded a flotilla in Nicaragua and set off for the Bay of Pigs, on Cuba's southern coast. This was the motley collection of soldiers, students, businessmen, farmers, and fishermen that the United States had been training since Eisenhower had set the plan in motion a year earlier. "You will be so strong," the chief U.S. adviser told them as they prepared to embark, "you will be getting so many people to your side, that you won't want to wait for us. You will go straight . . . into Havana"—as succinct a statement of U.S. self-delusion as any uttered on the eve of the invasion.

His overconfidence was shared by the CIA, and by the Joint Chiefs of Staff, and, apparently, by all the many advisers who had sold the idea to JFK and stiffened his spine when he wavered. Though he had doubts about the plan, he subscribed to its premise: that Castro was a cancer that must be excised from the hemisphere. He also accepted, without serious scrutiny, analysts' assurances that the "freedom fighters" (a term he himself had used) would spark an uprising without active U.S. intervention. It was "not a mere miscalculation but an absurdity," recalled Richard Goodwin, one of the few skeptics on the White House staff. This was made plain. The rebels were outnumbered, outgunned, and quickly cut down. On April 19, the morning after a tense, late-night meeting in which Kennedy rejected appeals by the CIA and the Joint Chiefs to launch a full-scale invasion, Salinger found him alone in his bedroom, in tears. "He was aghast at his own stupidity," Sorensen observed, and "angry at having been badly advised."

The shock and dismay registered widely. In Western Europe, Kennedy's election had been mostly welcomed; the new president was expected to be a reliable steward of the alliance and its values. Now he looked rash and fumbling and weak. His aide Arthur Schlesinger Jr. arrived in Europe that week for a conference and found the mood there unforgiving. The *Frankfurter Neue Presse* declared JFK "politically and morally defeated." The international press—and Communist propagandists—also took note of the police dogs that had been set upon peaceful civil rights demonstrators (among them, Medgar Evers, a field representative for the National Association for the Advancement of Colored People) in Jackson, Mississippi; of reports that a diplomat from Sierra Leone had been denied service at a restaurant in Hagerstown, Maryland, on the basis of his race; of the fact that in New Rochelle, New York, the home of Ghana's delegate to the U.N. had been defaced with racist epithets. The Soviet triumph in space, U.S. ineptitude in Cuba, and America's ongoing history of racial violence and hatred dealt a triple blow to the country's prestige.

From that point forward Kennedy would show less deference to the experts and rely more and more on his brother Robert to manage the foreign policy machinery. The debacle had implications for the manned space program, too, though it was not yet clear what they would be. On the one hand, the events of April heightened Kennedy's anxieties about Mercury-Redstone 3, which Shepard was preparing to fly. "Things had gone wrong on the Bay of Pigs and officials were worried. . . . They didn't want to get burned twice," Ed Welsh recalled. One of the members of the Ad Hoc Mercury Panel thought it best, on balance, to stick with the launch schedule but warned Sorensen that the chance of disaster was as high as one in ten. Some at NASA were even more pessimistic. "I shudder to think of that shot," John Hagen, who had run Project Vanguard, told a reporter from *Time*. "The allegedly reliable Redstone isn't reliable at all. It is going to be very iffy."

A successful flight, on the other hand, could shift the conversation—toward something exciting, something ennobling, an outlet for the energy that Kennedy had sparked in 1960 but that seemed to have weakened. White House advisers now sensed, in the president, a greater resolve—not clarity, yet, but resolve. On the morning of April 20, just before a cabinet meeting, Kennedy walked the grounds of the South Lawn with Lyndon Johnson, who asked him to issue a directive: to press all relevant parties, across the government, toward a decision point. Johnson would lead the review and shape the recommendations.

Sorensen drafted the memo that day and Kennedy signed it immediately. It instructed Johnson to review the space program and answer a series of questions: "Do we have a chance of beating the Soviets by putting a laboratory in space, or by a trip around the moon, or by a rocket to land on the moon, or by a rocket to go to the moon and back with a man. Is there any other space program which promises dramatic results in which we can win?" Kennedy asked LBJ to assess the cost of each option. He also wanted to know whether the country was "working 24 hours a day on existing programs" and "if not, why not? . . . Are we making maximum effort? Are we achieving necessary results?" He asked LBJ to report back "at the earliest possible moment."

The next morning Kennedy held a press conference. Some of the questions concerned Cuba, of course, but the furor had already begun to subside; Kennedy had been forthright in accepting blame for the Bay of Pigs, and he took this opportunity to underscore it again. "I'm the responsible officer of the government," he said. The reporters moved on, for the most part. One of them drew Kennedy into an extended back-and-forth about space. "Mr. President," he said, "you don't seem to be pushing the space program nearly as energetically now as you suggested during the campaign. . . . In view of the feeling of many people in this country that we must do everything we can to catch up with the Russians as soon as possible, do you anticipate applying any sort of crash program, or doing anything that would—" Here Kennedy cut in. Calmly, but jabbing his finger for emphasis, he delivered a remarkably candid assessment of the decision he faced.

"We have to consider whether there is any program now, regardless of its cost, which offers us hope of being pioneers in a project," Kennedy said. "It is possible to spend billions of dollars in this project in space to the detriment of other programs and still not be successful. . . . I don't want to start spending the kind of money that I am talking about without making a determination based on careful scientific judgment as to whether a real success can be achieved, or whether because we are so far behind now in this particular race we are going to be second in this decade." These questions, he said, were being considered by the vice president.

"Mr. President," the reporter interjected, "don't you agree that we should try to get to the moon before the Russians, if we can?"

"If we can get to the moon before the Russians, we should," Kennedy replied.

He had never said as much before. Not even during the campaign, an occasion for bold vision and big promises, had he set the moon as a goal. Now he had been backed into it on live television. Asked a yes-or-no question, Kennedy understood that no would have been a cry of surrender. Still, his yes was hedged: he favored going to the moon, but only "if we can" get there "before the Russians" do. This was not resolute enough for the reporter, who kept pressing. "Isn't it your responsibility," he said, "to apply the vigorous leadership to spark up this program?" But Kennedy had reached, perhaps exceeded, his limit. "When you say 'spark up the program,'" he snapped, "we first have to make a judgment based on the best information we can get whether we can be ahead of the Russians to the moon."

With that, the exchange ended. But behind closed doors in the White House, the Pentagon, and NASA headquarters, the search for a "program which promises dramatic results in which we can win," as Kennedy had put it, began now in earnest.

CHAPTER 10

Go or No Go

Glenn, Grissom, and Shepard—the First Team, as *Life* labeled them—spent most of the spring of 1961 living out of suitcases in a Holiday Inn in Cocoa Beach. They had chosen the motel over their crew quarters, which were clean and quiet but cramped: two wooden-frame bunk beds, a tiny desk, a television, and a surfeit of beige-colored furniture had been stuffed into a small room, S205, on the second-floor mezzanine of Hangar S at Cape Canaveral. Fluorescent lighting gave it the feel of a business park, which, in effect, it was. Next to the bedroom was a conference room, and next to that, the room where the astronauts were fitted into their pressure suits, and then, farther down the hall, the aeromedical room, full of "probing instruments," as Shepard put it. The motel, by contrast, had a pool and a bar, and it was close to other motels with pools and bars. It was also near the ocean, a hundred-foot walk through patches of beach grass. The Holiday Inn was their refuge, and they savored it.

The flight of Mercury-Redstone 3 had been delayed for so long—six months beyond what NASA had forecast—that military psychologists were sounding alarms about "overtraining," a deadening of the astronaut's instincts. But this, in part, was by design: the relentless rehearsals of every step of the flight plan and every conceivable contingency were meant to render the whole thing rote, to eliminate, as fully as possible, any element of surprise. For all involved—from the ground controllers to the astronaut himself—the launch was meant to feel like just another run on the trainer. Blockhouse communicators uttered, again and again, the phrases they would speak on the big day; the recovery ships and helicopters and frogmen and pararescue jumpers conducted maneuvers; medical attendants fitted each astronaut with sensors and the detested rectal thermometer and checked his vitals as he "flew" the simulator. Another team put the astronauts in the transfer van and practiced the ride that one of them would take from the hangar to the pad.

For Glenn, as the rehearsals grew closer, in every exacting detail, to the actual event, the exercise retained an air of unreality. Alone among the three, he was stuck in limbo. As the backup to both Shepard and Grissom, he had no guarantee that he would be the third or even the fourth or fifth American in space. Yet at the same time, if Shepard came down with a cold or twisted his ankle, Glenn would become the first American in space. He had to keep pace, therefore, with Shepard's training, step by step, while doing all the other things that Shepard, the focus of NASA's attention and hopes, no longer had time for: showing up at functions, returning phone calls, and attending weather briefings and "engineering meetings and planning meetings and operational meetings," as he said later. He reported back to Shepard; he spoke for Shepard. "I was now Al's alter ego, his virtual twin," he recalled. It took a supreme degree of self-discipline for Glenn to suppress the disappointment he still felt at being passed over; it took powerful self-control not to react when Shepard, needling him, called him "my backup."

The press—still in the dark, feverish with speculation—intensified

Glenn's struggle. He remained the presumptive choice, the odds-on favorite. Some newspapers gave the edge to Grissom, the only Air Force pilot in the running, but the smart money still seemed to be on Glenn. In late March the *Norfolk Ledger-Star* declared with authority that NASA had chosen Glenn back in February, even before the three finalists had been publicly identified. "With a mind bordering on genius," the article gushed, Glenn had outperformed the others. Much was made of the fact that he was the only astronaut to have attended all three recent test launches. And as the sole Marine among the astronauts, the paper added, Glenn would sidestep the rivalries between the Air Force and the Navy. The rumor was delivered with such brio that it acquired the authority of fact. Though NASA denied these reports, multiple sources—among them members of the congressional space committees—maintained that Glenn had a lock on the top slot.

Glenn, for his part, kept up the charade. He accepted, with grace, the warm wishes of friends and journalists and NASA technicians. He flashed them an opaque and indulgent smile and went back to work as Shepard's understudy.

PROJECT MERCURY PROCEEDED BY its own internal logic, moving inexorably, it appeared, toward its first manned spaceflight. But even with the launch on the calendar, everything remained open to question. With new urgency, the White House, Congress, the scientific community, and the press debated a question that was stated pointedly, in late April, by *The Nation:* "Is this trip necessary?" The President's Science Advisory Committee was not persuaded. Wiesner's phone rang with calls from scientists warning that a successful suborbital shot would look paltry compared to Gagarin's flight, while a failed shot would wreck what remained of U.S. prestige. Gilruth, meanwhile, was caught in a rolling, seemingly unending argument with NASA's biomedical advisers and the National Academy of Sciences, who insisted on six, even ten, more chimpanzee flights before they would back a single human flight. Gilruth, his affect flat as ever, said that if he was going to need that many chimps, he might move the program to Africa.

Why send a man to do an ape's job, anyway? That question had never really been put to rest, not even by the flight of Gagarin. James Van Allen, the physicist who, in 1958, had discovered the radiation belts that encircled the Earth, went on television to say that "a man is fairly useless in space." Whatever the Soviets had learned from Gagarin's orbital flight was not likely to be equaled, and was certainly not going to be exceeded, by what the United States might glean from a ballistic flight. It hardly seemed worth the risk.

And while the Soviets, it was rightly presumed, had been able to bury almost every piece of bad news, from delayed launches to botched tests to a dead cosmonaut, Gilruth had to step in front of the microphones and explain every embarrassment and near tragedy. Such was the price of a free press; but it was no wonder that some in Washington were asking whether the press ought to be quite so free to cover rocket launches. Senators publicly warned Kennedy against "another fiasco" so close on the heels of the Bay of Pigs, and called on the administration to conduct the upcoming flight in secret, lest it blow up on live TV. At the very least, they suggested, the networks should broadcast the launch on a tape delay.

"Many persons involved in the project have expressed anxiety over the mounting pressures of the press and TV for on-the-spot coverage," Wiesner informed Mac Bundy, the national security adviser, in March. As the launch date approached, those pressures increased. NASA had already consented to the steady encroachment of TV cameras; the whole facility was ringed with them. The networks had remote-controlled cameras and a crew near the launchpad, another crew in the press risers, additional crews on the recovery ships and at the tracking stations in the Atlantic—and still they complained that they were being curtailed. They demanded cameras in the command center. "The people in the blockhouse and in the control center are not professional actors," Wiesner cautioned Bundy, "but are technically trained people involved in a very complex and highly coordinated operation. The effect of TV cameras staring down their throats during this period of extreme tension, whether taped or live, could have a catastrophic effect." Walt Williams, in the end, stepped in to bar the door, but the incident served notice that the media would fight for every inch of access it could gain, and would not easily be walked back.

IF THERE WAS A JOYLESSNESS to the effort, a sense of simply plowing through, it fit the feeling in Washington. Though Kennedy's popularity rating had soared after he took responsibility for the Bay of Pigs, reaching 83 percent in late April ("My God," he said when he saw the polls), the talk in the capital was that he seemed outmatched by events. *Life* ran a photo spread of cabinet members, White House advisers, and congressional leaders looking disconsolate: grimacing, scowling, covering their faces with their hands. THE GRIM NEW LOOK OF KENNEDY'S TEAM, read the headline. *Life* explained that crises in Laos—where the rebels were steadily gaining ground—and Cuba were teaching the administration that "the Cold War is more total than they had thought." An editorial scolded Kennedy, urging him to abandon his "apathy," resume underground nuclear testing at once, and strengthen the defenses of Laos's neighbors South Vietnam and Thailand. Good reasons to be grim.

One of the faces in the spread, a long face with a deeply furrowed brow, was Lyndon Johnson's. But Johnson, possibly alone, was energized, invigorated. Previously, during his first few months in office, the role he had expected as vice president had been denied him by a distrustful and contemptuous White House staff; LBJ watched with mounting bitterness as Attorney General Robert Kennedy emerged as the second-most-powerful man in the nation. But the post-Gagarin panic put space—and, therefore, the chairman of the Space Council—at center stage. JFK's April 20 memo empowered him to shape a new approach. To that end, Johnson summoned Webb and Dryden, von Braun, Wiesner, Welsh, budget officials, generals, and businessmen to a meeting on April 24 in the Executive Office Building, next to the White House. The president, he told them, "wants the best hard-headed advice he can get—and he wants it now."

But first there was a recitation of grievances. The participants had many: that the United States had failed to clarify its goals for the space program, deferred tough decisions, and spent its resources haphazardly; the effort was fragmented; the press was intrusive; and private companies were working at cross-purposes. There were no fresh indictments here; each one had been made many times before. But the context was new. The Gagarin flight had shifted the balance decisively away from the voices of caution, giving weight to the impatient and indignant, the boldest thinkers and biggest gamblers, the most ardent Cold Warriors—who found, in Johnson, their champion. The nation, they argued, should not merely catch up with the Russians; it should surpass them.

This meant the moon. The meeting became, in NASA parlance, a "go or no go" discussion about a lunar landing. Johnson, whose answer was "go," moved around the conference table, collecting endorsements. There were holdouts: Wiesner hung back and said little. And there were reluctant conscripts, Webb among them. If JFK approved the mission, it was Webb who would have to carry it out. Opinion in NASA's senior ranks ran strongly in favor of a moon landing; it had been a stated goal from the start. Still, Webb doubted that his agency had truly faced up to the challenges of going to the moon. He was concerned that Johnson's group not push the president further than any of them were able to go.

But Kennedy, as his vice president understood, was looking to be pushed. Johnson moved swiftly to settle the matter. On April 28, four days after the meeting, he sent Kennedy a memo outlining his conclusions. It was a powerful piece of advocacy. Other nations, Johnson argued, "will tend to align themselves with the country which they believe will be the world leader—the winner in the long run." And like it or not, "dramatic accomplishments in space" were the mark of leadership. Johnson saw a reasonable likelihood that the United States could beat the Russians to the moon, possibly by 1966 or 1967—a more ambitious target than NASA had set. The nation had the resources to do this, he insisted. What it lacked, so far, was the determination to spend what it required, to break through the bottlenecks, to work around the clock in every plant in every part of the program. But time was short. Soon, he warned, "the margin of control over space and over men's minds through space accomplishments will have swung so far on the Russian side that we will not be able to catch up, let alone assume leadership."

Yet it was far from clear that the Johnson formula—"more resources and more effort"—would yield anything but more frustration. While it was true that future programs had been starved of funding, the administration had given Mercury an infusion of resources; and NASA, for its part, had already put many operations on a three-shifts-a-day, seven-days-a-week footing. Still the manned program continued to stumble.

On April 25, NASA launched what it called a "mechanical astronaut"—a package of instruments—aboard Mercury-Atlas 3, aiming to put the capsule into orbit. Gus Grissom and Gordon Cooper were assigned to fly "chase," to observe the Atlas in flight. Grissom, in an F-106A fighter, was to approach the rocket at 1,000 feet and climb alongside it; at 25,000 feet, Cooper would take his place. MA-3 never got that far. On leaving the launchpad it went dangerously awry. After forty seconds, a range safety officer hit the button to destroy it—forgetting, in that fevered instant, the two chase planes. As the Atlas exploded, it seemed from the ground that Grissom had been lost, consumed by the fireball. He was safe, miraculously, and, with Cooper, flew back unharmed. It was also a relief to NASA that the escape tower had done its work, carrying the capsule free of the Atlas just in time. It was recovered from the water, just two thousand yards offshore, and shipped back to McDonnell for refitting. "Today's flight will in no way change our plans to stick with the Atlas booster" for an orbital mission, Gilruth declared later that day—a vote of confidence that the rocket had not yet earned.

As Kennedy weighed the costs of a lunar landing, a more immediate question confronted him: whether to go forward with plans to put Shepard in space. There was too much at stake to leave the decision to NASA. On April 29, in the Oval Office, the president canvassed his advisers. Most urged postponement. Not even two weeks had passed since the Bay of Pigs, they pointed out; the nation could ill afford another disaster. Welsh, alone, waved off their concerns. The mission, he said, was no more hazardous than flying a plane in bad weather. "Why postpone a success?"

Kennedy decided against delay. The launch had been the goal for so long that it had achieved a kind of inevitability—or inescapability. Despite their own doubts, NASA officials were resolved to take the leap: the press had built a garrison of equipment trailers on a sandy knob at the Cape, and the drumbeat of expectation had grown so insistent that a postponement, while hardly a calamity, would embarrass

America in the eyes of the world. As anxious as Kennedy remained, he could not bring himself to pause the program. Weather permitting, the launch, set for May 2, was now "go."

As soon as Kennedy gave his assent, Webb released a public statement wishing "Godspeed" to the astronaut—whose identity, even now, he did not disclose. Webb called MR-3 "a most important step . . . a step that will lead on to man's ultimate conquest of this new and hostile environment." Yet he also stressed that it was "a serious step, for it cannot be taken without risk to human life. It is the kind of risk that Lindbergh took when he crossed the Atlantic, that Chuck Yeager took in the X-1's first supersonic flight." Webb expressed confidence that every reasonable effort had been made to protect the astronaut. Still, on May 1, he issued a cautionary note to the correspondents who were gathered at the Cape: "We must keep the perspective that each flight is but one of the many milestones we must pass. Some will completely succeed in every respect, some partially, and some will fail." And a failure, he warned, could have consequences around the world.

That afternoon, Webb and Pierre Salinger met with Kennedy in the Oval Office to review plans for television coverage of the event. Kennedy's secretary, Evelyn Lincoln, wrote in her diary that day that the president "is afraid of the reaction of the public in case there is a mishap in the firing." She added that Webb had said "he had tried to keep the press away from this and likewise the TV but they had been given the go sign long before he took over." NASA, with reluctance, had acceded to demands of the networks. "The space agency seems to have become trapped by its own desire to 'accommodate' the press," observed the *New York Times,* not without satisfaction. But the press had forced the issue—by monitoring whatever activity it could detect from the perimeter of the base, taking pictures with telephoto lenses, tracking the movements of Navy ships and the comings and goings of astronauts' wives, and, on that basis, jumping to erroneous (or sometimes accurate) conclusions, which were quickly and breathlessly reported as fact.

As Lincoln noted in her diary, Kennedy had tried and failed to reach a network executive in the vain hope of persuading him "to play down the publicity [of] this venture as much as possible." Now, conceding the issue, he approved a long-ignored proposal by Paul Haney of NASA's public affairs office to provide the press with real-time updates during the flight. When Salinger called Haney to relay the decision, reaching him at the MR-3 News Center in Cocoa Beach, Kennedy sought reassurance that the public was not about to witness, on live TV, the death of an astronaut. "Paul," Salinger said, "the president's wondering about that escape rocket on the Mercury."

He was wondering, specifically, how reliable it was. Haney replied that out of fifty-eight tests, the escape rocket had failed to lift the capsule to safety only once, maybe twice. Salinger turned and repeated this to Kennedy. Haney heard some chatter in the background, then laughter. "The president says go ahead," Salinger told him. "Give it a go. See if it works."

Kennedy meant the press policy, but he could just as easily have been referring to the launch itself: *Give it a go. See if it works.* At 8:30 that night, the countdown commenced.

AMID THE MOTELS and topless bars on Cocoa Beach were found, for the pious or the penitent, a surprising number of churches, and in the churches prayers were being said for the astronauts. The gravity of the moment had sunk in. The Holiday Inn, where Glenn, Grissom, and Shepard had been camped out for weeks, took down its roadside sign that read RIVIERA LOUNGE, ENTERTAINMENT and replaced it with ASTRONAUTS GOOD LUCK. The blanket benediction reflected NASA's success in keeping its selection a secret. Most outlets continued to assume it was Glenn, and as the cameras trailed him around Cocoa Beach, he gave no indication either way. His parents, watching the coverage from New Concord, worried about the toll the program was taking on their son, who was approaching forty. They had not seen him in person in some time, they told a reporter, but based on what they saw on TV, he seemed older; he was "balding more rapidly."

At 1:30 a.m. on May 2, the astronauts—having relocated from the Holiday Inn to Hangar S—were awakened in their bunk beds and began a choreographed sequence of events: a shower and shave, a meal of steak and eggs, a medical exam, a suiting up. But the slashing sound of the rain on the windows suggested that the launch would be scrubbed. By the time dawn arrived the storm had moved on, but where it had gone was three hundred miles to the southeast, near the Bahamas, where the capsule—christened *Freedom 7* by Shepard—was supposed to splash down. At 8:40 a.m., two hours and twenty minutes before liftoff, mission control canceled the countdown, postponing the flight for forty-eight hours or more.

No one outside the hangar had seen Shepard in his pressure suit, and NASA, revealing a flair for the needlessly dramatic, had made Glenn and Grissom suit up as decoys so they could maintain the mystery. (An earlier plan, roundly rejected by the astronauts, was to wear hoods all the way to the launchpad.) Still, Gilruth chose this moment to unburden himself and, at last, reveal his selection of Shepard. The correspondents were not cheered by the news. Shepard had never really caught the public imagination. Most reporters had been so eager for the choice to be Glenn that they demanded, now, to know why it wasn't. Some "openly expressed their strong displeasure to top NASA officials," one reporter recalled. Shorty Powers stammered that Shepard "had what all the others had, with just enough to spare to make him the logical man to go first"—a non-answer that would have to suffice. Headline writers, adjusting to the new order, readied their bromides—SHEPARD NEVER COULD PASS UP A CHALLENGE, one paper declared—and relegated Glenn to his secondary role. The *Washington Evening Star* identified Glenn's new "claim to fame: 'backup' astronaut."

Foul weather delayed the launch one more day, but by 8:30 p.m. on May 4 the sky had cleared and the countdown began again. That evening Shepard and Glenn ran together along a deserted stretch of beach, stopping to play idly, like kids, with sand crabs in the surf. The men had lived and trained in tandem for three months now and had, however grudgingly, grown closer than anyone could have expected. After dinner that night with the other astronauts, Shepard toasted Glenn: "John's been most kind." A rivalry remained, no doubt, but for now there was no outlet for it, nothing to compete for; Shepard had won the first round. Glenn's struggle, for the past few months, had not been with Shepard but with himself. Through force of will, through daily acts of self-abnegation, Glenn had made peace with his second-place role.

And his service to Shepard was nearly complete. At 1:10 a.m. on May 5, the astronauts' doctor, Bill Douglas, nudged both men awake. "Is there anything else I can do?" Glenn asked, after they'd had breakfast. Shepard said there wasn't, and went down the hall to get an exam and be fitted into his silver space suit. Glenn put on more modest attire: white coveralls and a white hat, which matched those of the technicians but made him look a little like a short-order cook. Ahead of Shepard, Glenn drove to the pad, rode the elevator up the gantry—the structure surrounding the Redstone—and stepped out on the platform that stood level with the capsule. He ran a series of checks. He'd also brought two items for Shepard—mementos, of a kind—which he now carefully taped to the instrument panel.

The first was a *Playboy* centerfold—a joke, obviously, at Glenn's own expense. Of course no astronaut was less likely than Glenn to encounter a *Playboy* magazine, much less liberate the centerfold from its staples for use in a prank. Coming from Glenn, who had lectured Shepard that the Cold War required him to keep his pants zipped, the photo was almost a sign of conciliation.

The second item was a handwritten sign, several inches long, clipped from a page of construction paper. In pale green pencil, Glenn had written a farewell message to Shepard, underlining each word in red or black:

JOSÉ, NO HANDBALL PLAYING IN THIS AREA DURING FLIGHT.
DO GOOD WORK, AL, AND HAVE A COOL APOGEE.

At 5:15 a.m. Shepard strode off the elevator and, in five minutes' time, began his insertion into the capsule. As he eased himself onto its fiberglass couch, he caught sight of Glenn's additions to the cockpit. He laughed—and then, through the hatch, saw Glenn laughing, too. Shepard said something about keeping the centerfold, but Glenn—mindful of the onboard camera—reached in and plucked it off the panel. About an hour later, he reached in again, this time to shake Shepard's hand. Shepard thanked him and managed, despite his pressurized glove, a thumbs-up. "Happy landings, Commander," Glenn said. The gantry crew cheered, and the hatch was closed. Through the fish-eye lens of the periscope, Shepard could still see Glenn's face—the last face he would see before being shot into the void.

The National Security Council meeting had just begun when an aide passed the president a note: "Slight hold T minus 6 and counting." When Evelyn Lincoln entered the room and handed him another note, this one reading, "Astronaut going up in 5 minutes," Kennedy pushed himself out of his chair and, trailed by LBJ, Bobby Kennedy, and several advisers, walked into Lincoln's office and stood, mute and uneasy, in front of a portable television atop a file cabinet. Johnson picked up the phone and got Webb on the line. Jacqueline Kennedy, on her way somewhere else, passed by the open doorway. "Come in and watch this," the president said, and she joined the group in front of the TV.

Freedom 7 was less a spacecraft than a projectile, flung upward like a stone. Still, there was a violent beauty to the effort, apparent even on a black-and-white set. At 10:34 the Redstone fired and rose from the pad. The live feed from the Cape was jittery—the rocket, within seconds, was little more than a white streak on a gray screen. For Shepard, the strange thing was how familiar it felt. All that conditioning had done its work. He was sitting atop a ballistic missile and yet the ride seemed smoother, on the whole, even quieter, than a run on the centrifuge. After two and a half minutes, the booster had done its work. Explosives cut it free. Soon Shepard was weightless, though he was strapped in so tight he didn't notice—until a washer, left loose in the capsule, drifted past his face. He switched off the Automatic Stabilization Control System and tested the manual controls, axis by axis—this, beyond surviving, was his principal task—and found that he was able to maneuver the capsule. And now, having been hurled 116 miles high, he began, as intended, to fall back to Earth. *Freedom 7* splashed down near the Bahamas. "Boy, what a ride!" he exclaimed. At the White House, the group cheered. "It's a success," Kennedy said quietly. The whole thing had taken fifteen minutes.

Aboard the recovery ship, the carrier *Lake Champlain,* Shepard was half an hour into dictating his report, recounting every detail of the flight into a tape recorder ("it was difficult for me to reach the filter-intensity knob with the suit on without bumping the abort handle"), when he was summoned onto the flag bridge to take a call from the president. "Commander," Kennedy said, "I want to congratulate you very much. . . . We watched you on TV, of course, and we are awfully pleased and proud of what you did."

"Well, thank you, sir," Shepard replied. "As you know by now, everything worked just about perfectly and it was a very rewarding experience for me and for the people who made it possible."

Pleased and proud, very rewarding—this undersold it. Across the country, the sense of relief and release was overwhelming. "He made it," a woman in Chicago gasped to a reporter, then began to sob. In New York City, people danced. The press hailed a new hero—finding in Shepard, in the span of fifteen minutes, so many of the qualities it had previously failed to see. He was, as James Reston put it in the *Times,* "the kid next door . . . with faith in the Lord and a glorious happy wife." Never mind that next door to Shepard's boyhood home in East Derry, New Hampshire, was his grandparents' mansion, with a ballroom as big as a tennis court, or that he belonged to no church, a fact he had acknowledged at that first press conference in 1959; this was not a time for splitting hairs.

NASA's concessions to the TV networks, so long a source of apprehension at the agency and the White House, now seemed a brilliant stroke. "Because the word 'freedom' emblazoned on the side of the capsule means what it says," the NBC correspondent Frank McGee declared proudly, "the launch had to be taken out in the open." This was well received across the non-Communist world. The contrast between Soviet secrecy and American openness—regarding not just the launch but also its results, the data that NASA freely shared—was clarifying. In a classified report later that month, the U.S. Information Agency observed that this, more than any other aspect of the flight, had made the "greatest impact on expressed opinion in the Free World" and had given the United States a "significant psychological advantage" over the USSR. "Candor and openness have their merits," Edward R. Murrow of USIA proclaimed at the National Press Club. ("They also have their demerits," he added, pointing to coverage of the Bay of Pigs disaster.)

The Soviets, for their part, took the posture that *Freedom 7* was almost unworthy of comment—except to dismiss it, with relish, as second-rate. Soviet scientists questioned whether suborbital flight counted as spaceflight at all. NASA declined to argue the point. It had never seen ballistic shots as more than a prelude to orbital flight. No one, in fact, had been more dismissive of suborbital flight than Dryden, who'd told Congress in 1958 that it was equivalent to "tossing a man up in the air"; it had, he said, "about the same technical value as the circus stunt of shooting a young lady from a cannon." None of this was lost on the astronauts. Compared to the Russians, Slayton said years later, "we . . . looked like weenies." (Of course, Slayton had competed hard to be that first weenie.)

Kennedy was looking ahead. At a press conference that afternoon, he again expressed pride in Shepard and NASA but pointed out that "we have a long way to go in the field of space. We are behind." The nation, he pledged, was "going to make a substantially larger effort in space," and he himself was going to ask Congress for more money; yet he never said, and the reporters never asked, what that money would do. But even as Kennedy stood before the press, his vice president was moving to convert the day's morale boost, which he knew would be fleeting, into a long-term commitment to put Americans on the moon.

"The Vice President just called," Webb wrote to O. B. Lloyd Jr., one of his public information officers, "and asked me to get with Mr. McNamara and prepare . . . a program for the President to send to Congress." Webb's task now was not, as he had planned, to frame the choice judiciously but instead, as he told Lloyd, to provide "a real build-up for the program." Well into the night, Webb and a small group of officials from NASA, the Pentagon, and the White House worked together to reconceive the space program. On May 8, the memo went to Kennedy's desk with a cover note from LBJ: "I am much impressed."

Sweeping in scope and ambition, the memo covered everything from weather satellites to large-scale boosters for military use. But its centerpiece was a manned lunar mission—"before the end of this decade." Webb and McNamara acknowledged that "this major objective has many implications. It will cost a great deal of money. It will require large efforts for a long time." But it urged JFK to accept the challenge, for "it is man, not merely machines, in space that captures the imagination of the world." They could offer no guarantee—only a chance. And "even if the Soviets get there first, as they may . . . , it is better for us to get there second than not at all."

AT TEN O'CLOCK that morning, three Marine Corps helicopters landed, one after another, on the South Lawn of the White House, carrying the astronauts, their wives, and others to a ceremony in Shepard's honor. The president, the first lady, and LBJ, standing in the driveway on a chalk-drawn rectangle marked PRESIDENT & MRS. KENNEDY, greeted Al and Louise Shepard and then the rest. It was the first time Kennedy had met the astronauts. He seemed energized, even a little awed, by their presence. In the Rose Garden, on a platform dressed up with bunting, Kennedy awarded Shepard the NASA Distinguished Service Medal. He called attention to "the fact that this flight was made out in the open with all the possibilities of failure, which would have been damaging to our country's prestige. Because great risks were taken in that regard, it seems to me that we have some right to claim that this open society of ours which risked much, gained much."

Behind them on the platform, facing the cameras and the crowd, stood the other six astronauts. "Our pride in them is equal," Kennedy said. He mentioned only one, besides Shepard, by name. That was Glenn—included in a list of people whose contributions, while unspecified, deserved "particular tribute." For Glenn, it was surely just as well that Kennedy never uttered the word *backup*. But that was still, for a few more moments, his role. In a NASA photograph of the event, Glenn occupies the middle of the frame—smiling warmly, applauding, bracketed on one side by his rival, on the other by the president, as those two shake hands.

With that, Glenn's time in Shepard's shadow was nearly complete. There was still a parade to get through, in which Glenn and the others would follow in Shepard's wake, like a train of attendants, looking at the back of the hero's head while he waved to the crowds along Pennsylvania Avenue. Glenn was assigned to "Car No. 6," ahead of the other five also-rans but trailing behind cars full of reporters and congressmen. When they arrived at the White House, Glenn made his way through the cluster of people in the Oval Office and found a seat on a couch opposite Kennedy, who was leaning forward in his rocking chair and asking the astronauts questions about the space program.

Some of Kennedy's questions were so basic—*Do we really need a manned space program? Can't machines do the job?*—that aides wondered whether he was just having fun with the astronauts, winding them up. But what he was doing was pumping them for information. He was preparing himself for the questions he knew he would face if he approved the lunar mission, which he now seemed determined to do. "I want to be first," he told Gilruth. "I want to go to the moon."

Two days later, on the afternoon of May 10, Kennedy gathered a small group in the Oval Office. Johnson, who had done more than anyone else to frame the decision Kennedy was making today, was absent: he was en route to South Vietnam, where he would reaffirm the administration's support for the government there in the face of an increasingly aggressive, effective campaign of terror by Communist Vietcong guerrillas. But Johnson could afford to miss this meeting. His position had prevailed. When Bundy voiced doubts about a moon shot, Kennedy seemed unconcerned—uninterested, even. He had made his decision, and at 4:30 p.m. the group split up and went to work fulfilling it.

CHAPTER 11

Holocaust or Humiliation

PRESIDENT KENNEDY HAD already given a speech to a joint session of Congress—a State of the Union address, back in January—but now, on May 25, to the surprise of the capital, he was back in the House chamber to deliver another. This one was billed as a speech on "urgent national needs." At 12:30 p.m., he began in grim fashion to recite them. Standing at the rostrum, he described the challenge of defending freedom at a time of "convulsive change." America, he said, was engaged in "a contest of will and purpose . . . , a battle for minds and souls as well as lives and territory. And in that contest, we cannot stand aside. . . . We do not intend to leave an open road for despotism."

A few weeks earlier, Wiesner had extracted a presidential promise not to justify the moon shot on the basis of science. This was probably unnecessary. Space Task Group spokesmen might talk about "the nature of the universe," but Kennedy was concerned with life on Earth and the system that would govern it. Space, he said now, was another battlefront in the Cold War. The Gagarin flight, he added, "should have made clear to us all, as did the Sputnik in 1957, the impact of this adventure on the minds of men everywhere, who are attempting to make a determination of which road they should take."

Therefore, Kennedy said,

> I believe that this nation should commit itself to achieving the goal, before this decade is out, of landing a man on the moon and returning him safely to the Earth. No single space project in this period will be more impressive to mankind, or more important for the long-range exploration of space; and none will be so difficult or expensive to accomplish.

He put the question to Congress and the country. Departing from his script, he urged every American to "consider the matter carefully . . . , because it is a heavy burden, and there, uh, is no sense, uh"—he began to arrange the pages of his speech, tapping the edges to align them—"in agreeing or desiring that the United States take an affirmative position in outer space, unless we are prepared to do the work and bear the burdens to make it successful. If we are not, we should decide today and this year."

To Ted Sorensen, who sat among the members of Congress, "the President looked strained in his effort to win them over. . . . His voice sounded urgent but a little uncertain." Kennedy had noticed the lack of applause. "Where are we going to get the money?" one Republican leader grumbled after the speech. It was a common refrain on both sides of the aisle. Clarence Cannon of Missouri, Democratic chairman of the House Appropriations Committee, said that the budget request for space was "wholly unrealistic and fantastic beyond measure," that it would cause enormous waste—and that none of this would stop Congress from giving Kennedy every dime he sought. This seemed fairly certain. Democrats held an advantage of thirty votes in the Senate and ninety-one in the House. Though some liberals worried that the lunar landing would undercut other priorities, such as aid to the elderly and the unemployed, they, like most other Democrats, were eager for a more assertive response to global communism. "There can be no price tag on survival in this dangerous world," said Robert Kerr.

It was far from clear that the country agreed. In late May, just before Kennedy's speech, Gallup reported that only a third of the public was willing to spend the $40 billion they were told it would cost to land a man on the moon; a large majority, 58 percent, was opposed. Shortly after the speech, when Gallup asked Americans to rank the issues for which they would be willing to make sacrifices, even if it meant paying more taxes, space came in fifth, behind retraining programs for the unemployed, military spending, civil defense—even USIA propaganda programs. Apollo was losing in a landslide to other objectives.

Kennedy had work to do. And the news media—while still behind the space program—was not going to do it for him. Network anchors and newspapers took a wary tone, conveying little sense that the announcement marked the start of some great adventure: coverage focused on the expense, not the excitement, of the mission. Even NASA was ambivalent. At the Marshall Space Flight Center in Huntsville, where von Braun and his colleagues listened to the speech, there were cheers of "Let's go!" Others felt proud that Kennedy would place that much confidence in the manned space program—despite the fact that, as one engineer recalled, "we'd made one 15-minute flight—that was it!" But as Gene Kranz, a flight controller, wrote in his memoir, "Those of us who had watched our rockets keel over, spin out of control, or blow up" found the idea of a moon shot "almost too breathtakingly ambitious." At Langley, upon hearing the news, members of the Space Task Group left their desks and stood around in disbelief.

The astronauts kept their own counsel. In May 1961, the prospect of landing an American on the moon—even if NASA met Kennedy's deadline—seemed remote. And at the end of the decade, Shepard would be forty-seven; Glenn would be nearly fifty. Even the youngest of the group, Gordon Cooper, would be forty-three—above the age limit of forty that NASA had set during its selection process. So the moon shot announcement, in an immediate sense, changed nothing for the astronauts. The work went on, same as before. On May 25, the day of the speech, Glenn was at the Cape, deep in the pages of McDonnell's Service Engineering Department Report 69—the section on the Mercury capsule's Automatic Stabilization Control System. The moon remained as distant as ever.

An ill wind welcomed Kennedy to Vienna on June 3. As he prepared to meet Khrushchev face-to-face—a summit both men sought, though for different and contradictory reasons—the problem of Berlin remained intractable. Khrushchev had let it be known that he had run out of patience with the United States and was determined to force the issue in the divided city—most likely by signing a treaty with East Germany, recognizing its "sovereignty" and, in this way, smashing the postwar arrangement of power in Berlin. One way or another, Khrushchev warned, "the German problem must be solved in 1961."

"The Soviet-American conflict cannot be easily settled by negotiation," Kennedy once said, but "it is far better that we meet at the summit than at the brink." He believed that by sitting across the table from Khrushchev, by establishing, as best he could, a personal rapport, he could lead the Russian to reason. *Time* quipped that it had identified the "Kennedy credo: when in doubt, talk." This overstated the case, but not by much. He did not imagine that he could talk his way past the impasse in Berlin. But he did think a nuclear test ban agreement was possible and could serve as a first step toward trust—and, ultimately, toward a peaceful resolution in Berlin.

Space, he thought, might present another opportunity. Since the start of his presidency, Kennedy had nurtured a not entirely idle hope that he could convince the Soviets to settle, in effect, for a draw in the space race—easing tensions and the enormous drain on both countries' budgets. In January he had made the pitch in his State of the Union address. "Both nations," he said, "would help themselves as well as other nations by removing these endeavors from the bitter and wasteful competition of the Cold War. . . . Nature makes natural allies of us all." Communications satellites, weather satellites, probes of Venus and Mars—all this, he said, could benefit mankind if approached in a spirit of common endeavor. He did not mention manned missions; the idea of a joint U.S.-Soviet trip to the moon, of two programs and two systems yoked together, was not the sort of fantasy to which a president, ten days into his term, might confess. But it appealed to him. Even more, he was eager to prevent outer space from becoming a theater of war.

Some of this was for show. In the contest for world opinion, there might be some advantage in putting forth these ideas before the Soviets did, and in greater apparent earnestness. At the same time, he was serious enough about the idea—and concerned enough about the cost of the moon shot—that in advance of the summit he had asked Wiesner to explore the possibilities. In April, Wiesner pulled together a group of experts to identify areas where the Cold War rivals could conceivably cooperate in space. There were no romantics at the table; there was no talk of a fellowship of nations; the group was engaged in a difficult, most likely fruitless exercise. Still, they did come up with a range of proposals—including a joint lunar mission. They did not expect the Soviets to embrace the idea. Nor, for that matter, Congress, the U.S. military, or the American people. Yet

Wiesner advised JFK to make the proposal anyway—lest the Russians beat him to it and reap the propaganda gains.

Overtures were made and were quickly rebuffed. On May 20, the Soviet foreign minister, Andrei Gromyko, said archly that cooperation in rocket technology was "inconceivable" unless the United States agreed to the disarming of Western Europe, the withdrawal not only of nuclear weapons but of all tanks and troops—a frequent Soviet demand. The curtness of the reply suggested that JFK's greatest hope for the summit—a nuclear test ban treaty—was probably beyond reach.

The start of the summit confirmed this suspicion. Khrushchev, confident that Kennedy could be cowed, made that his first order of business. At sixty-seven, he was twenty-three years older than Kennedy; at five feet, three inches tall, he was ten inches shorter; but virtually every exchange over their two days in Vienna revealed Khrushchev to be sharper, tougher, more vigorous, more tireless. He dominated their discussions, dominated the toasts and the banter during their luncheon at the U.S. ambassador's residence, dominated even a stroll they took, just the two of them, in the garden: Kennedy's advisers, from an upstairs window, saw Khrushchev "circling around Kennedy and snapping at him like a terrier and shaking his finger." Immune to Kennedy's charm, resistant to reason, Khrushchev lectured him about the historical inevitability of communism and berated him about Berlin while the painkillers and amphetamines in Kennedy's bloodstream—administered by his personal physician, Max Jacobson, known to his adherents as "Dr. Feelgood"—lost effect and the chronic pain in his back returned, dulling his wits. GIVE 'EM HELL, JACK, a placard at the Vienna airport had urged, but Kennedy was on the receiving end.

There was an element of sport to Khrushchev's performance, a competitive impulse, but it also had a serious aim: to bully Kennedy into conceding Berlin. Khrushchev's belligerence on the first day of the summit was only, it turned out, a warm-up for the second. His face went red and his arms waved as he shouted about "this thorn, this ulcer" that the Soviet Union was prepared to cut out on its own—if the Americans refused to recognize East German sovereignty and control of Berlin. "We fought our way there," Kennedy replied sharply, finding his voice. "If we accepted the loss of our rights no one would have any confidence in U.S. commitments and pledges." Khrushchev responded that the Soviet Union would sign its treaty with East Germany by year's end, and "if the U.S. should start a war over Berlin, let it be so."

Khrushchev's welcoming of war—even if the comment was only a shock tactic, and who could be sure?—made joint missions to the moon look like a delusion. Kennedy had raised the idea earlier in the talks, at the luncheon on June 3, when it still seemed possible to find some expression of mutual goodwill. Khrushchev, a dry martini in hand, was in an expansive mood, boasting about Gagarin—who, he said, had had such an easy time that he had sung songs in orbit. As for Kennedy's proposal, he greeted it coolly, noting that spaceflight had military implications. Then he shrugged—"All right, why not?"—a response too flippant to be taken seriously, but it was not a rejection. That came the next day. Having thought about it further, Khrushchev said over lunch, he was now of the view that the United States should go to the moon first, on its own. A lunar landing would be very costly, but the United States was rich; Russia would follow when it could. In any event, he added, the Soviet Union had seen few practical benefits from space exploration; the race was mainly for the purpose of prestige. For Khrushchev, it was a moment of candor amid two days of bluster and lies; for the two leaders, it was a rare point of concurrence.

"I've got a terrible problem," Kennedy confided to James Reston just before leaving Vienna. "If he thinks I'm inexperienced and have no guts, until we remove those ideas we can't get anywhere with him." And that was exactly what Khrushchev thought. Vienna had confirmed what the Bay of Pigs had revealed: that JFK was malleable, weak, "immature," as Khrushchev told his interpreter. He returned to the Kremlin even more determined to press his advantage.

KENNEDY HAD WELCOMED a debate on the moon shot, and now he was getting one. "We're obviously getting in trouble with the public," Dryden admitted in an interview. The tone of the discussion was captured by *U.S. News & World Report,* which asked, in a headline, WHY SPEND $20 BILLION TO GO TO THE MOON? The article went on—and on—in this vein: "Why bother? . . . Is the whole project anything more than an expensive stunt? What's the point? Scientific? Military? What?" Webb endeavored to answer each question, stressing the "practical value" of space technology, while Dryden, a bit too strenuously, tried to sell a vision of "a great variety of new consumer goods and industrial processes that will raise our standard of living." Senator Warren Magnuson, a Washington Democrat, warned that NASA had "a lot of missionary work" to do.

While headquarters applied itself to that, the Space Task Group remained focused on the next Mercury mission. The flight assignment would seem to have been settled back in January, when Gilruth had assembled the astronauts and told them that Shepard would go first, Grissom second, and Glenn would serve as backup to both. But somewhere along the way, in the months since, a "probably" had attached itself to Grissom's place in Mercury-Redstone 4. For Glenn, there was hope in the ambiguity. He now assumed he was next in line.

Clearly he had failed to learn that he should never make assumptions about his own fate. In June, Gilruth reaffirmed that Grissom was the prime pilot. "I was surprised and disappointed," Glenn said later—

an understatement. He had good cause for disappointment. With MR-4 scheduled for mid-July, he had a demoralizing stretch of duties ahead: attending meetings for Grissom, just as he had done for Shepard; taking notes and phone messages for Grissom and relaying Grissom's requests to the engineers, just as he had done for Shepard; following Grissom on the procedures trainer and mirroring his movements—another full dose of abasement. Worse still, the launch coincided with his fortieth birthday, July 18. Glenn felt he was crossing onto "the wrong side of some invisible barrier. . . . I could hear the clock ticking."

Not until July 17, the day before the scheduled launch, did the agency publicly confirm that its choice was Grissom. The news went down poorly. Reporters demanded to know whether Glenn's age was the issue; had he been disqualified? NASA denied it. Among the spectators on Cocoa Beach, the *Times* reported, "the most commonly held sentiment was one of sympathy for Colonel Glenn—a sharing of his certain disappointment at having missed the pilot assignment again. Tomorrow," the newspaper noted, "is his fortieth birthday."

Glenn celebrated, if that was the word for it, by waking up early and running on the beach with Grissom. It was an unlovely day. Late the night before, a heavy cloud cover had settled in, forcing a postponement. Glenn spent most of the day holed up in the crew quarters, doing little. There was a small birthday party, not an occasion for joy; and though Annie and the children had come to visit, Glenn was moody and withdrawn. Adding insult to injury, Gilruth told reporters that day that he had selected Grissom because he was the most fit to fly. Most likely, the statement was meant to tamp down rumors that Grissom had been awarded the slot out of deference to the Air Force, given that the first flight had gone to a Navy man. But the remark about fitness sounded like a dig at Glenn—as did a spokesman's comment that there was no guarantee that Glenn would even go third.

But perhaps Glenn had been spared a greater indignity. America's second suborbital flight, before it had even taken off, seemed a pale echo of *Freedom 7.* The crowds on Cocoa Beach were smaller and more subdued than they had been in May; and NASA, even more than last time, was on the defensive—explaining, then explaining again, why another ballistic shot was necessary when a Russian had already orbited the Earth. The honest answer was that a suborbital flight was the best NASA could do.

As the new launch date of July 21 approached, even NASA's Public Information Office was "acting rather ho-hum about it," Paul Haney later confessed. This was never the case in mission control, where the mortal danger of spaceflight was apparent at every instant; neither was it the case for the astronaut's family. Grissom's father, Dennis, who watched the coverage at home in Mitchell, Indiana, said that "pride ran second" that day; mostly what he felt was fear. But when *Liberty Bell 7* lifted off that morning, the nation did not appear to hold its breath, as it had when Shepard flew; the president and his aides did not huddle in anxious silence in front of a television set; and when the capsule splashed down fifteen and a half minutes later, ninety miles northeast of Grand Bahama Island, the feeling, on the whole, was of business as usual.

Then the escape hatch blew. It sailed five feet into the air and skipped across the waves. Grissom tossed his helmet aside and fought his way out of the capsule and into the sea. The capsule flooded and so did his suit, taking in water through its oxygen valve, leaking air out through the neck. He was struggling now—fighting to keep his head above the waves, gasping, going under the swells as recovery helicopters churned up the water with their rotors. After five desperate minutes Grissom was saved, pulled from the sea near a circling shark. His capsule was lost.

The next day, in that spirit of openness that Kennedy had lauded in May, NASA sat Grissom in front of the press at the Starlite Motel and had him tell his story of the flight, and the sinking, of *Liberty Bell 7*. Grissom explained that while waiting for the chopper, he had gotten ready by taking the safety cap off the detonator for the hatch, had been "laying there minding my own business, and *pow!*—the hatch went." Few believed him. For all the mechanical failures that had plagued Mercury from the start, there was an undercurrent of suspicion at NASA that Grissom had panicked. Some suggested that static electricity, generated by a helicopter, had triggered the accident. Others thought he must have bumped the detonator with his shoulder. The elusiveness of the truth, despite multiple rounds of what Grissom called "that investigation bullshit" (it did not help that Exhibit A was on the ocean floor, sixteen thousand feet down), allowed an impression to take hold that Grissom had "screwed the pooch," as test pilots put it.

"I hope it does something for you," Grissom told JFK after the flight, by radiophone from the recovery ship. He was alluding, however inartfully, to Kennedy's effort to build public support for the moon shot. "Oh, well, it's a big help, I'll tell you that," Kennedy replied. But in the days that followed, there was no White House ceremony for Grissom; no parade; no attempt to profit politically. Shepard had received his medal in the Rose Garden, from the president himself; when Grissom got his, it was given to him by Webb on the tarmac of an Air Force base.

NOT EVEN A FLAWLESS FLIGHT could have done much for Kennedy. An atmosphere of crisis had been building since Vienna. On June 11, Khrushchev made public the ultimatum he'd given to Kennedy at the summit: a six-month deadline to settle the fate of Berlin. Later that month, at a ceremony marking the twentieth anniversary of the Nazi invasion of Russia, Gagarin was brought out to praise Khrushchev as "the pioneer explorer of the cosmic age." Khrushchev—outfitted in a more generously tailored version of the uniform he had worn during World War II, when he'd served as chief political commissar at Stalingrad—grandly accepted another of the medals he had seen fit to award himself, before extending the honor to the seven thousand Soviets who had contributed to the success of *Vostok*. Then he turned to Berlin. Western leaders who refused to compromise would "share the fate of Hitler," he pledged. To a cheering crowd, he announced a buildup of weapons and troops—"everything necessary in order immediately to smash any opponent."

The move—and the bluster—was born of desperation. Khrushchev was growing more and more impatient at his inability to resolve the situation in Berlin. In record numbers, East Germans were making their way—by foot, by car, by train, by subway—to registration centers on the western side. West German officials had a word for what was happening. *Torschlusspanik,* they called it: panic "lest the door be shut in one's face." On a single weekend in July, more than four thousand people surged across the border despite heavy rains and a campaign of intimidation—beatings, abductions—by East German police. Nearly 120,000 had fled since the start of the year. Refugees were granted asylum and almost immediately flown out of the city to various parts of West Germany. On July 18, the day *Liberty Bell 7* was scheduled to launch, West German sources reported that Soviet and East German troops were setting up rings of tanks and anti-aircraft rocket bases thirty miles outside Berlin.

Since returning from Vienna, Kennedy had been weighing his options in Berlin, complaining that his advisers had given him two alternatives: "holocaust or humiliation." The hard-liners, led by Dean Acheson—secretary of state under Truman and one of Kennedy's "wise men"—sought military solutions: the declaration of a national emergency, the calling up of reserves, a massive buildup of forces. In the event that the Soviets cut off West Berlin, a move Acheson expected, he advocated a steady escalation of pressure. This included, possibly, the dispatch of a division down the autobahn, just to see what the Soviets would do. The endgame was nuclear war. Within the White House, the opposing faction favored negotiations, reinforced by a military buildup—less a plan, as yet, than an impulse. Kennedy demanded a more considered approach. This took time, and in the vacuum came a clamor for more aggressive action. Polls showed that a substantial majority of Americans would rather risk war than back down.

By late July, Kennedy had settled on a middle course. He had grown convinced that a long-term increase in U.S. troops in Europe would do more than any dramatic gesture to show that the cost of confrontation in Berlin would be high. "I hear it said that West Berlin is militarily untenable," he declared in an Oval Office address on July 25. "And so was Bastogne. And so, in fact, was Stalingrad. Any dangerous spot is tenable if men—brave men—will make it so. We do not want to fight," he continued, but he wanted it known that "we cannot and will not permit the Communists to drive us out of Berlin, either gradually or by force." Among other steps, he proposed to double and then triple draft quotas and called on Congress to increase the defense budget. At the same time, he said, the United States would not abandon its efforts toward a peaceful solution. Either way, there were "long days ahead."

It was a sober, even ominous appeal. Yet for now it quieted his critics—both at home and in Western Europe, where, after Vienna, Kennedy's determination had been in doubt. Khrushchev still considered Kennedy weak, still thought him a tool of warmongers and reactionaries, but no longer imagined that the Americans could be made to slink out of Berlin. Still caught in the vise, he thrashed about wildly. At a Warsaw Pact conference in Moscow, he railed against the United

States and promised to "meet war with war"; he vowed to make Kennedy "the last president of the United States."

On August 6 came another pointed statement. Early that morning, the cosmonaut Gherman Titov stood on the launchpad in Kazakhstan and read a speech over the public address system. He thanked the designers of his spacecraft, *Vostok II,* and thanked, for good measure, the Central Committee of the Communist Party. He vowed to succeed in his mission. It was markedly more ambitious than Gagarin's: Titov was to orbit for twenty-four hours, testing whether a person could fall asleep in a weightless state. If a man could sleep in space, the Soviets believed, he could live and work in space for longer durations. And that meant, in turn, that he could survive a trip to the moon.

Vostok II lifted off at 9:00 a.m., Moscow time. It circled the Earth seventeen and a half times. On his seventh orbit, Titov passed over Washington; a mere twenty minutes later, speeding through the void, he was above Moscow, where it was now early evening. "I beg to wish dear Muscovites good night," he said. "Now I'm going to lie down and sleep." This was not an idle comment but a boast, broadcast to the world. Titov did, eventually, fall asleep, although with difficulty; he struggled throughout his flight with vertigo, nausea, and headaches. But then, other than sleep, there was little that he was expected to accomplish in space. Soviet engineers had an even greater "love for automatic controls," as one of them put it, than their American counterparts did.

RED SPACEMAN LANDS! shouted the *New York Journal-American* the next morning. And by the time he landed, he had traveled farther than the distance to the moon and back. For every minute that either Shepard or Grissom had spent in space, Titov had spent more than an hour and a half—more than twenty-five hours altogether. "It leaves us almost speechless with admiration," the *New York Times* conceded.

Some simply refused to believe it: columnists and even some scientists indulged in conspiracy theories, postulating that the Soviets had sent a prerecorded tape into orbit on an unmanned satellite. But most Americans recognized that "unless the world changes drastically," as a news anchor intoned, the first lunar trip "will be taken by a spacecraft bearing a hammer and a sickle." NASA acknowledged that the soonest it could put a man in orbit was at the end of 1961 or, worse, in early 1962—assuming that it had, by then, managed to send a chimp into orbit (or a succession of chimps into orbit, depending on which NASA manager one asked).

In Moscow, Titov was feted in the now-familiar fashion: the motorcade with Khrushchev; cheering crowds of hundreds of thousands; models of spaceships in Red Square; declamations atop Lenin's Tomb. While Khrushchev beamed like a proud father, Titov proclaimed that *Vostok II* had proved that Russians could "land spaceship-satellites at any place on earth." (Soviet secrecy made it possible to maintain the fiction that Titov and Gagarin had gracefully landed their ships, rather than bailed out.) "The Soviet people have everything needed to crush the aggressor," Titov said, "should the enemies of peace launch another war."

At a Kremlin reception, Khrushchev gathered the foreign press and the diplomatic corps for a glass of champagne and a display of gloating. Standing before a bountiful table, he quipped that American spacecraft don't orbit, "they hop up and fall down in the ocean. We are glad," he added with a grin, "that the American flier did not get drowned." Then his remarks took a darker turn. "*Vostok II* did not carry atom bombs or other armaments for killing—but peaceful instruments," he said, making clear, by indirection, that future spacecraft might do otherwise. Again, however, it was Berlin that consumed his thoughts. He began to ramble with rising passion, insisting that nothing could stop his country from signing a treaty with East Germany. “Only lunatics would counter with war,” he said. “But you cannot exclude the possibility that such lunatics exist.” And in that event, “all Germany will be reduced to dust.”

That said, Khrushchev had other plans for Berlin. Four days later, just after midnight on August 13, convoys of East German troops, sirens screaming, fanned out across that divided city and turned the border between east and west into a barricade, choking off checkpoints with razor wire and stone blocks—the rudiments of a wall, a patchwork levee to hold back the human tide.

CHAPTER 12

Talking Our Extinction to Death

THE "SPACE GAP" was back. It had never been closed or even meaningfully narrowed; but it was clear that Titov's flight had wrenched it open wider, and the term began to pop up in newspaper articles and the speeches of congressmen. The charge that Kennedy had flung at Nixon in 1960 was now being leveled at him. Khrushchev's taunt that the U.S. manned program was capable only of hopping up and falling down was impossible to dispute; it would have to be disproved. But "at the rate we are going," a NASA scientist complained to the *New York Times,* "we will be lucky to reach the moon by 1975."

With haste, now, NASA retired Mercury-Redstone. A third and final suborbital mission, MR-5, had been planned for some time in the coming weeks, but in mid-August the Space Task Group went ahead and put an end to the embarrassment of the "short shots." It was time to give full focus to Mercury-Atlas—which needed it. Development of the Atlas had been limping along, beset by so many component failures and flaws in engineering and changes in design that NASA, Convair, and the Pentagon convened working groups to make sense of the trouble. On August 22, NASA had to postpone the flight of Mercury-Atlas 4, the first unmanned orbital test of the vehicle, one day before its scheduled launch; the rocket's transistors were defective.

The astronauts offered no public comment on the end of Mercury-Redstone. Some of them, privately, cheered the news. From the start of Project Mercury, orbital flight had always been the main event—"the big one," the seven called it. There was clearly no glory left in fifteen-minute flights. Glenn felt that way. Still, he was the one with the most to lose from further delays. His rivals were lobbying hard for the first orbital slot. The caricature of Glenn as the "sniveling" Marine, shrewdly advancing his interests, provided the others a scapegoat, or cover, as they shrewdly advanced their own. That included Shepard and Grissom, who had already flown. Both men—as Grissom wrote in a letter to his mother—"expressed our views" to NASA managers about the selection of a pilot for the first orbital flight. Those views, of course, were self-serving. The two—as yet only—experienced astronauts made the case that the best man to fly this new, more difficult, more dangerous mission was someone who had already withstood the rigors of spaceflight. The point was arguable; but no one could say it was being argued on principle.

Earlier that year, as Glenn later reflected, it had "seemed certain" that he would get the third flight. No longer: "The shift to orbital preparations threw the choice once again into confusion." For the other six, it was a moment of opportunity. For Glenn it was one more season in limbo—"a time," he wrote in his memoir, "of self-doubt and anxiety."

BUT SPACE AGAIN seemed a sideshow. "He's imprisoned by Berlin, that's all he thinks about," a cabinet member said of Kennedy that summer. The crisis had entered a new and more perilous phase, dominating virtually all discussion at the White House. What the papers, at first, called a "barricade" or a "border closing" was revealed, on August 17, to be the first step toward construction of a wall, as the jagged boundary of barbed wire and rope was rapidly replaced by concrete. No one could know what the next step might be. West Berliners and their mayor, Willy Brandt, grew restive while the United States weighed courses of action—options that struck Kennedy as either ineffectual (travel restrictions on East German diplomats) or unduly provocative (a ban on trade with the GDR). Addressing a rally outside the city hall, Brandt charged that Berlin might become "a new Munich," a casualty of Western appeasement. "Woe to us," he said gravely, "if through indifference or moral weakness we do not pass this test."

The wall was an abomination, Kennedy believed, but not worth a war. In a gesture—and it was mainly that, a gesture—of the U.S. commitment to West Berlin, he dispatched a small contingent of fifteen hundred soldiers to join the garrison there. He also sent his vice president. Accompanied by Lucius Clay, the Army general who in 1948 and 1949 had led the airlift of food and supplies to the Western sector during a Soviet blockade, Johnson arrived in West Berlin on August 19. A boisterous crowd of one hundred thousand lined the eight-mile route from the airport to the city hall; another three hundred thousand greeted him in the square outside the city hall, cheering and, in many cases, weeping at his words of solidarity. "This island does not stand alone," Johnson said. The United States is "prepared to carry the burden . . . to do our part to see that freedom is preserved."

As Kennedy had anticipated, the morale boost was short-lived, but it was even more fleeting than expected. Within days of Johnson's trip, the Kremlin charged the United States with flying "revanchists, extremists, saboteurs and spies" into West Berlin and openly questioned whether the Allies' access to West Berlin should remain so unfettered. Tensions simmered at the border: Communist guards fired machine guns and water cannons at Germans who approached the wall from the east; they fired warning shots to scare away a crowd that gathered along the west side. Khrushchev, in Moscow, gleefully predicted West Berlin's decline and death—its eventual abandonment by the Allies and even its own citizens. Few West Berliners voiced a contrary view. Few saw the wall as an end to Soviet encroachment—or to Western acquiescence.

Then, on August 30, McGeorge Bundy walked into the Oval Office carrying word—verified by the CIA—that the Soviet Union was about to announce a resumption of nuclear testing, ending the moratorium that had held, however shakily, since 1958. "Fucked again," Kennedy groaned. Since the start of his term, he had been resisting the pressure from nearly all quarters for the United States itself to resume atmospheric testing, a step he believed would enshroud the world with radioactive fallout and heighten the risk of nuclear war. Suddenly it seemed he had no choice—lest he submit to "atomic blackmail," as the White House called it. Khrushchev did not deny the charge. In fact he confirmed it, breezily, to a delegation from Britain, telling his visitors that he aimed to shock the West into negotiations. "All this makes Khrushchev look pretty tough," Kennedy said to Adlai Stevenson, the U.S. ambassador to the United Nations. "He has had a succession of apparent victories: space, Cuba," the closing of the border in Berlin. "He wants to give out the feeling that he has us on the run."

The first Soviet blast lit up the skies above central Asia on September 1. A second followed on September 4, and a third the next day, at which point Kennedy announced that the United States, seeing no alternative, would resume nuclear testing—underground, to prevent fallout, in contrast to the Soviet approach. When the United States conducted its first test, at a site in Nevada, on September 15, the White House again stressed that the president had been "forced reluctantly to make the decision" and had not, even now, abandoned his goal of a test ban of the "widest possible scope." Most of Kennedy's rhetoric was kept to this temperature—never overheated, never alarmist, but reflecting a stern resolve. He sought to preserve the possibility of talks—the only hope, he said, of getting the nuclear "genie back in the bottle." He knew he would have to be patient. "It's too early," he told his secretary of state, Dean Rusk. "They are bent on scaring the world to death before they begin negotiating, and they haven't quite brought the pot to boil. Not enough people are frightened."

Khrushchev set to work on that. By the end of September, the Soviets had tested fifteen nuclear weapons and gave no sign of relenting. By the first of November, the number of explosions had reached twenty-eight, the largest of which was estimated at fifty megatons (or more, Khrushchev insisted)—ten times the total power of all the weapons used by all nations during World War II, including the two atomic bombs that the United States dropped on Japan. That fall the United States conducted only four underground tests. Kennedy's restraint reflected the fact that America had the superior arsenal; while a fifty-megaton bomb was terrifying, explosive yield mattered less than the ability to deliver a nuclear payload with precision. Both sides were aware of the American advantage. "With respect to ICBMs," one Soviet commander complained to another, "we still don't have a damned thing."

The tests were shadowboxing: as an NSC aide wrote to Mac Bundy, they had to be considered "almost totally in a political and psychological context." By that standard, the Soviet Union was reaping mixed rewards. Politically, the so-called neutrals, mainly the postcolonial nations of Africa, Asia, and the Middle East, registered their unease; but psychologically, the explosions made their impact felt. In announcing the resumption of testing, the Kremlin declared that its bombs could be delivered anywhere by "powerful rockets like those Majors Yuri Gagarin and Gherman Titov rode to begin their unrivaled space flights." This was a lie—a compound lie built of smaller fictions, among them that the Soviets had a reliable ICBM program; that Titov had been able to maneuver his spacecraft at will; that the spacecraft could be armed; and that space-based weapons were in any way viable. But the Soviets' successes in space had bought them credibility. The claim went largely uncontested, even by experts. *Aviation Week,* for all its knowledge of weapons systems, saw the prospect of "a series of brightly painted *Vostok*s orbiting above every few hours and broadcasting messages to earth that this is 'your friendly Soviet astronaut number so-and-so with a 100-megaton warhead aboard ready to go any time.'"

In a poem titled "Fall 1961," Robert Lowell captured the national state of mind: "All autumn, the chafe and jar of nuclear war; we have talked our extinction to death." That overcast autumn, as the U.S. Public Health Service began monitoring fallout every twenty-four hours, as the nation acclimated itself to disquieting daily updates and obscure units of radioactive measurement, Americans sought refuge underground. A thirty-two-page booklet, *The Family Fallout Shelter,* had been much in demand since July, when Kennedy gave his speech calling for vigilance on the home front; inquiries rose sharply after the Soviets began exploding bombs in the atmosphere. One of the calls to civil defense headquarters came from a clerk to the House Science and Astronautics Committee, requesting copies for each of its members.

The mood was not panic but fevered preparedness. Newspapers printed lists of supplies to keep in a home shelter. The crucial quantities to remember, authorities instructed, were *twelve miles,* the outer boundary of the blast area of a one-hundred-megaton bomb, and *two weeks,* the point at which it would be safe, more or less, to step outside, the air largely clear of the worst radiation. It was also pointed out, as a practical fact, that the Soviets would probably strike at night, which meant that public shelters—in schools and office buildings and the like—mattered less than whatever fortifications Americans were able cobble together at home. There was an acute interest now in blueprints. *Life* did its part. The magazine's cover model on September 15 was not, as in most issues, a world leader or a movie star but a man in a radiation suit, his face half-obscured and his hand outstretched, waving or grasping. HOW YOU CAN SURVIVE FALLOUT, the cover promised. Inside were detailed plans for building family shelters. One required the use of a bulldozer and was given, by *Life,* an unfortunate if accurate title: "Big Pipe in the Backyard Under Three Feet of Earth."

On October 4—the fourth anniversary of Sputnik—the Soviets conducted another nuclear test, the seventeenth in the series, exploding a bomb of several megatons above the islands of Novaya Zemlya, in the Arctic. The United States marked the occasion with another setback. On what was meant to be a morale-building tour of U.S. space facilities, Lyndon Johnson stood in the heat on a desert hillside in southern California, at Edwards Air Force Base, waiting for a rocket engine to be fired on a test stand. He waited in vain. Mechanical trouble caused the test to be canceled. "You can see what can go wrong in the space program by what happened today," Johnson grumbled on his way out of Edwards. "But there's nothing Washington can do about that." He was not making excuses; he was, in frustration, simply stating a fact.

AT LANGLEY, October 4 was eventful, a day of decision.

Among the astronauts, it had been understood for weeks that Gilruth was weighing his choice for the first orbital flight. On September 13, the much-delayed, unmanned MA-4 had finally become the first Mercury spacecraft to attain a single orbit. "Atlas has the capability to fly a man in orbit," Gilruth pronounced at the postflight press conference. NASA still wanted to complete a three-orbit flight—

carrying either a chimp or a "crewman simulator," also referred to as a "mechanical astronaut"—before putting a man on board, but that day was approaching. NASA stepped up its training regimen. Now, in addition to long hours in the procedures trainer and egress training in open water, the astronauts were sent to a planetarium to review constellations, were told to brush up on Morse code, attended briefings by the U.S. Weather Bureau to get better at identifying cloud patterns, and more. Bob Voas sent them an extensive list of preparations, as well as an updated series of "failure training missions" they would undertake at the Cape. One night in late September, Carpenter came home "bone tired—6½ hours lying in the cocoon," Rene recorded in her diary. "He noted that after 2 or more hours little defects become demandingly apparent—his gloves are too short & hurt his fingertips; right foot too snug." On a happier note, while in the capsule "he used the urinal successfully."

On the evening of October 3, the seven flew from the Cape to Langley, arriving around 9:30, amid cold rain and gusts of wind. Gilruth had made his decision. This, despite the weather, had the group in high spirits. "Not only have they become my good friends," Scott Carpenter exclaimed to Rene, "what competition they're getting to be!" Scott brought Glenn home with him, since Glenn's family still lived in Arlington. "The boys" were "sunburned and bubbling" from their time in Florida, Rene observed in her diary. She, her husband, and their guest raised a glass and toasted the next day's announcement, whatever it might be; then they gave Annie a call.

In the morning Gilruth gathered the men in his office. Each took a seat. Gilruth, in his fashion, made no great speech; he moved quietly and matter-of-factly down his list of flight assignments. Glenn, he said, would go first. Carpenter was his backup. Slayton would fly the second orbital, with Schirra standing by. Shepard and Grissom, respectively, would "troubleshoot." Only Cooper was unaccounted for, a singular sort of humiliation. Eyes scanned the room for reactions. Everyone looked to Glenn; his face was impassive. There was a long silence.

"Well," said Grissom, breaking it, "I guess a handshake is in order for John." This loosened things up, and for Glenn there were not only handshakes now but claps on the back. Later, after the seven returned to their office, Shepard passed around drinks. Carpenter went home to tell Rene the news. He smiled with resignation—and a bit of bemusement. He felt he had outperformed Slayton in the training program, but there was no divining Gilruth's logic. "Guess he can't trust this old multi-engine pilot," Carpenter suggested. Still, as Glenn's backup, he had a shot at spaceflight, unlike Cooper (cutting him out, Rene wrote that day, was "a ghastly, searing thing to do"). "The selection still smarts," Carpenter added to the journal a few days later. "Although I wish it were for me, I am glad it's for John."

Carpenter was alone in that. "All of us are mad because Glenn was picked," Grissom wrote his mother on October 7, three days after the meeting. His graciousness in Gilruth's office had been a false front. The ongoing investigation of what had gone wrong with the hatch on *Liberty Bell 7* left him worried that he had fallen out of the rotation for good. "Neither Al nor I get one of the first two orbital shots," Grissom complained to his mother. "I've been feeling pretty low for the past few days," but "there isn't much we can do about it but support the flight and the program. Of course no one is to know a selection has been made," he cautioned, "so keep it under your hat."

They were all well practiced by now in keeping a lid on flight selections. But this news, for Glenn, was hard to contain. "I tried to take the news in stride," he later wrote in his memoir, but he was elated.

"When John found out about his selection," his mother remembered a few months later, "he acted just like a little boy."

Nearly five months had passed since Kennedy had announced his goal of sending a man to the moon by 1970, and—aside from a substantial budget increase for NASA—he had little to show for it. At a news conference on October 11, asked whether the space program was making progress, he refused to say yes. "Until we have a man on the moon," Kennedy replied flatly, "none of us will be satisfied. . . . I do believe a major effort is being made. But as I said before, we started far behind, and we're going to have to wait and see whether we catch up."

There were no follow-up questions. Reporters were far less concerned about space that day—or that fall—than in a host of pressing issues: Berlin; Soviet nuclear testing and the credibility of the U.S. deterrent; Kennedy's dispatch of General Maxwell Taylor, his closest military adviser, to South Vietnam on a fact-finding mission; and Americans' rising fear that war would break out. "They're naturally and quite correctly concerned," Kennedy said. "We happen to live in the most dangerous time in the history of the human race."

Given all this, it was perhaps inevitable that the military would make another play for control of the space program—or, at least, a role and budget on par with NASA's. The stakes, insisted the generals and planners and contractors, were existential. Walter Dornberger, a Bell Aerosystems executive who, with von Braun, had led development of the V-2 rocket for Hitler, predicted that soon the Soviet Union would strike against all U.S. space systems—reconnaissance, weather, communications, navigation—and that the United States could do nothing to stop it. Such concerns were echoed, emphatically, in *Aviation Week*, in effect the house organ of the aerospace and defense industries. "One of the worst policy decisions of the past decade was to try to label our national space program with the catchword of 'peace,'" declaimed an editorial. It called out the absurdity of dressing astronauts in business suits—business suits!—to stress the civilian nature of the U.S. program when the Russians had dropped any such pretense about their own purposes: on a visit to London to meet the queen and the prime minister, Gagarin had worn his Soviet Air Force uniform.

Frustrations flared into the open. For a full week in October, the American Rocket Society held an exposition at the New York Coliseum with the aim of educating the public about space. Among the model rockets and lunar landers, amid the talk of fly-by missions to Venus and Mars and the possibility of extraterrestrial civilizations, representatives of the military struck a discordant note. In a thundering speech, Trevor Gardner—the former Air Force official—argued that Titov's flight and the Soviet nuclear explosions were one and the same threat. NASA and the Pentagon, he said darkly, "must have their interface tightened." In a panel discussion, Arthur Kantrowitz—an engineer who had helped design a nose cone that could survive the heat of reentry—said that the notion of two separate programs, civilian and military, was crippling the space program. At a banquet the next night, Lyndon Johnson weighed in, insisting that the differences between civilian and military uses of space were "basic and not superficial." "There should be no confusion anywhere," he said, "about the central principles of American policy": to pursue both paths into space—civilian by choice, military by necessity. Johnson, it seemed, looked to end the debate by occupying both sides of it.

Either way, this marked, for Johnson, a brief return to center stage. In a bitter irony, his role in space policy had begun to recede the moment Kennedy had approved his recommendation of a lunar landing; the moon shot was Kennedy's mission now, and the job of fulfilling it belonged to James Webb. After the May 25 speech, the president had asked Webb to report directly to him, not through LBJ. Kennedy wanted straight talk from his advisers, that and skepticism; what he got from Johnson, invariably, was a sales pitch: “not whether, but how—not when, but now,” as Johnson had put it in May. As Johnson’s position—his hawkishness—became predictable, his advice became inessential. He continued to give speeches on space, visit assembly plants, and contrive to get his picture taken with the astronauts, but here, as in every other realm of domestic and foreign policy, LBJ was on the wane.

DURING A TYPICAL WEEK in October—as Glenn recorded in his calendar (a spiral-bound giveaway from Vought Astronautics, promising “365 opportunities in the challenge of space”)—he reviewed his flight plan at the Cape; went to Langley for meetings at the STG; practiced water egress off the coast of Virginia; and, less typically, spent a weekend at home in Arlington, timed to coincide with a visit from his parents. He was in constant motion, but Project Mercury, that fall, seemed stuck in neutral, gears grinding. On the first of November, after six months of redesigning and rechecking the equipment, Mercury-Scout 1—an important test of the global tracking network for an orbital flight—began to break up after liftoff and had to be destroyed. Then, on November 12, two days before a chimp was to be launched into orbit, NASA postponed the mission, disclosing only that the spacecraft had “some problems.” This, the last, crucial test before Glenn’s flight, was put off until the end of the month. In a memo to Mac Bundy, a NASA official conceded that it was unlikely that an astronaut would orbit the Earth before January 1962.

Lacking any fresh rumors about the pilot selection, the press turned its focus to the chimps. The chimp flight, MA-5, had been rescheduled

1. John Glenn, "go" for launch, February 20, 1962. "America needed this kind of hero," a *Time* reporter observed, "and got it."

2. Glenn at seventeen, 1938. As a small child in New Concord, Ohio, he went on a short flight in an open-cockpit biplane. After that, he dreamed of little else.

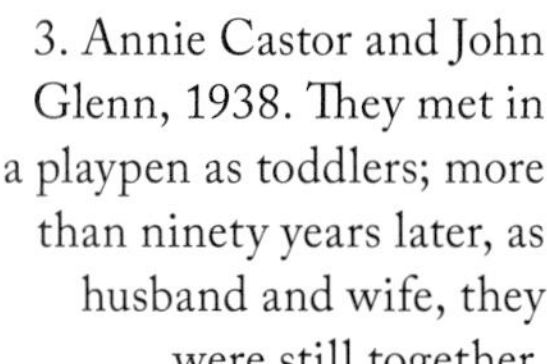

3. Annie Castor and John Glenn, 1938. They met in a playpen as toddlers; more than ninety years later, as husband and wife, they were still together.

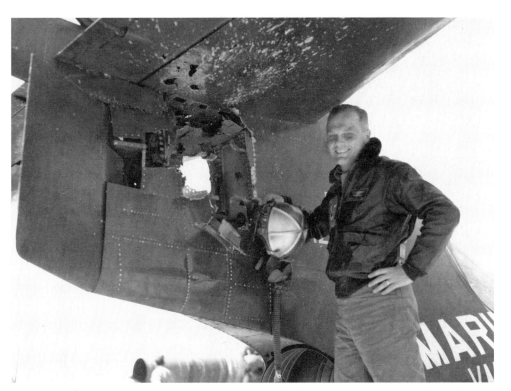

4. As a combat pilot in Korea, Glenn ignored orders against attacking a target and ended up with a hole in the tail of his plane. “The man is crazy,” said his wingman Ted Williams, the Boston Red Sox left fielder.

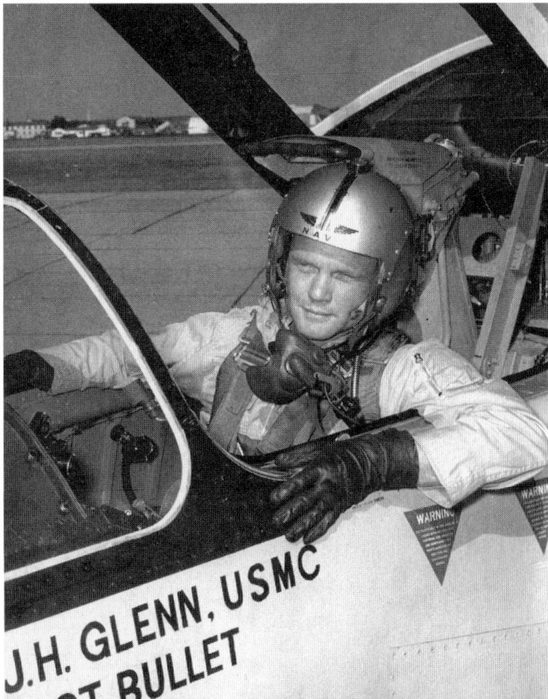

5. Project Bullet, July 1957. Glenn set a transcontinental speed record and was hailed, in headlines and newsreels, as a national hero.

6. At a press conference on October 9, 1957, Dwight D. Eisenhower shows his impatience at yet another question about Sputnik. The Russian satellite, he said, was no reason "to grow hysterical." Many Americans felt otherwise.

7. Lyndon B. Johnson, the Senate majority leader, looks at a spot believed to be the launch site of the second Russian satellite, Sputnik II, in November 1957. "Plunge heavily" into the space issue, an adviser told him; it might "elect you president." At left: Senator Ralph Flanders of Vermont; Admiral Hyman Rickover.

8. The Mercury Seven. It was inevitable, they believed, that one of them would be killed before Project Mercury was complete. Left to right: Wally Schirra, Gus Grissom, Glenn, Scott Carpenter, Deke Slayton, Gordon Cooper, Alan Shepard.

9. The "Chosen Three": Grissom, Shepard, Glenn. In February 1961, NASA identified them publicly as finalists for the first U.S. spaceflight. In truth, the contest was over: Glenn was the public's favorite, but Shepard had won the coveted slot.

10. Scott Carpenter, Glenn's only true friend among the astronauts and his backup for the orbital flight.

11. Nikita Khrushchev and his cosmonauts: at left, Gherman Titov; at right, Yuri Gagarin, the first man in space. Every Soviet "first," Khrushchev understood, struck a blow against American power and prestige.

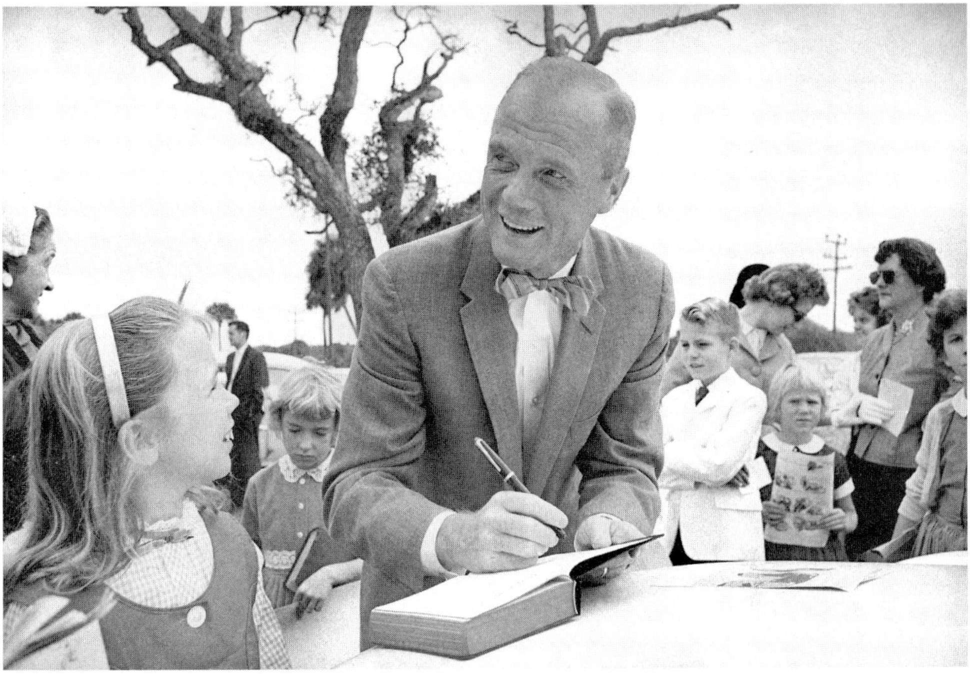

12. Glenn autographs a Bible after church services in Cocoa Beach, January 1962. The longer his launch was delayed—ten flights were scrubbed by mid-February—the more fervent the crowds became.

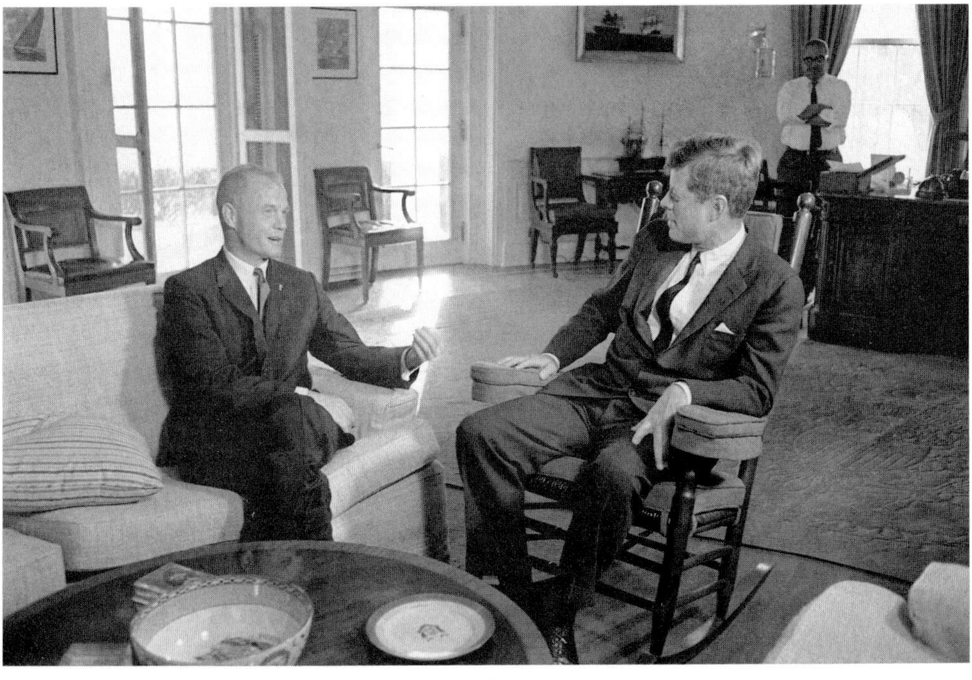

13. Glenn and John F. Kennedy, February 5, 1962—a relaxed conversation for the cameras at a time of mounting national tension.

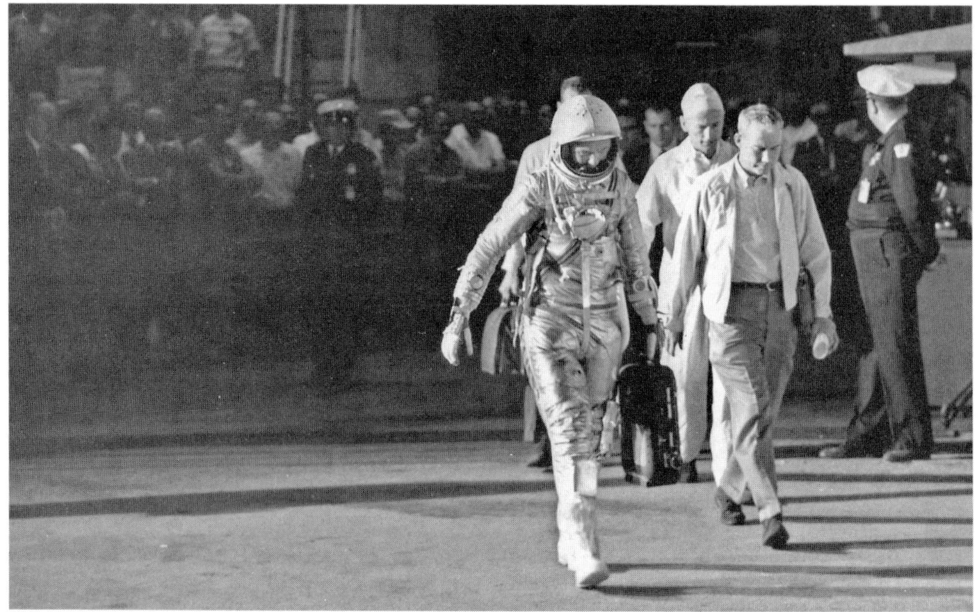

14. Arriving at the launchpad, February 20, 1962. At Glenn's side is Bill Douglas, the astronauts' physician; behind them is Joe Schmitt, the suit technician.

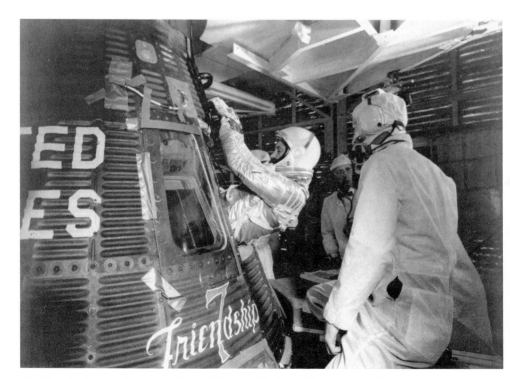

15. "You don't get in it, you put it on," Glenn said of the Mercury capsule.

16. Liftoff. "God-speed, John Glenn," said Scott Carpenter—an impromptu prayer.

17. (*Below*) Lyn, Annie, and Dave Glenn watch the coverage of the launch at home in Arlington, Virginia. Behind them is the family's minister, Reverend Frank Erwin.

18. Watching in silence at the White House. No one—besides Glenn himself—had more riding on the fate of *Friendship 7* than Kennedy. From left: Senator Hubert Humphrey, Johnson, Kennedy.

19. Grand Central Terminal, New York. For hours, thousands of commuters—cheering, praying—stood transfixed by the coverage.

20. Mercury Control Center, Cape Canaveral. On the map above, the orbital path.

21. Glenn in orbit. Ninety-five minutes into his flight, a warning light went on at Mercury Control. If the signal was right, Glenn's heat shield was loose; he would almost certainly burn up on reentering the atmosphere.

22. Chris Kraft, the flight director, ordered that Glenn be kept in the dark about the risk to his life. Kraft's word, in the control center, had the force of law.

23. If the heat shield was loose, Glenn's only hope of survival was that the titanium straps of the "retropack"—attached, at left, to this model capsule—would hold it in place.

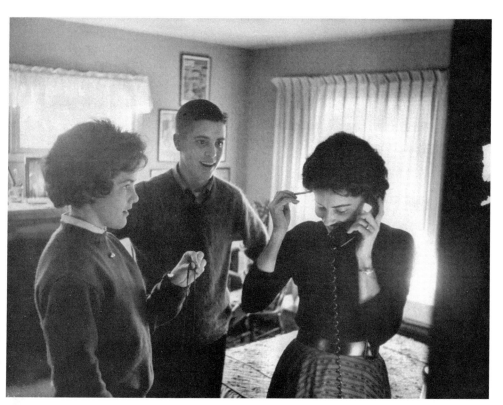

24. Lyn, Dave, and Annie get their first call from Glenn after his safe recovery. "You must be the most excited man in the world," Annie said.

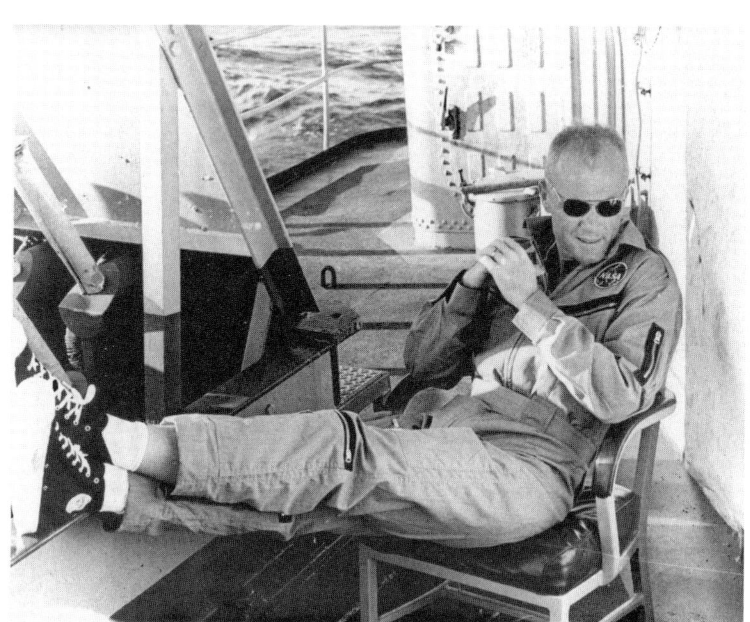

25. On the USS *Noa*, shortly after splashdown.

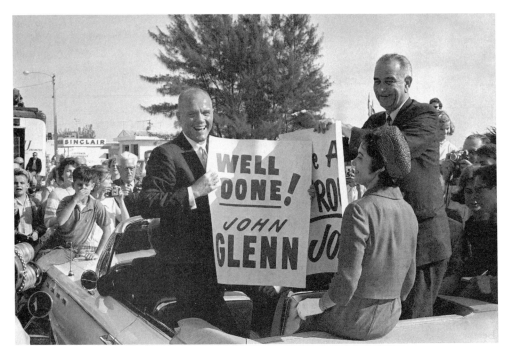

26. Glenn, Annie, and LBJ in the motorcade through Cocoa Beach, February 23, 1962. The parade route was "bedlam," a reporter said—less a celebration than a catharsis.

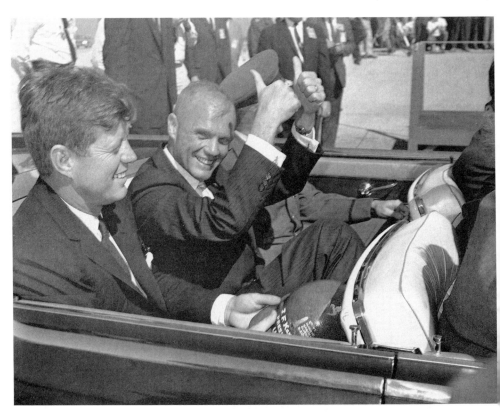

27. A tour—a victory lap—around the launchpad at Cape Canaveral.

28. Glenn puts an honorary hard hat—a gift from the launch crew—on Kennedy's head. Between the two men, looking down, is NASA administrator James Webb.

29. Glenn and Kennedy inspect the capsule. There was an affinity between the two men—a mutual esteem, even awe.

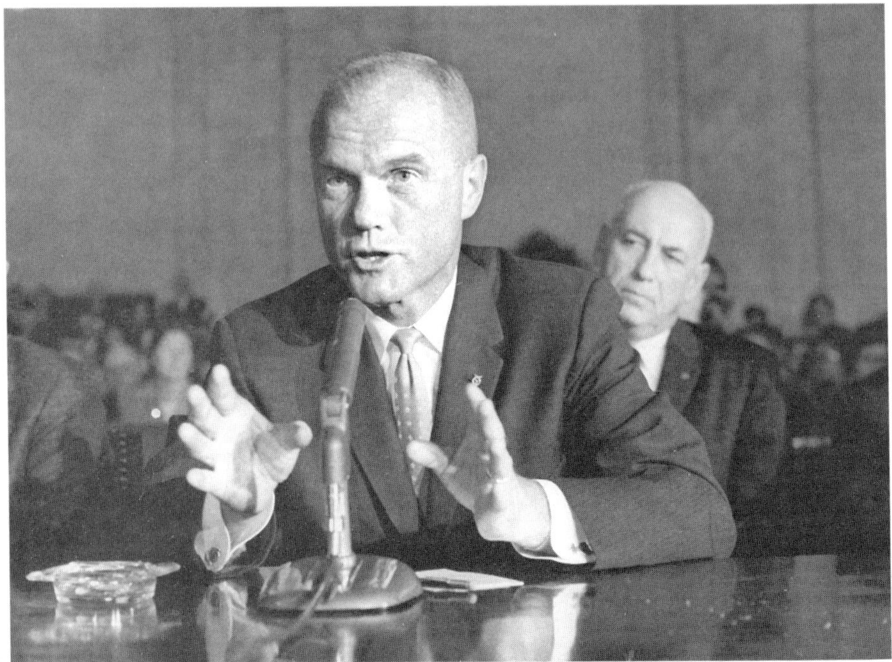

30. Robert Gilruth, director of the Space Task Group, sits behind Glenn at a congressional hearing after the flight of *Friendship 7.* Gilruth was inscrutable—as were his reasons for choosing Shepard over Glenn for the first spaceflight.

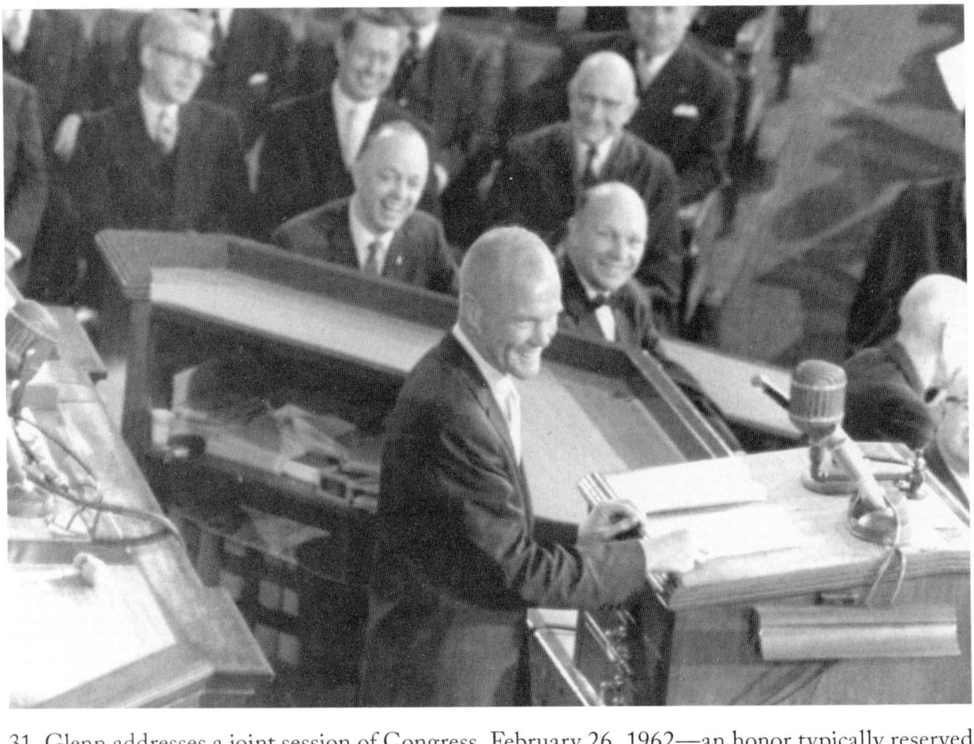

31. Glenn addresses a joint session of Congress, February 26, 1962—an honor typically reserved for heads of state. In 1974, he would join their ranks as a senator from Ohio.

for November 29. Ham, who had ridden MR-2 into space in January 1961, seemed to some the obvious choice; but the knives were out at the chimp training facility at Holloman. "Sources say . . . that Ham is going to be nosed out," the *Washington Post* reported; he had gotten "too fat." The slot for the three-orbit flight went to Enos—a risky choice. While Ham had proved easy to train, Enos had less patience for the system of rewards (banana pellets) and punishments (electric shocks to his feet) and was said to have a mean streak. To the Air Force cardiologist who examined him, he was "a little monster." But Enos was smart and effective, and the selection of chimps for flights had never been a popularity contest.

And no one faulted Enos for the trouble that followed. Early in the morning of November 29, as MA-5 sat, with its passenger, on Pad 14, a multitude of problems—including a defective transistor whose replacement was also defective—held up the launch by more than two hours. The flight that followed was a qualified success—from NASA's perspective if not, one imagined, from the chimp's. The spacecraft's systems worked well enough that Enos was able to perform his tasks, but the interior of the capsule overheated (it "merely got warm," a NASA spokesman said) and a lever malfunctioned, giving Enos repeated, unwarranted shocks. Of greater concern, the capsule had drifted out of attitude—its orientation in space—prompting the Automatic Stabilization Control System to fire thrusters to move it back into place. The cycle persisted: drift, correction, drift, correction, until Chris Kraft decided to bring the chimp back after two orbits rather than three. That order, relayed through a command post in California, nearly went unheard: communications cut out, stayed out for a while, and were restored just moments before Enos would have begun a third orbit. NASA was not to blame. Somewhere in Arizona, a tractor had plowed into a transcontinental telephone cable.

The safe return of chimp and capsule, despite their difficulties, raised hopes that an American might orbit the Earth before the end of 1961. "This year" had been the mantra in the papers ever since Gagarin had flown in April; during months of repetition it acquired an incantatory power, a significance wholly out of proportion to its practical import. "The reason for the hurry," chided the *Washington Post,* a rare voice of restraint, was "propaganda in its most trivial sense: the history books would note that both the Russians and the Americans put men in orbit during 1961." But NASA remained noncommittal. At a press conference that afternoon, Gilruth was asked whether the next flight would be made by a man. He shrugged. "Not necessarily."

But that message was largely overlooked in light of what Gilruth said next: that when the United States did send a man into orbit, that man would be Glenn.

Suddenly it all made sense. "I knew all along, John," a New Concord neighbor wrote him, "they were saving you for this." Instantly the idea took hold—not just among Glenn's friends but among newspaper editors and politicians and the public—that the NASA managers had passed Glenn over previously because, as one Marine Corps officer declared with authority in the *Times,* they had been "holding [him] back to ride the big one." Was it not self-evident? The news seemed to confirm what so many people had believed all along: that he was the first among equals, "cut from a different stone" than Shepard or Grissom, as a *Time* correspondent told his editor; that he had been blessed with special, ineffable qualities, possessed of a certain greatness that made him the obvious choice, the only choice, for this role. It seemed fated, destined, predetermined.

Yet Gilruth never said or even implied as much—not at the time or in the decades that followed. Neither did Williams or any other participant in the flight assignment process. In the discussions that yielded the selections of Shepard and Grissom, there appears to have been no talk of keeping anyone in reserve for a later, more important orbital flight. An astronaut was not a pitcher who needed to rest his arm before the big game. On the contrary, it might have been argued—indeed it was argued, even if the position did not prevail—that the best way to train for an orbital trip was to take a suborbital trip, not to log extra hours in the simulator. Besides, if NASA managers, back in January, had been looking to "save" someone for future use, it would probably have been the man they judged their most proficient pilot: Al Shepard.

The fact that Glenn ended up with "the first major flight," as Bob Voas reflected, was a product not of planning but of "chance or . . . happenstance." Glenn's bad luck had become his good fortune—his exceedingly, if inadvertently, good fortune. His selection to fly Mercury-Atlas 6 ended—in large measure erased—the humiliation of the past year. Glenn had been the backup to the first and second Americans in space. But now the rightful order of things had, it seemed, been restored, and the first and second Americans in space reclassified, in retrospect, as warm-up acts for the third.

CHAPTER 13

Contingencies

THE ARGUMENT REMAINED unresolved. In 1959, a good many engineers and space officials had made the case that the principal task of an astronaut, when he someday rode a rocket into space, would be to keep his gloved hands to himself. The automatic system would fly the spacecraft; the pilot, said the research chief at Bell Labs, should avoid interfering lest he "louse things up." Taking the opposite view, John Glenn had told reporters that he had no plans to "just sit there" in the capsule but would, instead, be "working the controls." But now, two years on, as Glenn prepared to be sent into orbit, this was still mostly an aspiration—and a source of conflict.

On November 29, the day his selection was publicly announced, Glenn sat in a meeting room at the Missile Test Annex at Cape Canaveral and reviewed the flight plan—the list of responsibilities he was expected to fulfill. Six minutes and fifteen seconds after liftoff, for example, he was to "report initial 'zero-g' sensation"; at the twenty-minute mark he was to observe Lake Chad, in Africa; two minutes later, he would conduct a yaw maneuver, slowly shifting the capsule 60 degrees to the right and back. The plan accounted for every check of every system and switch, every test of the high and low thrusters. Little would be left to chance—another way of saying that little would be left to the astronaut's discretion. As a member of NASA's Flight Activities Section presented the plan, Glenn marked up his own copy with a pencil. He numbered a series of points from one to thirteen. Then he delivered, in effect, a rebuttal.

"We have had troubles" with ASCS, the Automatic Stabilization Control System, on every flight, Glenn began. Walt Williams had been "sold a bill of goods": automatic control was not safer than putting an experienced pilot in control. No test pilot at Edwards would ever fly an experimental aircraft on autopilot, Glenn said; why should an astronaut? All seven, he said, had the "impression that some feel we're irresponsible and have the 'I can fly anything' attitude," and that NASA managers were taking a "180° opposite tack to try to keep us in bounds." NASA's disregard for its astronauts ran so deep, Glenn complained, that changes to the flight plan were presented to the pilots as holy writ; the paper trail of revisions often bypassed the astronauts' office. "We not only have felt that we were not part of [the] team," Glenn's notes stated, "[we] have had [the] feeling people felt they wouldn't even bother with our opinions, and intentionally avoided us."

The reference to Edwards was telling. The role of the test pilot, as the saying went, was to push the envelope—to press the equipment to the very edge of its capabilities. Here Glenn was the equipment. MA-4 and MA-5 had tested the capsule in orbit. The main objective of MA-6 was to gauge man's ability to function in space—to determine "how much work we can let a man do up there," as Dr. Stanley White, chief of Mercury's Life Systems Division, put it. With longer, more demanding missions ahead, this was becoming a pressing question. The Soviets, presumably, had answered it to their own satisfaction, but NASA knew almost nothing about the assignments Gagarin or Titov had performed or how well they had fared.

In Glenn's view, the flight plan left him no room to test the most essential human capacity of all—judgment, decision-making—in a hostile environment. Back in May, Shepard had been given the latitude to switch the control system briefly to manual, allowing him to roll the capsule and adjust it up and down and left to right. But the suborbital flights of 1961 were too short to have raised, in any serious way, the question of what an astronaut was capable of doing. Glenn would spend nearly five hours in space, and the idea of having every move predetermined was unacceptable to him; it was a waste of his abilities, a squandering of what he and his flight could achieve. He let this be known.

When Glenn had entered the program in 1959, he'd looked like a man who could not be denied a thing. This had not proved true. In an argument, Glenn was more formidable than any of his peers—but no more likely, in the end, to get what he sought. On December 6, one week after the flight plan review, Helmut A. Kuehnel, head of the Flight Activities Section, noted in a memo that "as a result of this presentation, the flight plan has been modified to be . . . extremely conservative." It called for the ASCS to remain in control for most of the mission, while the astronaut acted in a "backup capacity to the automatic systems." Glenn was finally the prime pilot. Yet he remained, all the same, a backup.

"One politico has decided we will get MA-6 off on time or else," Scott Carpenter wrote in his journal that fall. "Launch operations have been ordered to do it. That's a hell of a way to run this affair. The politicos should leave the engineers alone."

There is no evidence that any politician, even the president, issued such an order; but then, no one needed to. The pressure to put the first

American in orbit "on time"—a fluid term that had once (long ago) meant January 1961, before it slipped to May, until it was put off to July, then October, and now December—was felt throughout the program. In the rush to launch before the end of the year, the flight crews were being pushed so hard that NASA officials began to fear they would slip up, cut corners, "do something wrong," as Shorty Powers put it privately.

In early December, Glenn and Carpenter left the relative comfort of the Holiday Inn on Cocoa Beach and settled in, for the duration, to their hermetic quarters in Hangar S—the blue walls and metal bunk beds familiar to Glenn after his weeks here with Shepard and Grissom. He welcomed the isolation. One of his greatest worries now was the common cold. "I could see, in my worst dreams, coming up to flight day and catching some kind of bug that would get me bounced off the flight," he later recalled. So could a sprained wrist, a wrenched back—the list of disqualifying injuries and ailments seemed to grow longer as the launch date approached.

He did, however, keep up his regimen of running five miles a day—a calculated risk. It gave him a sense of peace. It also kept his weight down, which took work for Glenn, who was naturally thick; he was well aware that every extra pound he weighed at liftoff would require an additional thousand pounds of fuel to propel him into space. So almost every evening, after his last check of the capsule or last session on the simulator, Glenn put on white sneakers, black shorts, and a white T-shirt, clipped his plastic ID badge to the V-neck, and headed for the restricted stretch of beach that lined Cape Canaveral. He hammered his way along the hard-packed sand, often alone among the gulls and crabs, sometimes joined by Carpenter.

"I have resolved to train myself as thoroughly as John," Carpenter had vowed in his journal after their selection. At thirty-six, he was younger and fitter than Glenn, but he hated running; and he soon found that Glenn's relentless pace extended to every area of conditioning, physical and mental. "John, I must say, is wearing me out," Carpenter confessed. "I am not capable of the continued dedication to this thing that he is—I just have to break now and then and think of something else—have a drink, listen to music, talk about a good book, play the uke[lele] . . . , but not John. Plug, plug, plug—it kills me. It amazes me. It makes me a little proud and a little sorry."

The pair maintained an easy camaraderie—a good thing, since they were about to spend more time in close quarters. On December 6, Gilruth and Williams announced that there would be no orbital flight in 1961. "Early next year" was the most they could say about the new launch date. The capsule sat in Hangar S; the booster stood inert on the pad—hostages, both, to technical problems. The disappointment was widely felt. "Long-held hopes . . . vanished today," mourned the *New York Times* after NASA announced the delay.

Over the course of the year, in addition to the Mercury-Redstone flights, the United States had launched three times as many scientific satellites as the Russians had since the start of their program. But what Ted Sorensen said of JFK—"he thought of space primarily in symbolic terms"—was true of most Americans, and since Sputnik there had been no symbol of space supremacy more powerful than the image of Gagarin, in his Soviet Air Force uniform, standing with Khrushchev atop Lenin's Tomb. The sight of Glenn returning home for Christmas was a symbol, too.

YET IT WAS GOOD to be home. During the past three months, Glenn had spent only eight days with Annie and the kids. With his launch postponed until the new year, he was given a full week off. On December 21 he'd flown north to Virginia and thrown himself into last-minute shopping, visiting with neighbors—enjoying vestiges of the life that he was no longer living. Everything was the same, but nothing was normal. No one could shake the feeling that this might be their last Christmas together. Lyn, now fourteen, was the only one willing to say it out loud. When she had first been told about her father's selection for the flight, she had blurted, "It's wonderful—if he comes back."

This was not a new worry. Annie had been living with that fear since World War II. The orbital flight, she told a reporter, was "just another step in the program." She paused, fidgeting with the little blue replica of the capsule on her charm bracelet. "No, that's really not right," she corrected herself. "It's a big event in our life and I'm anxious."

Just after Christmas, the Glenns took a short drive north to Great Falls Park, where the Potomac widens into a concourse of rock and churning water. They lit a campfire. It was Lyn, as Glenn remembered it, who raised the subject: whether he would survive his flight. "This was something I probably should have brought up before," he reflected years later, but in his eagerness to "keep the best outlook," he hadn't. He had brought home models of the capsule and cheerfully walked his children through the flight plan, the engineering, the way the systems worked. But he had never discussed the risks except to dismiss them; he had never acknowledged that he might die. Now he told Lyn and Dave that he expected to get back safely, but if he didn't, if something went wrong, they shouldn't blame God, or NASA, or anyone. "I wanted this," he said. And he had done it not just because he wanted it but because he thought space exploration was important for the country, even if it cost him his life.

"I was just so stunned that I couldn't even respond," Lyn later recalled. No one could. They sat in silence for a while, staring at the fire, then hugged each other and left it at that.

"It's been a tough first year," President Kennedy said, "but then they're all going to be tough." Gagarin, the Bay of Pigs, Vienna, Berlin, the nuclear tests—each had been a blow to Kennedy's expectations, each a rebuke to his sense of what it would take to prevail in the "long twilight struggle" he'd described in his inaugural address. The certitude of those early days had given way, painfully, to a recognition of realities in their maddening complexity. In November, in a reflective speech at the University of Washington, Kennedy seemed to be in dialogue with himself: "We have learned," he said, "that reason does not always appeal to unreasonable men. . . . No one should be under the illusion that negotiations for the sake of negotiations always advance the cause of peace. . . . It is a test of our national maturity to accept the fact that negotiations are not a contest spelling victory or defeat." Hard-earned lessons, all.

But in losing his illusions, he discovered his resilience. Having passed through a perilous year, Kennedy emerged with a new confidence—not the brash variety of January 1961 but "a wiser, more mature kind," as *Time* discerned in January 1962, naming him its Man of the Year. Whether his decision, in August, to send troops to fortify West Berlin had indeed been the turning point, as the magazine judged, he seemed more secure in his footing. He was no less intent on talks with the Soviets but was less willing to overlook their obstructions. "You have offered to trade us an apple for an orchard," he reproved a Russian diplomat. "We don't do that in this country." He strengthened his hand by dramatically increasing the defense budget. And the American public was solidly behind him. Gallup had his approval rating at 78 percent, higher than Eisenhower's at the one-year mark.

On January 11, Kennedy rang in his second year with one of the longest speeches of his public life: a fifty-three-minute State of the Union address that rambled across the landscape of concerns. But for all its breadth, it had almost nothing to say about the issue that AP newspapers and radio stations had voted the top news story of 1961, above Berlin, above the Bay of Pigs: the space race. Before the address, Ed Welsh of the Space Council had sent suggestions to Sorensen; Webb and Dryden had drafted a passage recounting the steps NASA had taken in 1961, adding that "progress in our manned spaceflight program has been gratifying." Kennedy used none of it. Instead he tucked the discussion of space into a section on the United Nations, committing to share the benefits of the U.S. program with the entire free world. In passing he vowed—if "vow" was not too strong a word—that the United States would be "among the first" countries to land a man on the moon.

Such was his state of confidence in NASA. In 1960 he had attacked Nixon for sugarcoating the U.S. effort in space; he was not going to let himself be accused of the same. Every day of delay for Glenn's flight was a day when the Soviets might make one-man orbital shots a thing of the past. At Cape Canaveral, rumors made the rounds that the Soviets were about to put a two-man spacecraft into orbit around the Earth. Drew Pearson, in his syndicated column, wrote that the Russians were weeks away from launching a cosmonaut into orbit around the moon. His source, he said, was the CIA. So until John Glenn had been sent into space and had come back alive, Kennedy was not going to stand before the nation and declare himself gratified.

AT THE CAPE, the year began with another postponement. No sooner had Glenn's capsule been "mated" to its booster, Atlas 109-D, than NASA announced that it was delaying the launch by a week, maybe more, to fix a problem with the booster's fuel tanks. Walt Williams returned to the Cape after the holidays in a funk. Driving alone across the causeway from the mainland, he caught sight of the launchpads and tried, and failed, to will himself into a state of excitement. He was all too aware that when he saw his family next, it would "either be a time of national triumph or of bitter failure and perhaps tragedy."

The mood was brighter, determinedly so, at "Astro," the Astronautics division of Convair, which had built the Atlas booster. At the plant in San Diego, Convair executives led the press on a tour, projecting throughout "a bracing air of confidence," as a *Time* reporter described it. There was a good deal of buoyant talk about Astro's quality assurance program, its painstaking process of stress-testing every sheet of thin-gauge steel and X-raying every tiny transistor and putting every missile through computer simulations of the maneuvers it would perform. Only then, engineers explained, did an Atlas receive its seal of approval: a round, white decal bearing the astrological symbol for Mercury and an *R* for "Reliability." "There's not a shadow of a doubt about any component," one test manager said. Nor was there room for it: "We're working with Colonel Glenn's life." But pressed by reporters, Astro's president, Jim Dempsey, put the reliability of the Atlas at 80 percent—a one-in-five chance of failure. No matter how many inspections this missile had passed, all it would take was one bad transistor or valve to turn it into a fireball. The longer the briefing went on, the more apparent this became. "In no case," protested Dempsey, "has there been an Atlas failure in which an astronaut could not have successfully escaped."

A dubious claim, but even if one took it at face value, there was plenty else to worry about. Another concern, if not a new one, was that Glenn would crash-land at some remote spot along his orbital track: in the Kalahari Desert of southern Africa, possibly, or in the jungle of Papua New Guinea. Glenn was well prepared for the possibility of being lost in the wild; but now, in the cloisters of the Space Task Group, white men in white shirtsleeves were suddenly seized by the worry that Glenn would be not lost but found—by indigenous tribes. Glenn himself imagined the scene: a sonic boom; a speck falling from the sky; the plume of the parachute; then, Glenn said years later, "the hatch blows off the side and out steps this thing in a silver suit."

Calls were made; cultural experts were consulted. NASA's Office of International Programs supplied Glenn with a statement to make in the event of an emergency landing in a non-Western (non-white) country. He would carry typewritten copies with him on the flight—translated into French, Spanish, Hausa (spoken mainly in Nigeria and Niger), and Swahili—and was told to hand one to whomever he encountered first. He was encouraged to memorize key lines (underlined by NASA to aid in the process). "I am an American," the statement began.

> I have returned to the earth in this craft/vehicle after a test flight in space for peaceful purposes. I ask your help. Please tell your officials that I am here. Please inform the nearest American Government officials that I am here. I will stay with this vehicle until arrangements can be made to return me and the vehicle to my country. Thank you.

In other words: "Take me to your leader."

Gallows humor was rife at the Cape. Technicians worked together on an "Ode to the Astronauts," which did not mince words:

Some like it cool, some like it hot,
But no choice has the Astronaut,
If on return he's not awake,
He'll start to feel like well-done steak.

And so on in this vein, stanza after stanza. Glenn was given a copy.

It did not escape comment that Glenn's spacecraft was Capsule No. 13. The number had been assigned to the machine by McDonnell more than a year earlier, well before it was slated for the first orbital flight, and nobody said a word about it. But aviators, as a group, are superstitious: in wartime, especially, the pockets of pilots jingled with good-luck charms and coins and religious medallions; cockpits carried horseshoes, baby shoes, an assortment of talismans. Glenn, putting his faith in a higher order than magic, shrugged off suggestions that he change the number. Still, his capsule needed a name—a good, appealing name, something catchy for public consumption, like Shepard's *Freedom 7* or Grissom's *Liberty Bell 7.* Glenn had a few ideas, but he saw the naming process as a means of making his children feel a part of his mission. He gave Lyn and Dave a single guideline: the name should say something about America's role in the world, since the world would be watching.

The children went about their work with earnestness: consulting a thesaurus, canvassing friends at school, jotting ideas in a notebook. They and their father culled a list of favorites that ranged from the patriotic (*Independence, Republic, America*) to the steadfast (*Defender, Resolute*), the intrepid (*Columbia, Endeavor, Voyager*), and the ingratiating (*Brother, Partner, Companion*). By Glenn's telling, he let Lyn and Dave decide, and they happened to settle on his own top choice: *Friendship.* In Lyn's recollection, "We each had a vote, and I'm sure Dad had more of a vote than we did." Either way, Capsule No. 13 now had a name. On the list he'd been keeping, Glenn added a "7" and a check mark after *Friendship.* He informed officials at NASA of his selection. Then he told them he wanted the name hand-painted on the capsule in script—the kind of personal touch fighter pilots added to their aircraft—not in little block letters, sprayed through a stencil, as *Freedom 7* and *Liberty Bell 7* had been.

He might have expected the reaction he got. Shepard and Grissom let it be known in the halls of Hangar S that Glenn thought he was *too good* for a stencil, that he wanted an *artist* to paint him a *logo.* That job went to Cecelia "Cece" Bibby, a contract artist, on the grounds that, her supervisor said, women had better handwriting than men. Bibby sketched out three options, which her boss carried to the astronauts' office to discuss with Glenn—returning, in short order, with a sour expression. He tossed Glenn's selection onto Bibby's drawing board. "He said that Colonel Glenn wanted the person who made the design to paint it on the capsule," Bibby recalled. Which was a problem, because that person was a woman. And women, they all understood, were not permitted on the gantry, where the capsule now sat atop its booster. "But she's a woman"—the art director's objection—had gotten nowhere with Glenn, so Bibby was given the assignment and was sent to meet the astronaut.

Bibby had never met any of the seven; she didn't have clearance to enter Hangar S. Wally Schirra escorted her upstairs. Glenn excepted, they treated Bibby as a curiosity, a novelty act: a girl (at thirty-three) with a paintbrush. "Well, you're not what we expected," Gordon Cooper said. "You don't giggle when we talk to you." The pad crew was less welcoming. Stepping into the "white room," a clean room that surrounded the capsule, Bibby was confronted by one of NASA's German imports, Guenter Wendt, the pad leader—"Pad Führer," the astronauts called him, with a certain affection. Wendt told her that a woman had no business on the gantry. Bibby replied that he could take it up with Glenn. Then, amid wolf whistles and taunts from pad workers, she went to take her first look at the capsule.

The astronauts came to treat Bibby as one of their own. She had withstood the hazing of their fraternity and—no small matter—she had a better sports car than any of them did, an A.C. Ace, and could repair it herself. They enlisted her in their pranks. One day Bibby ran into Gus Grissom on the stairs in Hangar S, and he asked her how her paint job for the "Boy Scout" was going. What the capsule really needed, Grissom offered, was "naked ladies" on the hull, like the painted pinups on the noses of fighter planes; that would shake up the straitlaced Glenn. Bibby replied that it wasn't worth her getting fired. Grissom called her chicken, clucked to underscore the point, and headed up the stairs.

Bibby found a way to take the dare. On paper—not on the titanium skin of the capsule—she drew a woman, naked except for her pom-pom slippers, propped up on her elbows, running a finger along the word *Friendship*. "It's you and me, John-Baby . . . against the world," the caption read. Bibby enlisted a white-room engineer to attach the image at the end of the capsule's periscope when the countdown began. She went back to painting the logo and finished on January 19, an occasion to pose for photos—Glenn in his space suit and Bibby, holding her paintbrush above her handiwork, in a white lab coat and white head scarf.

Pranks did little to lighten the mood. The tension sought release and found it in contentiousness. Shorty Powers grumbled that Glenn had been staging "impromptu press conferences on his front lawn" (an exaggeration) and that, contrary to past practice, the astronaut had invited a *Life* reporter and photographer into his house to watch the launch with his wife and children, insisting that the reporter, Loudon Wainwright, was a friend (this was true). Powers was also angered by Glenn's request to bring his family to the Cape and let photographers record the visit. In Powers's view, as Williams described it, "this is too much of a 'last supper' type exercise, and he doesn't think these are the kind of pictures to have around in the event of a fatality."

For Glenn, the flashpoint was the flight plan. He and Voas were dressed down by Chris Kraft over their request to loosen the restrictions on the use of manual control. The argument was ongoing, and nerves remained raw. Glenn readied himself for another meeting—or altercation—about the plan by preparing a litany of complaints, scrawl-ing them at all angles on either side of a large note card. He had just happened to learn, by accident, that the flight plan had been changed again—no one had mentioned it to him. He was curtly told that all changes were for his safety. "No one's better able [to] judge that than I am," he wrote. All he was asking, he insisted, was to be kept in the loop, but he had run afoul of Gilruth for having too many opinions: "People seem to think [I'm] trying to run everything." At the bottom of his note card, Glenn added a last plaintive note: "Felt we'd have help. Never felt so alone & so many against. Why?"

THERE WERE MANY REASONS why: among them, management's exasperation with Glenn for his refusal to fall in line and the failure of the astronauts—riven by jealousies and petty antagonisms—to present a united front. But if Glenn felt alone it was, above all, because he was about to be shot into the blackness of space, at the mercy of machinery that had just been invented. Although thousands of engineers were focused on his fate, the fact remained that what Glenn was about to undergo, he would undergo alone.

On January 19, four days before the scheduled launch, Glenn was interviewed by CBS about the dangers of spaceflight and about the strength that his religion provided him and his family. "We don't look at this flight as being a big dangerous thing that we have to suddenly go into all kinds of divine help to accomplish," he said on the broadcast. "We try and live our religion day in and day out, more than just call on it in an emergency. I dislike intensely the people who make a sort of fire-engine religion, I guess I like to call it, in which they only call on their Maker when the chips are down." At the end of the program, Walter Cronkite pronounced Glenn to be "spiritually prepared to face his next test next week." Perhaps he was. Privately, though, Glenn wrestled with his faith. "Any time a person goes through a big, rather dramatic experience," he said later, it invites reflection about one's "relationship with the rest of the world and with God and whatever God may be to you. And I certainly did my share of that before the flight."

He spent time thinking about his family's conversation in the cold at Great Falls—about what had been said and what he had not found words to say. In the isolation of the ready room, as the launch date approached, Glenn wrote a letter to the "troops," as he had often called his children in the notes he sent home. But he had never written a letter like this. "There have been so many things I have wanted to say to you before the mission that I hardly know where to start," he began.

> I guess it is natural to wax a little philosophical with a big event coming up; I hope that this letter may be applicable for whatever happens on the mission. If the mission is a huge success, then I will be glad I had this chance to let you know how I feel and we can discuss it when I get home. If, by chance, the mission turns out otherwise for me, then I will be glad I took this opportunity to "talk" to you. Let me add that I certainly have no premonition of anything but a very successful flight for Friendship 7. . . .
>
> We know we should each live every day in such a way that we'd be proud to have it be our last day. This requires some thinking about what you want your life to be like. I suppose I could title this letter my "Legacy to the Troops" because I would like to let you know what I hope we have instilled into you while you have been growing from children to young adults.

That legacy, Glenn believed, would find expression in "an honest mind. Candid intelligence. Loyal spirit. An understanding heart. . . . A firm confidence and pride in your own personal abilities. . . . A strong religious faith, for it is our overall guide and gives meaning and purpose to our lives." But the focus of his letter was courage—what it meant, why it mattered above all else.

> Courage is only present when there is also fear. In a dangerous situation we all have fear. It would be foolish to not be afraid. I have never gone into combat, for instance, without being afraid, but the important thing is what we do about being afraid. . . .
>
> I can tell you that I will be afraid when the booster is getting ready to fire because I know there are dangers involved much greater than normally experienced. What I do at that time is the important thing. . . . Human progress has never been fostered by the cowards who have let fear rule their lives.

Whatever his children thought of the letter, it brought Glenn no evident peace. He had meant his words to be helpful "whatever happens," but as he continued to brood about the possibility that he would not make it home alive, his letter to the "troops" seemed somehow insufficient. So he wrote another—not a letter, exactly, but a script. Over the years, during his long stints away from his family, Glenn had made a practice of sending them tape-recorded messages. And now, on the eve of his flight, he took out a yellow legal pad and wrote himself a script for a recording he hoped his children would never have to hear.

> If you hear this, I've been killed.
>
> Made peace with God a long time ago before this happened. Felt good to live with that feeling of not being afraid to die. Didn't always live like it. Kept trying.
>
> Always believed life continuing after & it does. I can tell you that. Just diff. form. I have a secret from you because I know what

it's like now and you don't, but you have something to really look fwd to. Grave not goal of life. I'm better off right now than you are.

Now let's get practical and discuss funerals. You're aware of my feelings regarding them. Please keep it simple. The father and husband you knew and know is not in that body, even if you get it back, remember I'm not in it. No need to fuss over something like that, is there? Little like fussing over an engine when all fuel has evaporated from tank. . . .

There will be a lot of attention on you for a while. Don't let it affect you too much. Rather than being so sad, look at the other side. I was able to contribute some very worthwhile things during my lifetime. I'm proud of that & so should you. That's how progress is made—little contributions from many. Mine was maybe a little more spectacular but was certainly not as much of a contribution to progress as many people make in their fields each day. Just resolve to always do your best in whatever work you do. So be glad, as I am, that my life was not wasted.

I like to think of my life continuing in you.

Back to immediate problems—the funeral. If that body is there and is put into Arlington, here's what I'd like for you to do. Whenever things get a little rough for one of you, just say to each other "DW" and that means "Dad's waiting," and I want you to think of me laughing and joking as we've done so many times, because that's exactly what I'll be doing while you're so unhappy. Big smile, then, from each of you in return to me, OK? Now, at Arlington, if they lower that body into the ground, rather than watching that, I want you three to step outside the canopy. Dave, you pick the highest tree you can see, point it out, and I want each of you to look at the highest little branch in the tree, and when that little branch waves in the wind, that will be a sign just for us that I'm waving to you,

OK? Maybe that sort of symbolizes the fact that we tried hard, and got to a high point. Now it's up to others to get a little higher. . . .

I'm proud of each of you, more than you can know, and rather than feeling sad at my departure, let's be thankful we had so much time together and look forward to what's ahead. I love you very much and I'll see you later. "DW," troops—go look at that tree 'cause the top branch is waving at you, I can see it from here.

Glenn flipped the page and wrote Annie's name at the top of the next sheet. He jotted a few notes, scattered words and phrases, for a message he would later record for her:

What if—
Had many years together
Interlude
Remarry
Wonderful person & mother
Vibrant & alive

Here his notes trailed off, and Glenn went back to waiting for the countdown.

NASA, too, was at work on a script. On January 16, in a confidential memo with a decidedly bland title—"MA-6 Contingencies"— O. B. Lloyd Jr., of the Public Information Office, warned Webb that the "death of the pilot would likely provoke an enormous public reaction critical of the entire United States manned space effort." He listed several catastrophes that could befall Glenn—none of them likely, Lloyd judged, but "such possibilities must be reckoned with." The booster rocket might explode on the pad, for example, or fail in flight; the spacecraft might break up in orbit or on reentry; the parachute might fail; the capsule might sink with the astronaut aboard. But "the most bizarre fatal situation," in Lloyd's view, would be a failure of the retrorockets or the stabilization control system, which "could leave the spacecraft literally 'stuck' in orbit for several days. The onboard oxygen systems required to support life would last only 10 to 12 hours." In the event of Glenn's death, the memo advised, remarks should be made by the president and vice president, Webb, Gilruth, and one of the astronauts, probably Shepard. Lloyd proceeded to draft most of these statements—each one expressing, in its own distinct way, sorrow at the loss of a friend and pioneer.

"There is a strong rumor at Canaveral that [Glenn] isn't going to make it," a *Time* reporter wrote his editor. He had been making the rounds, collecting anecdotes about Glenn for a cover story that would run after the flight. "I fervently hope," the reporter confided, "this is a biography and not an obituary we're working on."

CHAPTER 14

The Big Scrub

THIS TIME IT WAS the life-support system. On January 22, the day before the scheduled launch of *Friendship 7*, NASA announced that it was postponing the flight until, at the earliest, January 27. Pad technicians, conducting their final checks of the capsule, had found that under certain conditions the system would pump so much oxygen into Glenn's space suit so quickly that he would run out of air three hours into his five-hour flight. "This cause[d] a lot of head-scratching," Walt Williams recalled. The fault, tests revealed, was in the space suit: the seal rings that attached Glenn's gloves to his suit had been installed backward and were leaking oxygen. The B. F. Goodrich Company, which had manufactured the gloves, quickly sent a new pair to Cape Canaveral, and the matter—this small but all-too-typical embarrassment—was kept out of the papers. Still, there was nothing NASA could do to stop the press from tallying the number of postponements. This was the fourth since December.

Glenn drew a neat line through the words "MA-6 LAUNCH" on his calendar and went on with his training, trying not to be dulled by the routine. In his hours outside the simulator, he busied himself by studying astronomy and maps of the Earth—not an idle pursuit. If the automatic control system failed, could he align himself by the stars? Could he orient himself by landmarks on the ground when his field of view spanned nine hundred miles, the arc of the horizon he would see from that height? This had big implications for longer spaceflights.

There was a scientific element as well: for months now, under the direction of Jocelyn Gill, who chaired NASA's Ad Hoc Committee on Astronomical Tasks for the Mercury Astronaut, Glenn had been briefed on the observations they wanted him to make. "Main interest is in appearance of corona," he recorded in his notebook, doodling a little sun. The committee was especially eager to know about comets: their precise positions, the length and direction of their tails. Among the astronomers, excitement was building. "There may be surprises!" Gill wrote.

Bob Voas, who was responsible for most of the scientific agenda, had proposed in the fall that astronauts on orbital flights be given a camera. Gilruth had liked the idea, but other managers didn't, so four months later Voas and Glenn were still trying to convince them. The official view, delivered in the same peremptory manner that had been driving Glenn crazy, was that a camera would be a distraction for the astronaut, who had more important things to do than take snapshots. Besides, they said, it would be hard to operate a camera while wearing gloves. Engineers did retool a Leica 35mm camera to allow Glenn to take a single roll of ultraviolet spectrographic images of the stars. But this was not what he had in mind, as he later wrote in his memoir: "The slide-rule-and-computer contingent lacked the imagination to see the value of photographs that would help translate an astronaut's experience for anyone who saw them. They had their checklist, and that was all that was important."

Finally Glenn pleaded his case to Gilruth, who overruled the slide-rule types and put the machine shop to work on a proper camera—one that was simple enough to operate with a single gloved hand. None of their inventions worked very well. One day in Cocoa Beach, on his way to get a haircut, Glenn stopped at a drugstore and happened to spot a camera in a display case. He asked to take a closer look. It was a forty-five-dollar Ansco Autoset, made by Minolta and equipped with automatic exposure—a must for Glenn, who was not going to be able to adjust his aperture setting in flight. He paid for the Ansco and brought it back to the Cape. (He was not reimbursed.) The engineers flipped it upside down and stuck a large viewfinder on what was now the top, making it possible for Glenn to set up a shot while his helmet was on; at the bottom they attached a pistol grip and trigger, allowing him to snap a picture and advance the film one-handed. He was prepared, now, to photograph his journey, the first astronaut to do so.

Sightings of Glenn in Cocoa Beach—at the barber, at the drugstore—became increasingly rare. In December it had still been possible for Glenn to make his way around town unmolested, to show up at the Kontiki Village with Carpenter and listen to the lounge singers. By January this was impossible. Correspondents were everywhere. They kept watch on the gates at the NASA facility, went looking for Glenn at his old haunts, crowded into the back pews of Riverside Presbyterian Church, where he attended services on Sundays, or ambushed him outside it, demanding to know if the delays were making him restless or scared (no to both). Except for the weekly service and his weekly haircut—both sacrosanct—Glenn holed up in Hangar S, day and night.

If anyone was growing restless, it was the press. An unruly battalion of more than six hundred correspondents, photographers, and TV cameramen, foreign and domestic, had been accredited by NASA; hundreds of others hovered around the perimeter of the base or spread out across the beach. As the long wait grew longer, reporters were reduced to reporting on themselves. The care and feeding of the press corps, principally Powers's job, was becoming more contentious. Not only were there more of them to manage, but the gloss had gone off the relationship. Powers had set up a news center at Cocoa Beach and kept it well stocked with fact sheets, but none of this was as interesting as whatever the astronaut might be saying or doing. And on that question—actually, on most questions—Powers could no longer be trusted. His answers were often inaccurate or cryptic—cloaked, reporters said, in a "fog of half-information." Worse, word had gotten round that he was fabricating many of the colorful facts and quotes he attributed to the astronauts—including a story that Grissom, the day before his flight, had gone fishing and cooked the catch for dinner. The loathing was mutual and deep. Powers raged about reporters' ignorance and "flamboyant writing." As a correspondent for *Space Digest* complained, "Newsmen were . . . lectured on their own responsibilities, admonished not to sleep in the sun during important activities, warned archly against rumor-mongering, and . . . were advised also (and we for one found it rather hard to believe) that Col. Glenn had said something about doing his job and hoping the newsmen were doing theirs."

NASA was losing control of the storyline. Everything was "proceeding according to plan," Williams cheerfully insisted on January 24—but this had the authenticity of a forced smile. That day, a *Time* reporter told his editor that sources put "the possibility of a three-orbit, completely successful shoot [at] about fifty-fifty. The pessimism is due partly to . . . the recurring trouble with the spacecraft; the latest difficulty in the life support system [which] required extensive dismantling of the delicate instrument panel; the failure yesterday of the Polaris"—a much-hyped missile that, in a test on January 23, fell short of its planned trajectory—and the "blow-up this morning of the Navy's . . . attempt to launch multiple satellites from a Thor rocket." What could have been a big week before the "big one" looked instead like a breakdown. The Thor's failure to send its five satellites into orbit—it dispatched them, instead, into the sea—was followed by an even deeper disappointment. Ranger 3, a spacecraft with a large dish antenna and a pair of solar panels that flanked it like wings, was meant to crash-land a seismometer on the moon and take close-up TV images as it approached. Instead, it went off course, missed the moon by nearly twenty-three thousand miles, and sailed off to orbit the sun. On the way, in a final indignity, its camera cut out.

News coverage took a darker turn. Excitement about Glenn's flight was giving way to resignation and fear; the nation appeared to be steeling itself. "Many things could go wrong," the *New York Times* pointed out, providing, for reference, a list, and hastening to add that the spacecraft had thousands of parts and, therefore, thousands of ways it "could mean death for the red-haired test pilot." Newspapers described a debate among NASA engineers over the length of time that Glenn's capsule would remain stuck in orbit if its retrorockets failed (eventually its orbit would degrade and the capsule would fall back to Earth without a push; estimates ranged from twenty-four hours to ten days) and recounted another argument, among critics of the space program, about who or what was to blame for the run of recent failures. On January 26, one day before the scheduled liftoff, a reporter asked Powers whether the astronaut would be carrying "a cyanide capsule or anything like that." "No," he replied, "nothing like that." Glenn, he said, "wants you to understand that he and we have reduced the risk as far as humanly possible but that . . . there could be a malfunction, that something could happen to him. He pointed out that pioneers have faced risks many times before, that Admiral Richard E. Byrd almost died in the Antarctic but that didn't stop polar exploration."

David Bell, the White House budget director, was neither a space official nor even a supporter of manned spaceflight. Yet on January 26 it was Bell's misfortune to be in front of a microphone in the New Senate Office Building, slated to testify about the Economic Report of the

President for 1962. He was quickly drawn into defending the space program and, in particular, the plan to land a man on the moon. Much of the criticism came from Senator Paul Douglas, an Illinois Democrat and prominent liberal. "A great many people," Douglas said, "and I am one of them, who do not know very much about this program, wonder whether it is worth it. . . . Some of us have doubts when we think what $20 billion could do in the form of schools, in the form of health, and in the form of education. If it is purely a stunt to get there before someone else, that raises a question."

"This is not embarked on by the Government of the United States as a stunt," Bell said crisply. "This is embarked on because it is the opinion of the responsible officers in the executive branch, headed by the President, endorsed by the Congress, after full hearings and debate—"

"After very cursory debate," Douglas interjected.

"Nevertheless endorsed by the Congress," Bell continued, "that the United States should embark on this very costly and very major effort to achieve a capability of moving about in space, of being able to explore space . . . , of being able to use it to the same extent that any other country can."

With the flight of *Friendship 7* only one day away, this goal, somehow, still seemed remote. Glenn himself, amid final preparations, drafted a statement to the American people and asked Powers to release it if something went terribly wrong with the flight. Glenn's note urged the nation, just as he had urged his own children, to stand behind NASA and not to stop exploring the heavens.

At two o'clock in the morning on January 27, a Saturday, Glenn was awakened by Bill Douglas, the astronauts' physician, and was told his flight was a go, despite concerns about the weather. Glenn showered and shaved, as he might any morning. Soon he had visitors: Gilruth and Williams; Slayton, who had been assigned the next flight; G. Merritt Preston, chief of Cape Operations; and David M. Shoup, commandant of the Marine Corps. The group joined Glenn and Carpenter for a "low-residue" breakfast, the genteel term of art for food—in this case, filet mignon, scrambled eggs, and toast with jelly—that might spare the astronaut from moving his bowels while in space. After the meal, Douglas attached EKG and biosensor leads to the tiny dots that had been tattooed onto Glenn's torso to ensure proper placement; an equipment specialist helped Glenn into his pressure suit. At a quarter to five, like a workman with a lunch pail, Glenn strode out of the crew quarters toting a gray metal box—an air-cooling unit, connected by a white tube to Glenn's suit, just below his rib cage. He stepped into a van for the short ride to Launch Complex 14.

Cloud cover shrouded the coast and the sun had yet to rise; but a seven-mile stretch of Cocoa Beach was alight with driftwood bonfires and charcoal grills. About ten thousand people had spent the night on the sand, drinking beer, cooking steaks and chops and hot dogs, playing transistor radios. Over the next few hours another sixty-five thousand joined them. They abandoned their cars along the strip: in front of the used-car dealer offering "countdown specials," or the Starlite Motel, home to both the NASA press operation and a dance revue called "Girls in Orbit," or the Holiday Inn, whose marquee now read, HOPES AND PRAYERS OF FREE WORLD ARE WITH COL. JOHN GLENN. Henri Landwirth, the manager of the Holiday Inn and a friend of the astronauts', prepared to unveil, in Glenn's honor, a nine-hundred-pound cake in the shape of *Friendship 7*. A truck in the parking lot held the sections of the cake; bakers planned to assemble them while Glenn was in space.

As dawn broke, the nation turned on its television sets. In Grand Central Terminal, in Manhattan, hundreds of early-morning commuters stared at a waiting-room screen. A man walked down Connecticut Avenue in Washington carrying, and watching, a battery-powered set. President Kennedy, spending the weekend at his family's estate in Palm Beach, Florida, got up early and clicked on the set in his second-floor bedroom. In Glenn's hometown of New Concord, a crowd of more than a thousand, giddy and tense, assembled in the Muskingum College gymnasium, where TV monitors had been set up.

Glenn's parents watched at home. In their small living room, they had lined up four TV sets: one for each network plus a spare. Later, if the flight was a success, they would go to Muskingum to meet the press. John Glenn Sr. had bought a pocket handkerchief and a "brand spanking new tie" for the occasion, he'd told reporters the day before, much of which he and Clara Glenn had spent in that same living room, fielding questions. "The delays are all a part of the space business," he acknowledged with a shrug. For a week now, ever since a detachment of feature writers had descended on New Concord, Clara had been trying to keep up a brave face. "I don't think I really worry," she told William Shelton of *Time*. "I know Bud is optimistic. He explained to us all the safety precautions that are taken. He has always felt that the Project Mercury people have done everything they could to be sure everything worked out as planned. If my son has faith in the program and is optimistic about what he is going to do, then I am, too."

In Arlington, Annie Glenn had gone to bed the night before with a migraine headache—she was prone to these—and woke up the next morning with another. At least she had the support of her neighbors, a contingent of ex-Marines and their wives who called their block "the barracks." Early that morning, friends dropped by to put a pot roast and potatoes in the oven for Annie, the children, and Annie's parents, who had arrived the previous week. Soon Loudon Wainwright of *Life* showed up to watch the launch with the family—or, rather, to watch the family watching the launch and to take notes discreetly while a photographer snapped pictures. The group settled into sofas—facing four TVs, just as Glenn's parents were doing. As the clock counted down, Annie stepped out of the room to take a call from John, who was already strapped and plugged into the capsule that, in the image on her TV screen, was too tiny to distinguish from the booster rocket.

After Glenn was inserted into the capsule ("You don't get in it, you put it on," he once joked), there was a moment of levity: Cece Bibby's "naked lady" made her planned appearance through the periscope, and "John-Baby" relished the joke. But nearly everyone at the Cape was on edge. Behind the blastproof doors of the blockhouse, 250 yards from the launch stand, Walt Williams found the engineers "tremendously keyed up," their eyes bloodshot. Then, at a quarter to eight, forty-five minutes before liftoff—T-minus 45—Powers came on the public address system to announce that the countdown was being paused. Technicians had to "clean up some details." The trouble was minor—an ill-fitting gasket on the hatch—but took several attempts to resolve. All the while, the bank of clouds above the Cape thickened and darkened.

It was a "peculiar situation," Powers said over the loudspeakers at 8:40, but the weather, really, was typical for that time of year. What Powers probably meant, but was not prepared to say, was that it was a troubling situation, because weather was a far greater concern for an orbital flight than for the short hops Shepard and Grissom had made. A fifteen-minute ballistic flight needed only a small window of good weather over a narrow area, extending a few hundred miles to the Bahamas, where *Freedom 7* and *Liberty Bell 7* had splashed down. But Glenn's flight, set to circle the globe, required clear weather all the way to the coast of northwestern Africa—the part of the path where trouble with the spacecraft, during those dangerous first minutes after liftoff, was most likely to occur. Across this vast expanse, spanning multiple weather systems, NASA had identified several emergency recovery areas. Unless the wind and sea conditions were moderate at each, unless rescue crews would be able to spot the capsule's main parachute at 10,000 feet, unless the cloud ceiling was sufficiently high and the time of day sufficiently early that search aircraft could operate for at least four hours before nightfall, MA-6 would not be cleared for launch. "The combination of variables," *Aviation Week* observed, "is formidable."

And today, it appeared, insurmountable. The countdown began again, was paused again, was resumed again; and by 9:00 this stutter-stepping had gone on long enough that there would be time for only two orbits, not three. Meanwhile the cloud bank, sitting gloomily in place, would make it impossible for NASA's cameras to track the spacecraft on its path through the atmosphere. "It was one of those days," Williams said later, "when nothing was wrong but nothing was just right, either. I welcomed that overcast." A "gut-feeling thing" had been troubling him all morning, and now he had reached his limit: he would scrub the launch, shut it down. But first he had to tell the president. A call was made to Palm Beach; two minutes later Williams issued the order. The countdown clock read T-20. It took another ninety minutes to remove the hatch cover. By then, Glenn had been inside the capsule for five hours and eleven minutes—half an hour longer than it would have taken him to orbit the Earth three times.

In the Muskingum gym, the audience gasped. Up the road, Glenn's parents sat numbly in their living room. "Too bad," they said—their only comment before the press was ushered out. On Cocoa Beach, some of the spectators, reluctant to leave, kept the bonfires burning and drank their last beers with breakfast. Most others—sand in their hair, eyes bleary from lack of sleep—packed up their tents and beach chairs and marched a sullen retreat to their cars, many of which, they found, were stuck in the sand. It took all morning for tow trucks to arrive.

Traffic on the strip took more than three hours to clear. NASA's decision to delay the launch five or six days—until February 1 or 2—was a blow to anyone who had taken time off from work to travel to Florida.

"Well, there'll be another day," Glenn said back at the crew quarters, but as he was helped out of his suit, he did not appear as sanguine as that. He was sweaty and tired and clearly unhappy. He put on a terry-cloth robe and headed for the showers. As he was on the way, NASA officials pulled him into a conference room. They must be eager, he thought, to see how he was holding up, to chuck him on the shoulder and give him a pep talk. Instead, sharply, they asked him to call his wife in Virginia and tell her to allow Vice President Johnson, who was sitting in a limousine a few blocks away from the house, to come pay his respects. Also, she should let him bring much of the Washington press corps in with him. Also, she should ask Loudon Wainwright to leave.

Annie Glenn—having endured months, indeed years, of anticipating this moment; having spoken to John that morning for what could have been the last time; having sat through more than five hours of speculation by TV anchors about the outcome of the launch; having the day of reckoning deferred once again; having spent all day in the blinding fog of a migraine—was buckling. She was, Lyn recalled, "as tense and angry as I've ever seen her." Annie's stutter, which got worse when she was upset, made it hard for her to be understood. All she wanted was to get back into bed, but first she had to contend with Johnson and, trailing in his wake, an aggrieved group of correspondents and cameramen who for three years had been denied access to the astronauts' families by the terms of their exclusive agreement with *Life*.

For the Mercury Seven and their wives, the *Life* contract was never just about money: it was a buffer against the kind of frenzy now unfolding on the lawn. Only two weeks earlier, in a letter to Glenn, the editor of *Life,* Edward K. Thompson, had promised to "keep the wolves away from Annie's door (and these include those working for our dear sister magazine, *Time*)." LBJ knew well enough to align himself with the pack and not *Life,* the lone wolf. And now, with his sanction, the wolves amassed outside the door, ready to displace Wainwright from his position at Annie's side.

Glenn was unyielding—unmoved by the argument that LBJ's needs or sensitivities should take precedence over his family's or should override a contract he had signed in good faith. The very idea was offensive to him. He reminded the group in the conference room that the astronauts had an agreement with *Life* and he said he was going to honor it. This went down badly. So did his refusal to call his wife and convince her otherwise. Instead, on the spot, he dialed Annie and told her that if she didn't want anyone else in the house—and that included the vice president and ABC, CBS, and NBC—it was all right with him. Then he turned to the NASA managers, announced he was going to take a shower, and stalked out. In Arlington, meanwhile, Wainwright—his station secure—listened as Annie placed a call to Johnson. Despite her halting speech, she deftly put the matter to rest: "You're so nice to call, Mr. Vice President, and you surely understand how it is. We'd just like to be together in the family at this time."

For the press corps, the siege of Arlington was wasted energy. They already had their front-page story. Nearly every newspaper ran a photograph of Glenn exiting his capsule glassy-eyed, open-mouthed, exhausted. The *New York Times* captured the indignity: FOILED BY CLOUDS. "It had come so close," sighed the writer. The Soviets, of course, savored the setback. A Russian science commentator blamed "feverish haste" for the postponement of Glenn's flight, as well as for the "complete fiasco" of the Ranger 3 lunar probe earlier that week. It was clear, he said, that "even the trodden path is not easy."

Gagarin, on a similar note, mused about the "serious psychological and moral pressure exerted" on Glenn by his long wait in the capsule. Yet if Glenn felt the pressure, he gave no sign of it. The morning after what he called "the big scrub," he strode cheerfully up the front path of Riverside Presbyterian Church, bantering with the photographers who'd followed him there. "You'd be better off inside," he joked. (Declining the invitation, they hung around outside and, when the service let out, jumped into their cars and trailed him back to the gate at Cape Canaveral.) Decades later, Glenn acknowledged that "it was a lot of stress," having the launch canceled after "you . . . get ready and get yourself psyched up to go." And then he would have to do it again. That is, unless he caught a cold and got bumped from the flight—a continuing concern. (Earlier that week, the astronauts' secretary, Nancy Lowe, had picked up a slight cold and, by order of Bill Douglas, had been forbidden from getting near Glenn.) And every new delay, Glenn reflected, gave his family more time—and more cause—to worry about "the safety of this whole thing."

The Glenns had at least two more weeks of that. On January 30, NASA called a news conference and announced—to groans—that it was postponing MA-6 a sixth time, until February 13, due to "technical difficulties with the launch vehicle." Powers read a statement from Glenn: "Sure, I'm disappointed, but this is a complicated business. I don't think we should fly until all elements of the mission are ready. When we have completed all our tests satisfactorily then we'll go." But the press was long past being mollified by words that Powers had probably scripted. *What technical difficulties?* reporters asked. *And why two weeks?* Because, the spokesman said, it would take two weeks "to do all the things that need to be done"—a non-answer that inflamed tensions further, if not as much as another comment he made: "Of course we know just what the trouble is, but I'm not telling you." Powers kept to his word, letting a full day pass before revealing that the tank of rocket propellant, RP-1 kerosene, was leaking.

By this point, his audience had thinned. Most news crews had been called home, their bosses unwilling to let them keep running up motel bills and bar tabs when there was nothing to report. It had been costing TV and radio stations $50,000 a day, collectively, to cover what the *Times* called "the most extensively and expensively reported single news story" in history. On top of that, the networks, in the name of bragging rights, had been engaged in an extravagant arms race: when NBC built itself an observation deck at the end of its equipment trailer, blocking ABC's view, ABC built itself a bigger structure, two stories high, claiming it was the tallest building on the Cape, and provided tours. But for now, there was nothing for anyone to see on the launchpad but the Atlas—a "sick bird," as Walt Williams put it—being slowly emptied of oxidizer and propellant.

Glenn, too, went home. Again he drew a line through the words "MA-6 LAUNCH" on his calendar, adding "SCRUB" in block letters and, more happily, "TO D.C." He spent most of February 2 with family and friends, and running errands in Annie's maroon station wagon. The next day, despite the frigid temperature, he held a news conference on his front porch. "It looks like Hangar S was not such a bad place after all," he said, scanning the crowd of newsmen on his snow-covered lawn. Wearing a white winter coat and flanked by Annie, Lyn, and Dave—all silent but smiling—Glenn struck an upbeat note. Reporters told him he had appeared tired in photos of the scrub. "I didn't feel that bad," he countered. Wasn't the launch date of the thirteenth bad luck? they asked. On the contrary, Glenn said: it "can only bode good for success." And further delays, he added, wouldn't trouble him a bit. Glenn's nonchalance was widely noted. One correspondent wrote his editor that "even if he falls on his face . . . , Glenn will somehow look good doing it. Like Ruth striking out. Or Joe Louis taking a long count from the canvas."

Yet the longer the launch vehicle sat on the pad, the more likely it was that some component, somewhere, would fail. At the Pentagon, the members of a secret task force considered that possibility in light of their ongoing effort to overthrow Fidel Castro. On February 2, the group sent a memo to General Edward G. Lansdale of the CIA, proposing a wide range of covert actions in Cuba. The memo does not seem to have received serious attention. Indeed, some of it was comical: a plan to air-drop one-way plane tickets good for passage to neighboring countries ("not the U.S."), a plan to distribute doctored photographs of "an obese Castro with two beauties in any situation desired." But it was developed in earnest. Among its more practicable proposals was a suggestion that if Glenn's flight failed, the United States should concoct "irrevocable proof" that it was Castro's fault. This could be achieved "by manufacturing various pieces of evidence which would prove electronic interference on the part of the Cubans." The task force called it "Operation DIRTY TRICK."

Over the weekend, Glenn got word that President Kennedy wanted to see him Monday morning. Thus did Glenn, with the punctuality one might expect from a Marine Corps officer, find himself standing at attention at the White House on February 5, waiting for the president's helicopter to arrive from Middleburg. Glenn wore civilian clothes: dark suit, white shirt, narrow red-and-blue rep tie. The red in his tie picked up the color in his face; he was sunburned from running on the beach. This would be the first time he met JFK without the other astronauts present, though he had a chaperone in Powers. A conversation between an untroubled president and an untroubled astronaut might serve, it was thought, as a balm to the nation—or at least a boost to the program as its critics continued to circle.

At 9:40 the two men sat down in the Oval Office: JFK in his rocking chair, Glenn on the couch. Powers kept watch from a respectful distance, standing near the door. Before the press was brought in, Kennedy asked Glenn a rapid-fire series of technical questions. "He was interested very much in the anticipated G level during launch," Glenn later recalled, and "what kind of sensations we expected during launch; what kind of control we had over the booster during launch; were we actually going to drive it like we did an airplane or were we pretty much at the mercy of the guidance systems until we were in orbit. . . . He was interested in real detail." And he was concerned about safety. Kennedy wanted to know whether NASA had done everything in its power to protect its astronauts. Glenn explained that at the start of the program, Gilruth had given them veto power: if they ever, for any reason, felt unsafe, they wouldn't fly. This satisfied Kennedy; it seemed to him a sensible approach.

Now the doors were opened and the press pushed its way in, the usual commotion. Kennedy turned to Glenn and said he was probably used to all this attention. Glenn replied that he was, that he knew it was part of his role, inviting the public to "live through this experience in Walter Mitty style"— a vicarious thrill. "I don't mind," he continued, "but I think our emphasis is on the wrong point sometimes. There is a great deal of publicity, but not enough emphasis on the scientific aspects of Mercury"—as opposed, he said, to "how Annie is getting her hair fixed." Yes, Kennedy replied, but the science was complicated; publicity humanized the program. With that, the two men got up, moved to the center of the room, and stood there awhile, smiling broadly for the cameras.

The next morning NASA announced that the date for the launch had slipped yet again—one more day, until February 14—to address minor problems with the booster. Sources told UPI that the new timetable was optimistic. "As I've said from the beginning, we've been behind," Kennedy told the press that day. "And we are running into the difficulties which come from starting late." Still, he insisted, "we're making a maximum effort." Later that week he received a thank-you letter from Glenn, who had included some maps of his orbital path. "I enjoyed our visit very much," he wrote. "Knowing of your interest in our orbital mission, I thought perhaps these maps might enable you to more closely follow the flight. . . . My best personal regards to you and I hope we can successfully complete the manned orbital flight with no more delay."

Predictably, it was not to be. February was living up to its billing as the worst month of the year, weather-wise, in the region. Early on the fourteenth, clouds blanketed the Cape at 7,000 feet—too low for NASA's cameras to track the spacecraft. Another day, another delay. NASA would try again tomorrow. "This is the eighth time," a reporter needlessly pointed out at JFK's regular news conference that morning, noting "the considerable ordeal on Colonel Glenn." Wouldn't it be better, the reporter asked, to put the whole thing off until springtime, when the weather would have improved? "I know it strains Colonel Glenn," Kennedy replied. "It has delayed our program. It puts burdens on all of those who must make these decisions as to whether the mission should go or not. I think it's been very unfortunate." Still, he trusted NASA to make the right decision; he, for one, suspected that "they would be reluctant to have it canceled for another three or four months, because it would slow our whole space program down."

"We've had bad luck," he added; and it was not done yet. As the next day dawned, the central Atlantic convulsed, whipped up by gale-force winds—right where Glenn would splash down if he had to abort after liftoff. The flight was put off a ninth time. Twenty-four hours passed. A front moved eastward across the Gulf of Mexico. The flight was put off a tenth time. Glenn was told at 12:50 a.m. “I guess it was to be expected,” he said in a statement. “We all knew the weather was marginal.” It was now Friday, February 16, and a Saturday launch appeared a “lost cause,” in the frank assessment of the U.S. Weather Bureau. NASA, losing in a rout to the winter skies, made uncertain noises about Tuesday, February 20, and left the matter to be decided later. The calculus would have to account not only for Glenn but for the more than eighteen thousand people—flight controllers, pilots, sailors, rescue swimmers, engineers, physicians, others—spread across the world, manning sixteen remote tracking stations and a recovery fleet of twenty-four ships and more than sixty aircraft. These were well-trained units, and in late January Gene Kranz had informed Chris Kraft that they were in peak condition—not altogether good news, because their morale and performance could only decline as they sat idle. Since then, three weeks had passed. Still they remained at their posts, hostages to the weather report.

The press, meanwhile, had filtered back to Cocoa Beach, checked into the familiar motels, and renewed their complaints about NASA’s close-fisted information policy. The reporters’ mood, clearly, had not improved during their time away. They bemoaned the fact that they had nothing new to say. “The battery of temporary teletype machines is chattering out to the world the same old pre-flight stories,” *Aviation Week* observed. “Only the dates are new.”

After the tenth postponement, there was a flicker of news when a psychiatrist named Constantine Generales—billed, somewhat grandly, as coordinator of the Space Medicine Program at New York Medical College—claimed that as a result of the delays, “anxiety had undoubtedly built up in Glenn’s subconscious.” Glenn’s performance, the doctor said, would suffer; NASA should send Carpenter instead. Glenn was livid. It was ridiculous, he told the NASA managers, to be diagnosed at a distance of a thousand miles by a doctor with no connection to the program. Still, it rattled him. In notes for a talk at Langley a few weeks later—notes he then scratched out—he admitted that his "main concern . . . was that someone might listen to one of these false self-styled experts." To put the matter to rest, as Bob Voas recalled, Powers called in the cameras and prodded Voas, a psychologist, to "tell the psychiatrist he didn't know what he was talking about."

Glenn was "taking it very well," Voas told the press. "There is no evidence that he is building up any frustrations or annoyance." The most that Powers would concede was that Glenn was disappointed. Asked if Glenn had greeted news of the delay with any four-letter words, he was noncommittal. "Well, he's a red-blooded American, and he knows some, ah, slang," he said. ("How refreshing it would have been to have heard him say 'damn!'" one reporter remarked.) If Glenn felt stymied, he gave no sign of it in public. On February 18, standing outside his church, he urged Americans to "stay relaxed. I've been at this thing for three years now," he explained. "I feel fine. Sure, we regret the delays. But as Scott Carpenter said, it gives us a chance to 'hone our capabilities.'" JOHN'S NERVES ARE A-OK, judged the *Zanesville Times Recorder*, the paper in New Concord's neighboring town, adding, "Glenn Urges Folks to Take It Easy as Blast-Off Looms."

In truth he was tense, increasingly so. For two months now, Glenn had been grounded—pinned down by the elements—and it was wearing on him. Word at the Cape was that if the flight had to be delayed once more, NASA would wait a month before trying again. By now Glenn had come to expect it. "I was not exactly hopeful," he said later. On cue, a cold front moved across central Florida on February 19, the day before Glenn's eleventh attempt. The mid-Atlantic storms had subsided, leaving relatively calm seas in the three key recovery areas; but the Weather Bureau, looking at that local front, predicted broken cloudiness and gave Glenn fifty-fifty odds that his flight would go. He called Annie that evening and told her to expect another postponement. Still, in the event that the weather cleared, he needed a full night's sleep. He went to bed around seven o'clock. Before he turned out the light, however, he took one more look at his operations manual, especially the section on the Automatic Stabilization and Control System. He had meant what he'd said outside the church. He would hone his capabilities to the last.

The pad crew at Launch Complex 14 retired early, too; a second shift took over the preflight checks. One young engineer was told to monitor the rocket's central engine: the "sustainer," which sat between the two booster engines and was meant to burn the longest. The problem with the sustainer was that it had a tendency to leak fuel. Not much fuel, but enough that the quantity had to be checked hourly. That night, while Glenn slept, the engine kept leaking—a slow, steady drip of kerosene. "It was determined to be an acceptable seep," the engineer recalled; the countdown continued. The leak was assigned to a category of risks, a crowded class of imperfections that had to be accepted if there was ever to be any chance of getting airborne. At a press conference that day, Chris Kraft had put it bluntly: "If I thought about the odds at all," he said, "we'd never go to the pad."

CHAPTER 15

—— ✦ ——

Godspeed, John Glenn

At a quarter past five in the morning on Tuesday, February 20, John Glenn, space suit on, helmet sealed, was sitting in the back of a small white trailer truck on the launchpad, waiting, still waiting, for the skies to clear. He reclined in his seat, opened the blinds on the small window, and peered up at the Atlas rocket, shining silver and emitting vapor in the glare of the arc lights. It looked, Glenn thought, just as it had that night in 1959 when the astronauts, only a month into their training, watched a missile like this explode above their heads.

Glenn was waiting not only for the clouds to pass but for the missile's guidance system to be replaced; its radio beacon was malfunctioning. He was calm, relaxed. Bill Douglas, the astronauts' physician, knew this because Glenn's bioconnector, trailing from a port on his thigh, was plugged into a unit that revealed that his heart rate was normal. The countdown clock stayed frozen at T-120 while Glenn, Douglas, Deke Slayton, and Joe Schmitt, the suit technician, sat for twenty minutes, then thirty, then forty. From the moment he had woken up—1:30 a.m., half an hour early—Glenn had been expecting another scrub, and now, as the hold lengthened, that feeling grew stronger. A crowd of technicians in white hard hats stood idly around the truck at the base of the gantry. Fastened to a fence was an exhortation: "Complex 14 has worked 57 days without a serious accident."

To Glenn, it all seemed routine. This was the fourth time he had suited up, the fourth time he had gotten "psyched up to go" while suspecting he wouldn't. "This time you're going," Walt Williams had told him after breakfast, shaking his hand. "This time, boy, you're going to take a ride!" But as six o'clock approached, the only ride Glenn had taken was in the truck with the words NASA TRANSFER VAN on the side. The men in the blockhouse were confident that the flight would go, but they had no particular cause for that; it was just a feeling. The astronaut himself still felt otherwise—even when the word came down, at 5:58, that mission control was ready to resume the countdown. He exited the van in a hurry, stepping briskly down to the pad, holding his air-cooling unit in one gloved hand and waving with the other to the technicians, two hundred of them, who clapped and cheered as he walked to the elevator. They watched as its doors closed behind him, watched as it climbed the exoskeleton of the gantry.

At the top, workmen in white caps and white smocks greeted Glenn and patted him on the arm. Scott Carpenter, playing the backup role that Glenn knew well, had been in the white room for hours, inspecting the capsule. It all checked out, he told Glenn now. While they talked, Schmitt, the suit tech, leaned into *Friendship 7*, arranging cables and hoses on Glenn's form-fitting couch and laying out the restraints that would bind him to it: shoulder harness, lap belt, knee straps, chest straps. Within a few minutes Glenn was cleared for "crew ingress." Feet-first, he slid like a contortionist through the two-foot-square hatch. Once inside, he had little room to maneuver—the cabin was about as small as the cockpit of a fighter jet. The instrument panel wrapped around him in segments, color-coded by function. Built into it was a new navigational tool: a revolving model of the Earth, not much bigger than a baseball. Once Glenn was in orbit, he would position a tiny capsule along its orbital track, turn a knob to wind the gears, and set the globe in motion.

Above his knees was the periscope display, and on the display, as Glenn settled in, was another cartoon by Cece Bibby. Not, this time, a "naked lady," but a dowdy woman in a housedress, her hair in a scarf. She held a mop and a bucket. The bucket bore the *Friendship 7* logo. Bibby had added a caption: "You were maybe expecting someone else . . . JOHN BABY??" When the drawing was removed, a technician dangled a *Playboy* centerfold in its place.

Small problems kept the engineers occupied and on edge. A respiration sensor attached to the microphone on Glenn's helmet got knocked out of place, making it impossible to keep accurate track of his breathing rate. Fixing this would require prying open his suit—a delicate operation, and not worth the effort. Then the bracket supporting the microphone broke. This could not be shrugged off. If Glenn couldn't communicate with mission control, he couldn't fly. There was a backup helmet in the van parked below; a tech was sent down to extract its working bracket. This took ten minutes. Finally the countdown restarted—then stopped again, this time for forty minutes, while engineers replaced a broken bolt on the hatch door. All the while the clouds hung stubbornly in place and Glenn, tipped backward in his couch, reviewed his checklist. On his headset he listened to the control center chattering about surface winds in the recovery areas and the hydrogen peroxide levels for the thrusters and other things that mattered only if the flight was a go. As directed, he checked the switches on the panel, making sure they were in the proper position, a process he later described as completely unnecessary, "designed to keep the astronaut busy." But then, he reflected, "I didn't have anything else better to do at the time."

Now the nation was waking up. On Cocoa Beach, thousands had again spent the night in the back seats of cars, under pup tents, on cots, in blankets on the sand. It was chilly out; a light wind was blowing, and people stayed under wraps as morning broke. The crowds were smaller than they had been in January, and more subdued. "It's really no carnival atmosphere," NBC reported, panning across the scene in the half-light of dawn. Across the country, bars opened early and turned on TV sets. In Manhattan's Grand Central Terminal, commuters began to fill up the mezzanine beneath a massive screen; it was framed by the words CBS NEWS: FOR GLENN IN ORBIT. President Kennedy and Jackie were watching the same broadcast in their second-floor bedroom in the White House. Downstairs, a portable TV was wheeled into the Family Dining Room, where, shortly, the president would be hosting a weekly breakfast for legislative leaders. It was not a day for talking shop.

Annie Glenn had awakened in the dark at 5:30 to the sound of TV crews setting up on her front lawn. She had slept badly. The city of Roanoke had sent her husband a twelve-foot tall, six-hundred-pound wooden valentine, adorned with plastic roses, and there had been no place to put it but to prop it up in the carport, by the station wagon. Annie spent a restless night worrying that the wind would smash it into the car. Now she made coffee, set out sweet rolls and coffee cake, started a fire, put a new reel of tape in a recording machine. Her parents emerged from a bedroom. So did Lyn and Dave, neatly dressed for the cameras. Loudon Wainwright and a photographer from *Life* were there already: at Annie's invitation, they had spent the night at the house so they wouldn't have to push their way, in the morning, through a hostile crowd of less-favored reporters. Frank Erwin, the Glenns' minister, arrived shortly, as did their closest friends, the Millers. By 6:30 everyone had filled up plates with food and ate breakfast in the living room while Wainwright padded quietly around the house, peering into rooms and open closets, taking notes on what he saw: a rack of brightly patterned bow ties; a scale model of *Friendship 7* on a desk in the den.

They filled the time as best they could. Tom Miller moved from TV set to TV set, adjusting the picture on each and turning the sound off, save for CBS. Dave started reading a book, then brushed the family cat, a Siamese. Lyn polished her loafers and stood up to sweep the fireplace. On the coffee table, beside a vase that held a single red rose, a substantial hourglass ran out; now and again, someone turned it over. Mostly they all sat and stared at the screens—a seemingly endless loop of anchors at desks, old footage of Glenn in his space suit, and live images of the house they were in, curtains drawn, revealing nothing. Then, around eight o'clock, a weather report brought the house to life. "Suddenly we have almost completely blue skies here," an announcer said from the Cape. "I told you! I told you!" Lyn shouted.

Glenn could see it himself through the periscope. "Looks like the weather is breaking up," he said. Electric horns confirmed it, blaring a warning that all personnel should clear the gantry. Then the whole of it—that entire exoskeleton, almost 150 feet tall, painted a bold and now radiant red—rolled back from the spacecraft, leaving it bathed in bright light. Sun streamed into the window and got in Glenn's eyes. The bonds of Earth were being released: the last tie to the tower was an "umbilical cord," tethering the rocket to its electrical supply. There was a brief hold as the booster's liquid oxygen tanks were topped up, stretching the metal skin of the Atlas and causing it, 93 feet from its base to its top, to rumble and shake. Up until the removal of the gantry, every step and every feeling had been familiar, a recapitulation of last month's "big scrub"; but this was new, this sense he had that the booster was alive. It screeched and growled. When he shifted back and forth it moved, just slightly.

The count resumed at T-45. A few minutes later, Scott Carpenter, now in the blockhouse, placed a call to the house in Arlington.

Annie picked up the phone in the bedroom and waited as Carpenter patched her husband through. The radio link ran through the blockhouse; to keep John and Annie connected, Carpenter had to stay on the line, silent witness to what might be the last conversation between the two. The weather was fine, John brightly told Annie; the sky, he could see, was clear and blue. He was tense with excitement—"keyed up," in his phrase. Annie understood his mood but in no way could match it. She was rattled by the noises that filled the spaces between her husband's words: a loud squeal, a low rumble. It was just the booster, Glenn assured her, nothing unexpected. "I can tell you," he had written to his "troops" back in January, "that I will be afraid when the booster is getting ready to fire. . . . What I do at that time is the important thing." Now, as that time approached, he said nothing of his fears. But he did ask Annie about the recordings he had made—one for her, the other for their children—in the event that the worst happened. He had sent the tapes home; had Annie received them? She said she had.

Now she called Lyn and Dave to the phone to say good-bye. Then she did the same. Glenn could hear the catch in her voice. He told her not to be scared, told her he loved her and that he would call when he landed. "Remember," he said, "I'm just going down to the corner store to get a pack of gum." It was his old wartime sign-off, meant to lighten the mood and ease Annie's fears. Annie knew her line in this routine, but could not say it with any conviction: "Well, don't take too long." With that, she picked herself up, clenched her jaw, and walked back to the living room, tears in her eyes, for the final countdown. She sat on a cushion on the floor and pulled her knees up, locked her hands around them. She was ringed by chairs and couches filled with family and friends.

"What must be the thoughts of John Glenn at this moment?" an anchor intoned on NBC. (The house had, for the moment, forsaken

Cronkite.) "You can't scare us," Annie's father retorted. "I bet he has a big smile on his face," Annie said. "Certainly he must know," the anchor continued, "that the fervent prayers and hopes of a nation are with him, as most of the nation is sharing this experience . . . as best we earthbound human beings can do." The TV screen showed the rocket—which still stood on the pad but was sending forth such a rush of vapor that it almost seemed in motion. "T-minus four minutes. And the countdown continues, moving toward a moment of zero, a moment of ignition."

THE CALL AND RESPONSE was rapid-fire. Sixty seconds were left before launch. Thomas J. O'Malley, the test conductor, sat in the blockhouse before a steel-blue panel of switches and blinking lights, a black telephone at his left elbow and a well-filled ashtray at his right, and barked into his headset. He ran though his checklist like a drill sergeant calling roll:

"Communications?"

"Go!" a ground controller answered.

"ASCS?"

"Go!"

"Aeromed?"

"Go!"

"Status check: pressurization?"

"Go!"

"Mercury capsule?"

"Go!"

Glenn took hold of the abort handle. This was standard procedure.

"The ready light is on," O'Malley shouted. "Eject Mercury umbilical." Glenn watched it fall away before his periscope retracted. "T-18 seconds and counting, engines start," O'Malley said.

"May the wee ones be with you, Thomas"—an Irish invocation from his boss, Byron MacNabb, in the control center. O'Malley crossed himself. "Good Lord ride all the way," he said.

Then Carpenter, beside him in the blockhouse, said his own impromptu prayer:

"Godspeed, John Glenn."

It was his farewell to a friend. But his friend hadn't heard it: all Glenn could hear now above the roar was the voice of Al Shepard, counting down from ten.

THREE ENGINES ERUPTED with a force that shook the ground and made the air vibrate for miles. For the first two seconds there was no movement, only noise. Amid a surging mass of smoke the rocket stood still, building the thrust it would take to lift off. Then, heavily, it rose. It kicked free of the restraints at its base and shot upward, with gathering speed, riding a column of yellow-and-pink flame—a column nearly as long as the booster itself. "The clock is operating," Glenn said calmly. "We're underway." Above his right knee was a counter labeled TIME FROM LAUNCH. It read 00:00:03. On the Florida coast, now below him, it was 9:47 a.m.

On the beach, people jumped and cheered. Some waved their hands, as if Glenn could see them. In Grand Central Terminal, someone shrieked, "Go, go, go!" But for many, the tension of all those months of waiting found no outlet, at least not yet. The public address system of the New York City subway carried a somber benediction: "Colonel John H. Glenn has just taken off in his rocket for orbit. Please say a little prayer for him." The message was repeated every ten minutes. It was echoed, in multitudes of ways, around the world. In Rome, a Vatican spokesman announced that Pope John XXIII was praying for Glenn's safe return and would be kept up to date throughout the flight.

The mood at the White House was similarly grave. Kennedy and his guests that morning—LBJ, House and Senate leaders, a few presidential aides—left the breakfast table three minutes before liftoff, pulled chairs up to the TV set, then stood in front of it: the president was too restless to sit down. "Go, baby!" they heard Cronkite shout. "Looks like a good flight. Oh, *go,* baby." The men watched in silence. It was far too early for exultation. At the house on North Harrison Street, Lyn Glenn covered her face with her hands as the networks showed the rocket disappearing from view. Tears ran down her cheeks. Annie, still seated on the floor, rested her head on her knees and wept.

"We got a pitch!" Cronkite said as the rocket, according to plan, rolled slightly to the northeast. "That was a critical moment." It was one of several over the next five minutes, the time it would take to find that "keyhole in the sky," as Glenn put it, and enter orbit. "Little bumpy," he reported. Then, as expected, it got a great deal bumpier. After forty-five seconds of flight, at an altitude of 35,000 feet, MA-6 hit "maximum Q," the most intense point of aerodynamic pressure—the point where the spacecraft was at the greatest risk of exploding. In 1960, an unmanned flight had done just that, in this very same zone. Now, as Glenn's spacecraft plowed through the air like a ship through high seas, it shook him so hard that it threatened his grip on the abort handle. "Coming into high Q a little bit," he said coolly, but his voice vibrated, and he wondered how close he was to the limit—the threshold when the escape rocket would fire automatically and pull him free of disaster. And then, in half a minute, the worst of it had passed. "Feels good," he told Shepard. "Smoothing out real fine."

The rocket was also speeding up—burning two thousand pounds of fuel a second, growing faster as it got lighter. The g-forces shoved Glenn into his seat with the pressure of six times gravity. There was a kind of satisfaction in this: not only was it almost laughably short of the 16 g's the astronauts had endured during their spins on the centrifuge, but as Carpenter had predicted before the flight, Glenn was glad to accelerate in a line, for a change, instead of in circles; he felt that he was getting somewhere. And he was: forty miles high now and climbing. When his clock read 00:02:09, the two outboard engines, having served their purpose, were cut loose in a flash of smoke; one remaining engine pushed the capsule onward. Then the escape tower was dispatched, too: here, outside the stratosphere, it was only dead weight. Glenn watched as it was shot like a dart into the void. Now the Atlas, guided from the ground, adjusted its aim to ease the spacecraft sideways through the "insertion point," into orbit. Glenn got his first glance at the Earth: clouds dotting the Atlantic beneath a deep black sky.

He was nearing the end of powered flight. In orbit, momentum would propel him. Nearly emptied of fuel, the Atlas was little more than a "stainless-steel balloon," as Glenn once described it, and its thin walls, to his surprise, began to oscillate; he seemed to be at the end of a spring. On cue, just past the five-minute mark, the engine cut off. Suddenly Glenn felt he was tumbling forward—another sensation he knew from the centrifuge, though the effect, like the g-forces, was milder here. Then explosive bolts blew and flung the Atlas away. In the same instant, three tiny rockets, attached to the base of the capsule, fired—a last little kick to move the spacecraft farther clear of the missile. Glenn was through it now, the keyhole. He was in orbit. He was a hundred miles high, traveling 17,500 miles per hour, nearly 26,000 feet per second: frictionless flight.

The weight left his chest. Every pressure point eased. He felt he was floating inside his suit. The restraints held him in place but loosened just slightly; the ends of the straps floated up as if dancing on a breeze. In 1959, after his first experience of weightlessness on a plunging parabolic flight, Glenn had effused that he had "finally found the element in which I belong." He felt that way now: "comfortable . . . natural." And elated. "Zero G and I feel fine!" he told Mercury Control. "Capsule is turning around." *Friendship 7* had started along its path with its narrow end forward, like a bell that had tipped over; now the capsule swung 180 degrees to lead with its blunt end—its heat shield—better positioned in the event of an unplanned reentry. For Glenn this meant facing backward, watching continents recede as they passed below. If facing forward was, as he said later, "like sitting up in front of a Greyhound bus watching the world come at you," this was like staring out the back window. He showed no sign, though, of disappointment. "Oh," he said, "that view is tremendous!"

For now, there was little time to enjoy it. Glenn had a substantial list of tasks to do, systems to test, switches to check, observations to make, readings of instruments to report. For reference, the astronauts' secretary, Nancy Lowe, had typed the flight plan onto a scroll, about five inches wide and four and a half feet long. It was attached, at either end, to a bobbin; Glenn advanced it, segment by segment, with his thumb as he moved along the orbital path. It identified landmarks he should look for on Earth ("BLOWING DUNES"; "CONGO RIVER") and in space ("MOON SET"; "CASSIOPEIA: COUNT STARS") and where to spot them; it reminded him when to "CHNG FILM—COLOR" and when to do his exercises (to assess the effects of weightlessness, he would pull a handle attached to an elastic cable thirty times in thirty seconds). Tucked beneath the panel he had a series of laminated star charts, corresponding to his position in space at given times, and a spiral-bound map book, annotated with yet another checklist and with a guide to the location of Navy recovery ships; a pencil was attached to a spring-coiled tether, lest it float out of reach.

But in these first moments of his first pass around the Earth, Glenn's main job was to determine whether the stabilization control systems, which oriented the capsule, were working. If there was a problem, it was best to know it now, before Mercury Control committed to a second orbit. In the spirit of redundancy—of "backups to backups," as Bob Voas put it—there were three systems: the Automatic Stabilization Control System (ASCS), which was the default mode, preferred by flight controllers and resented by astronauts; a manual system, controlled by a stick at Glenn's right hand, which used its own set of thrusters and a separate fuel supply; and fly-by-wire, which also used the stick but drew on the fuel and the jets of the ASCS. One by one, Glenn tested each system on all three axes: he pitched the capsule up and down, yawed it left and right, and rolled it, like a drill bit. As the thrusters expelled little blasts of vapor, he could feel, more than hear, a high-frequency buzz; all sounds were muffled. Oxygen hissed faintly as it flowed into his helmet; electrical coils clicked as they engaged. After two minutes of maneuvering, Glenn had determined that all systems worked well—or, in the case of the manual mode, well enough. "It felt mushy," he said later, more difficult to operate with precision than it had been on the procedures trainer. He much preferred the fly-by-wire system: "That," he said, "really gives you fine control. . . . I liked it very much." But as decreed, that pleasure was denied him. He reverted to autopilot and his role as observer.

In a few minutes' time he was most of the way across the Atlantic. "The horizon," he reported, "is a brilliant blue"—not much different, he thought, than it looked from a high-altitude airplane, but more vivid for its contrast with the blackness of space. Glenn could still make out the shell of the Atlas, below and behind him, tumbling like a baton, reflecting the sun. The sun itself, as its light streamed through the window, was intense, a luminous white, yet not as blinding as NASA scientists had expected. Much of the ocean was covered with clouds, but where the skies opened up, the water appeared as bands of blue: the currents, he assumed, of the Gulf Stream. Through the periscope, the Canary Islands came into view. Glenn reached for his equipment pouch, fastened just below the hatch by strips of Velcro. Inside was the camera he had bought in Cocoa Beach. There was also a surprise, stowed there by Al Shepard: a toy mouse. It floated into Glenn's field of vision—until its long tail, knotted to something, stopped it from drifting. The mouse, as Glenn knew, was a reference to José Jiménez, the comedian Bill Dana's anxious astronaut ("They put that mouse in the nose cone . . . then they closed the door . . . on that little mouse . . . *I don't want to talk about it!*").

"This is *Friendship 7*, still on ASCS," Glenn said. Twenty-one minutes after liftoff, he was passing over Africa. "I can see dust storms down there blowing across the desert, a lot of dust; it's difficult to see the ground in some areas." With his jury-rigged camera, he snapped pictures of the Sahara. This was more, however, than a travelogue. One of Glenn's key tasks was to determine how well a human being could see from space: how much detail he could discern on the Earth's surface, how accurately he could gauge the distance between objects in orbit (the capsule and the empty Atlas, for example), how well he could identify lakes and rivers and mountain ranges. Some of this mattered for future missions, some for the sake of scientific discovery; in other cases, the military implications were clear, if unstated. Glenn applied himself to all of it, assiduously. That included his biomedical tests. He raised his visor and popped a nickel-sized sugar pill ("Standby . . . taking xylose") for a boost of energy and to see whether weightlessness affected his digestion. He shook and nodded his head to see if the movements made him dizzy, as Titov had been while in orbit ("Feel fine," Glenn said). He easily read the vision chart mounted on his instrument panel, assuaging the doctors' concern that an astronaut's eyeballs, in zero g, would lose their round shape and fail to focus. And he duly did his exercises, pulling the elastic cord and then pumping the blood-pressure cuff on his left arm. Telemetry sent the data to the tracking station in Zanzibar. "*Friendship 7,* this is Zanzibar Surgeon," came the response. "Blood pressure 136 systolic . . . just under 90 for diastolic. . . . Everything on the dials indicates excellent aeromedical status. Over."

As Africa rolled behind him, the Earth began to darken; Glenn was flying into shadow, toward the night. He "dark-adapted" the cockpit—clicking off certain lights, placing red covers over others—to make it easier to see the stars and to judge whether the horizon, at night, was visible enough to use as a reference point for the capsule's attitude. About 150 miles above the Indian Ocean—near the apogee, the highest point, of his elliptical orbit—Glenn rolled *Friendship 7* to get a better view of the setting sun. Given the spacecraft's speed and elevation, the sun sank quickly: eighteen times faster than when seen from the ground. To protect his vision, he watched it through the polarizing filter of a photometer, held to his eye like a kaleidoscope. What he saw was breathtaking. As the sun began to set, it seemed to flatten into the horizon, almost to melt, pooling liquid light across the curve of the Earth. Moments later the ribbon of white broke out in color: orange at the base, then red, then purple, then a deepening blue. For a few minutes more this thin band, more radiant than a rainbow, traced the boundary between Earth and space, then dimmed and went out. The world, now, was so dark Glenn could see nothing of it at all; looking down, he thought, was like staring into a black hole.

On the East Coast of the United States, it was still morning; only forty minutes had passed since liftoff. The phone rang on North Harrison Street: Scott Carpenter was calling to assure Annie that everything was going as planned. Annie, no less than Mercury Control, had been more worried about those five minutes between liftoff and insertion than any other phase of the flight, so she had been breathing a bit easier since her husband had entered orbit. "Gosh, isn't that wonderful!" she had said, and finally smiled, when the news anchors reported that the escape tower had been jettisoned. The networks, excitedly, began broadcasting Glenn's transmissions from the capsule, but not quite in real time; out of concern for emergencies, NASA had insisted on a few minutes' delay. When they began to run the recordings, which would play all day long, Annie laughed out loud with joy. Lyn said, "That's Dad!" Now, after Carpenter's call, Annie sat at her kitchen table with Wainwright, sipped a glass of cranberry juice, and reflected for a moment. For now, Annie Glenn, like the rest of the nation, was a spectator—a member of one of the more than forty million households, the largest daytime audience in history, fixed in place in front of their televisions, watching commentators point to orbital maps and listening to meteorologists review the weather in the recovery areas. "It doesn't seem like he's my man," she said. "Now he's everybody else's hero."

In New York, the crowd at Grand Central had swelled to nearly ten thousand. How long they intended to linger there, as the workday began, was unclear, but no one seemed in a hurry: they stood shoulder to shoulder, transfixed by the screen. At National Airport in Washington, the public address system apprised passengers of Glenn's progress. Across the Potomac, at the Washington Monument, tourists and guards carried transistor radios. The National Gallery of Art, which did not permit them, was virtually abandoned. Schools canceled classes and let students listen to the news. In Albany, the New York State Assembly stood for a silent prayer, then listened to the radio coverage over the loudspeakers. A judge in Grand Rapids, Michigan, suspended a trial so he and all involved could watch the coverage on TV—specifically, a stolen TV that had been introduced as evidence against the accused.

Life stood still at the White House as well. Around 10:30 a.m., as Glenn flew in darkness, President Kennedy led a solemn procession of legislative leaders out of the residence, where they had been eating—or not eating—breakfast, and brought them into the Oval Office to continue watching the coverage on television. Throughout the White House, TV sets were on, and they were, it seemed, everywhere: in the president's office; in his secretary's office; in the Fish Room, where reporters had huddled; on the second floor of the residence, where Jackie Kennedy sat; and on the third floor, where four-year-old Caroline and her play group were watching. None of the legislators—who were soon displaced by a delegation from the state of Minnesota—made any pretense that business had been transacted: Kennedy was preoccupied all morning by the flight. Between meetings, he turned the TV in the Oval Office back on; he scrutinized the orbital maps Glenn had sent him, maps his naval aide, Captain Tazewell Shepard Jr., had mounted on boards and annotated. He was so caught up that he forgot to eat lunch.

Down the hall, in the press office, Pierre Salinger had a direct line open to the Mercury Control Center. The word from the control room was no different from what Powers had been telling the public: the mission was going according to plan. Still, the mood at the Cape remained wary—not apprehensive, exactly, but watchful. That, of course, was their role, these shirtsleeved men in headsets, seated on swivel chairs in rows: they kept a vigilant eye on the dials and gauges and illuminated buttons on their consoles and followed Glenn's progress on the world-wide map that filled the front wall. A red icon of the capsule, suspended by wires above dark green continents and pale green seas, moved shakily along the first of its prescribed paths, each orbit enumerated, 1, 2, and 3. Flight controllers were monitoring about ninety kinds of telemetered information, reporting to Chris Kraft on everything from the control stick position to the pressure in Glenn's space suit; but it was all, in Gene Kranz's view, "very rudimentary." To Kranz, a former flight test engineer, Mercury seemed almost "a step back from a standpoint of the technology of command and control."

Indeed, the name of the facility—"Mercury Control Center"—was still, at this point in the program, aspirational. For all the data they had flowing in, for all the expertise concentrated in that glass-paneled room and elsewhere in the complex, there was no getting around the fact that the controllers were, like JFK, like Annie Glenn, observers. They had no illusions about their ability to determine Glenn's fate. Just after liftoff, when the head of the Air Force Space Systems Division had approached Walt Williams, offering congratulations, he'd brushed them off. "Well," Williams said, "now all we've got to do is get him down."

CHAPTER 16

A Real Fireball

THE NIGHT, FOR GLENN, was forty minutes long and as busy as the day had been. He clicked on the lights built into his gloves—tiny bulbs attached to the tips of the index finger and middle finger of each hand—and extracted his star chart from the case below the instrument panel. The Earth was pitch-black; the horizon marked a silhouette against a vast field of stars. They didn't twinkle: with no atmosphere to refract the light, the stars appeared as clear pinpoints. Glenn picked out the constellations on his map. He felt confident that, if need be, he could navigate by them. Still, the thick glass of his window had an obscuring effect; he was disappointed to find that the lab tests had been correct, that he was no better able to see the stars than he could from, say, the desert on a clear night. Soon, however, there was a light show below: two massive storm systems sent sheets of lightning across the tops of clouds; thunderheads popped like flashbulbs. He noticed a faint, then intensifying glow behind him. "I think the moon is probably coming up," he told the tracking ship in the Indian Ocean. "Yes, I can see it in the scope back here and it's making a very white light on the clouds below." The moon was nearly full, and now it brightened his path.

Over his headset Glenn heard a familiar voice: Gordon Cooper, at the tracking station in Muchea, Australia, just outside Perth on the southwestern coast. "How're you doing, Gordo? We're doing real fine up here. Everything is going very well," he said. "John, you sound good," Cooper replied, and he did: he sounded exuberant, in fact, punching phrases (*"real fine"*) for emphasis and rolling the first *r* in *"Roger!"* Glenn ran through his checklist like an announcer calling a baseball game: "Control fuel is *nine-zero one hundred* [90 to 100 percent]; oxygen is *seven-five one hundred*; amps are *two-two*; all systems are *still go*. Having no problems at all. Control system operating fine." Though he had little time for it, he felt like making conversation. "That was sure a short day," he told Cooper.

"Say again, *Friendship 7*?"

"Now, that was about the shortest day I've ever run into."

"Time passes rapidly, huh?"

"Yes, sir," Glenn said.

As HE PASSED ABOVE PERTH, he could see a small illuminated patchwork: as a greeting to Glenn, the mayor had asked the city's residents to turn on their lights. British Petroleum, for its part, lit oil-fired torches. "The lights show up very well," Glenn told Cooper. "Thank everybody for turning them on, will you?"

"We sure will, John."

He flew eastward, toward the rising sun. Nearly an hour and a quarter after liftoff, Glenn had his first meal—applesauce squeezed from a metal tube—and shut his visor. As the sun came up, it shone through the scope, a brilliant red. "You are very lucky," said the CapCom—the capsule communicator—on Canton Island, midway between Australia and Hawaii. "You're right, man; this is beautiful," Glenn answered. Then, through the window, he saw something unexpected: "I'll try to describe what I'm in, here. I am in a big mass of some very small particles that are brilliantly lit up like they're luminescent. I never saw anything like it." They were minuscule—some as small as the head of a pin, the largest no bigger than three-eighths of an inch—and glowed a yellowish green. They looked like fireflies, he thought, a whole pasture full of them, or a shower of little stars. They extended as far as he could see, thousands in every direction, swirling around the capsule and slowly trailing away.

"Roger, *Friendship 7*," the CapCom replied. "Can you hear any impact with the capsule?"

"Negative, negative. They're very slow . . . they're going at the same speed I am, approximately." It occurred to Glenn that they might be ice crystals, frozen water vapor from the thruster nozzles. He tapped the controls to see if he could create more, but none appeared, only steam. Once the sun had fully risen, most of the particles had vanished. Fifteen minutes later, Glenn was still talking about them, trying to make sense of them. The next CapCom along the path, in Guaymas, Mexico, was more concerned with a "scope retract" light flashing on consoles at the Cape, indicating that the astronaut, for some reason, had pulled in his periscope. He hadn't. The signal was wrong.

But a few seconds later, Glenn had an actual problem to report. The blunt end of the capsule had begun skating to the right, like a car out of alignment. It drifted about 20 degrees, at which point the autopilot, slow to react, fired the thrusters to bring the capsule back in line. Then the two-step repeated itself—slide right, swing left, every correction a waste of fuel—the same two-step that, last November, had caused Chris Kraft to cut short the flight of Enos the chimp. "Have had troubles [with] ASCS . . . every flight," Glenn had written at the time, in his notes for a meeting on his flight plan, and this, now, made it a clean sweep. The capsule was at no risk of slipping off course; its velocity was too great, its inertia too powerful. But if the ASCS continued to spend this much fuel correcting its own errors, Glenn would be going home early, just as Enos had.

Except that Enos had had no recourse. Glenn did. He shut the automatic system off, took hold of the control stick, and switched to fly-by-wire. "The fact that the pilot is aboard," Powers told the press, "and has the ability to analyze the problem and take judgmental corrective action, is a demonstration of the value of man in space flight." NASA, clearly, still had something to prove in this regard—as did Glenn. He had wanted to fly *Friendship 7,* and now he was flying it. Any kind of technical trouble was unwelcome, and controlling the capsule's orientation manually would take the place of other planned activities: experiments, observations, photographs. But the principle of "pilot in command" had been reaffirmed. "*Seven*, we concur here," Al Shepard said as Glenn approached the Cape to begin his second orbit. "Recommending you remain fly-by-wire."

"Roger," Glenn replied, "remaining fly-by-wire."

"*Seven*, this is Cape. The president will be talking to you."

"Ah—Mr. President—"

"Go ahead, Mr. President," Shepard said.

Silence. Forty seconds passed: only static. "This is *Friendship 7,* standing by," Glenn chimed in. "Roger, *Seven*," Shepard answered quickly, "we're having a little difficulty. Start off with your 30-minute report."

"Roger," Glenn said, and cheerfully went down the list, everything normal except the ASCS, every switch in its proper position, and the view out the window, the view of Florida, was beautiful. He paused. "I thought if the other call"—the call from the White House—"was in, I would stop the check."

"Not as yet," Shepard said, "we'll get you next time." But Shepard already suspected that this would not be so.

THE "LITTLE DIFFICULTY" was a warning light reading "Segment 51." A technician named Bill Saunders, conducting a routine scan of his console, was the first to see it flashing. "I've got a valid signal on Segment 51," he said, and Shepard, hearing this on his headset, made a note on a pad of paper and then sat and blinked in disbelief. Numbly, he walked over to Kraft and told him the news. On the other side of a glass panel, in an observation room, John Yardley, the lead engineer at McDonnell, picked up the phone and started making calls: to the factory that had built the capsule; to Max Faget, its chief designer; to "anybody with any knowledge of this," as one of the men in the room recalled. Someone was sent scrambling to round up, at great haste, as many experts as possible. Their discussions for now concerned a single question: whether to believe the signal. Because if the signal was correct, Yardley said later, "it's sure death."

What the signal said was that Glenn's heat shield had come loose. It was designed to come loose—but not before it shielded the capsule and the astronaut from the ferocious heat of reentry, temperatures of more than 3,000 degrees Fahrenheit. A ring of twenty-four locking pins was supposed to hold the shield in place—securing it to the blunt end of the spacecraft—while the capsule plowed through the atmosphere. Only after the main parachute had deployed would the pins release and the shield drop down—attached by cables but dangling, as its rubberized cloth cushion extended and filled with air, like the bellows of an accordion. This landing bag, or "impact skirt," would buffer the capsule (and its occupant) in case it hit land; if it splashed down, according to plan, the landing bag would keep it steady. All this was hours away—that is, unless a malfunction caused the pins to unlock in space. Even a slight gap between the spacecraft and its shield would become a fatal breach. Glenn would burn up on reentry, the capsule his crucible, a streak across the sky.

The debate, hushed but heated, began in the control room, spread into side rooms, was carried into back offices. Clusters of men went over wiring diagrams. A circle formed around Kraft, who struck a skeptical note. Check the signal again, Kraft told Saunders; the connection was probably spotty or weak. Saunders checked: the signal held steady at 80 percent strength. The blockhouse picked it up, too, as Glenn passed above. The possibility—the hope—remained that the problem was not in the capsule but at the Cape: a bad radio relay; a bad transmitter. Unless the remote sites around the world were receiving the same indication. Conscripting a teletype operator, Gene Kranz sent an urgent message to all of the tracking stations: "Report the readings on Segment 51, ASAP." Urgency, though, was poorly conveyed by teletype. While some of the sites had dedicated phone lines, most did not, and the Cape's queries went out on UHF single sideband, like a ham radio signal—passing first through the communications system at the Goddard Space Flight Center in Maryland. The process of transmitting a message took ten to fifteen minutes. In the meantime, the argument intensified. "We were desperate to find a solution," Kranz remembered, "without being sure we knew the problem."

Decades later, it would remain unclear who first mentioned the retropack. Surely it had been on many of their minds. The retropack—a unit of three reverse-propulsion thrusters attached to the blunt end—was designed to act like brakes at retrofire, slowing the spacecraft enough that gravity could pull it back to Earth. But the pack did one other thing: it held the heat shield in place. Three titanium straps bound the pack, and with it the heat shield, to the rim of the capsule's blunt end. And presumably—nothing now was taken for granted—the straps would continue to prevent the shield from coming loose, even if the locking pins did not. Don Arabian, one of the engineers, made this point emphatically: the landing bag could not possibly have deployed, he said, because the retropack prevented it. Even so, this was cold comfort. Once the retrorockets were fired, the whole apparatus, straps and all, would be jettisoned automatically into space.

Unless, perhaps, it shouldn't be. The source of this idea, too, was unclear; but maybe it was self-evident, or quickly became so, that the only conceivable way to save Glenn's life—if, again, the signal was right—was to leave the pack attached as *Friendship 7* plunged through the atmosphere. Faget, on the phone from Langley, declared this a fine solution. He and his engineers, out of concern that the retropack might fail to jettison, had tested a capsule—pack attached—in their wind tunnels, and deemed it stable for reentry. Yet few of the controllers were convinced. "There is one hell of a difference," Jerry Roberts of McDonnell said later, "between a wind tunnel test and an actual reentry, with this thing coming back in at more than 17,000 miles an hour. We had no wind tunnels that could simulate that!" Faget's assurances aside, no one was prepared to say that a three-foot-wide package of retrorockets, igniters, straps, electrical wires, and explosive bolts, sitting directly on the center of the otherwise smooth heat shield, would not throw the capsule off its very precise course through the upper atmosphere and send it spinning into what the engineers, in clinical fashion, called a "fatal attitude."

There were related fears. Even if the straps held long enough for g-forces to take over and press the heat shield in place, the retropack was not designed to withstand the heat of reentry. Its base was made of aluminum, its rockets of steel. Most of it would melt or burn off.

The danger with that, as Yardley and others saw it, was that the pack would burn off unevenly, destabilizing the capsule. And as it burned—breaking apart into blazing fragments, molten metal streaming along the skin of the spacecraft—what damage might it do? Worse, if any of the retrorockets had residual propellant, the heat would make them explode. The debate had reached a stalemate.

Then, as if on cue, the reports began to come in from the tracking stations. The ship in the Atlantic was first: affirmative. The landing bag signal was at 84 percent. Kano, Nigeria: affirmative, 84 percent. Zanzibar: affirmative; according to the telemetry, the landing bag was deployed throughout Glenn's pass over eastern Africa. Not every site picked up the signal; but most did, and strongly. "There is no doubt about it," Walt Williams concluded. "This is a valid signal that the heat shield is loose."

Glenn himself had said nothing about it. But then, he knew nothing about it. No one had told him a thing. "He's got enough to do without worrying about this." That was Kraft's stated view; and Kraft's view, in the control center, had the force of law. Never mind that even the most cramped conception of the astronaut's role had given Glenn responsibility for monitoring the capsule systems; Mercury Control was now attempting to do that without him. For forty-five minutes, while controllers agonized about what might or might not be happening in space, they refused even to ask Glenn whether the landing bag warning light just above his left knee was illuminated. Neither did they ask if the landing bag switch at its left was set at AUTO, OFF, or MANUAL. They did not want him to panic. In this, though, they misunderstood their pilot, and test pilots in general.

"Even if a plane got completely out of control," Glenn once wrote, "I do not think an experienced pilot would panic. There would be no point in that. He would still be sitting there trying to figure out what to do right up to the last minute." This could have been the test pilot's credo, or the combat pilot's: *No point in panic.* Glenn had been selected in 1959 for many reasons, but one of the most important was precisely this, that he didn't panic—not in Korea, when an enemy shell put three hundred holes in his plane, including its cockpit; not when an anti-aircraft gun blew a hole the size of a basketball in his plane's tail; not when he ran out of fuel while searching for a downed pilot and had to glide, without power, across the span of North Korea to get back to base while frost crept across his field of view. The thing that made pilots like Glenn anxious was not danger, not even mortal danger, but a lack of information and a loss, therefore, of control.

The control center's approach had exactly that effect: to arouse Glenn's suspicion that he was being kept in the dark. In the tiny teletype room—reverberant with the clatter of machines, reeking with the oil that kept the gears and bearings moving—Kranz and other flight controllers scripted questions for the remote stations to ask Glenn as he came into range. Two hours and nineteen minutes into the flight, three-quarters of an hour since the warning light went on at the Cape, the Indian Ocean CapCom gave the first, cryptic indication that something was wrong: "We have message from MCC for you to keep your Landing Bag switch in 'off' position. Landing Bag switch in 'off' position. Over."

Glenn looked at the switch. It was off. Of course it was off. In fact, he had just told Zanzibar it was off, as part of the standard report he issued every half hour.

"Roger," Glenn replied, and left it at that.

Seven minutes later, he was over Australia and back in contact with Gordon Cooper. "Will you confirm the Landing Bag switch is in the 'off' position?" Cooper asked. "That is affirmative," Glenn said. "Landing Bag switch is in the center 'off' position."

"You haven't had any banging noises or anything of this type?"

Banging noises. A leading question, but leading where?

"Negative," Glenn said.

"They"—the Cape—"wanted this answer," Cooper added, by way of explanation or apology, though the comment revealed nothing that Glenn hadn't known. The question wasn't who wanted to know about banging noises, but why.

Kraft had been right about one thing: Glenn was too busy to worry about it. For most of his second orbit he was preoccupied with the autopilot. Rather than give up on the ASCS entirely, Glenn turned it on and off, trying to diagnose the problem, trying to pin down what had triggered it, but there was no apparent pattern. At first the small thruster on the left ("a delicate piece of jewelry," as Williams described it) had seemed to be sticking; then it was the one on the right. Glenn was still able to use fly-by-wire without incident: over the Atlantic, he had turned the capsule to face forward again, just for a moment. "I like this attitude very much, you can see where you're going," he said. But when he returned to flying backward, the automatic system went seriously awry, far worse than before.

In every axis now—yaw, pitch, and roll—the instruments were at odds with what Glenn could see out the window. He was also now struggling with other issues, none of them serious, but all requiring his attention: the oxygen supply in his secondary tank was dropping, even though he wasn't using it, suggesting a leak; the EXCESS CABIN H2O light had gone on, indicating a buildup of humidity, possibly from the water coolant in his suit; the FUEL QUAN warning went on, too, unsurprisingly. The quantity was not dangerously low; but Glenn was depleting his supply more quickly than expected, given the work he was doing to compensate for the autopilot. When the Indian Ocean CapCom asked whether he'd identified any constellations, now that the sun had set again, he replied that he just hadn't had the time. The white tube containing his second meal—puréed roast beef and vegetables—remained untouched.

As DAWN APPROACHED, the fireflies returned. This time, somewhere above the Pacific, Glenn tried to take pictures of them. He parked his camera in front of his visor—he found he could simply let go and leave it to float in place—while he fished in the pouch for another roll of film. One slipped from his grasp while he was reloading the camera; attempting to grab it, Glenn batted it farther out of reach and watched it drift behind the instrument panel. He was narrating the experience when the Canton Island station broke in. "We . . . have no indication," the CapCom there said, apropos of nothing, "that your landing bag might be deployed."

Finally Glenn asked the obvious question: "Did someone report landing bag could be down?"

"Negative," Canton Island replied. "We had a request to monitor this and to ask you if you heard any flapping, when you had high capsule rates—"

"Negative," Glenn said. Now it all made sense, this line of questioning. It was about the fireflies. Mercury Control, he deduced, must have a theory. "They probably thought these particles I saw might have come from that"—from the heat shield, like a loose rear bumper, banging into the capsule and knocking flakes of dust, frost, something from its surface. Except that the theory didn't fit what Glenn was seeing. "There are thousands of these things," he pointed out, "and they go out for, it looks like, miles in each direction."

The CapCom did not correct Glenn's misapprehension. The issue, Glenn believed, had been put to rest. In short order the Cape appeared to confirm it: as the flight neared the three-hour mark, they gave Glenn the go-ahead for a third orbit. If there was trouble, he concluded, they'd have brought him back early. "You look real good up there," Wally Schirra told him from the tracking site on the California coast. "I feel real good, Wally," Glenn responded. "No problems at all."

IN THE CONTROL CENTER, the mood was one of cold panic. Kranz saw "faces drained of blood." Yardley, in the teletype room, had tears in his eyes. More than ninety minutes after the warning light had blinked on, the conversation, drawing in more and more controllers, had become such a distraction that Kraft ordered all but a few back to their seats. Absent any new information, absent any new notions of how to save Glenn, positions hardened and were put forward with vehemence; the debate was getting heated. On one side were the capsule's designers and manufacturers—Faget; Yardley and his McDonnell engineers—who argued, most of them, for leaving the retropack on. On the other was the team that analyzed, and in this case doubted, the telemetry: Arabian and others, echoing Kraft, continued to insist that the safest approach was to leave well enough alone. But "there was no right answer that day," Kranz recalled, "nothing in black and white. . . . The last orbit was a stalemate."

One way or another, it would have to be broken. Glenn's third orbit gave them more time to keep arguing, but they could not leave him up there indefinitely. Even if his oxygen supply held out for a fourth pass around the Earth, his fuel might not, given the issues with the autopilot. And, of course, the longer he spent in space, the later his reentry: assuming, as no one could with confidence now, that the capsule splashed down near the Florida coast, recovery crews would have little time to find it before darkness set in. "Either you give me a decision or

I'm going to make one myself," Kraft snapped at Williams, whose presence in the room—rather than behind glass, where he had been sitting with Gilruth and the NASA hierarchy—was, increasingly, an irritant to Kraft. As operations director, Williams outranked Kraft, the flight director; but now that men and not monkeys were being sent into orbit, Kraft, from his seat at the center of it all, tended to regard this as a technicality. "While the mission is underway," he wrote in his memoir, "I'm Flight. And Flight is God."

With ten minutes left before retrofire—when Glenn would fire the three rockets, one after the other, to begin deceleration and reentry—Williams told Yardley that he needed the engineers' final recommendation. Yardley knew his own mind, but he scrambled for the phone to hold one last round of consultations with St. Louis and Langley. Then he walked to the floor to give them the word. In a huddle with Williams, Kraft, Shepard, and Slayton, he said that leaving the retropack on seemed safe enough to take the risk. That view, Yardley added, wasn't unanimous. But it was his view. With that, Williams went back into the viewing booth to talk it over with Gilruth, who sat, saying nothing, listening to the conversations between the astronaut and the remote sites. Williams asked whether he had any objections to the idea of leaving the retropack attached. Gilruth simply shook his head.

But even that didn't settle the issue. Kraft, of his own accord, made one last attempt to determine whether the heat shield was loose. Just before giving Glenn the go for a third orbit, the Cape had told the remote sites to stop reporting on Segment 51; no clarity had come of it. But now, as Glenn made his final pass above the Pacific, Kraft countermanded his own order: "Monitor Seg. 51, note time signal on or off, duration of signal, percent of deflection. Forward ASAP." Then Kraft upped the stakes. At his direction, the Hawaii station, high above

Waimea Canyon on the island of Kauai, radioed Glenn a new directive. "*Friendship 7*," the CapCom said, "we have been reading an indication on the ground of Segment 51, which is Landing Bag Deploy. We suspect this is an erroneous signal. However, Cape would like you to check this by putting the Landing Bag switch in auto position, and see if you get a light. Do you concur with this? Over."

Glenn's silence was an answer of sorts. He let the line crackle awhile. No, he did not concur, not that his opinion was actually sought. Glenn understood that if he clicked the switch from OFF to AUTO, and the light then flashed green, it would be almost certain that his heat shield was loose. It was better to know than to remain in the dark. This was always his view. Yet given the chance—the likelihood—that the switch itself was malfunctioning, that its circuits were bad, then clicking it back and forth risked the very thing that could cost him his life: it might well trigger the landing bag to deploy.

In the years before this moment Glenn had spoken many times—to the press, to Annie, to their children—about his faith in the program. Here, now, was a profound test of it. If the risk occurred to him, he reasoned, then it had occurred to Mercury Control. "Okay," he finally replied. "If that's what they recommend, we'll go ahead and try it."

He flipped the switch to AUTO and quickly back to OFF.

"Negative," he reported, "in automatic position did not get a light."

"Roger, that's fine," the CapCom said. "In this case, we'll go ahead, and the reentry sequence will be normal."

As opposed to what? Kraft had told the Hawaii station that if no light went on, Glenn should be instructed to reenter as planned, but if a light did go on, indicating a loose heat shield, he should reenter with the retropack attached. None of this, however, had been communicated to Glenn. (Neither was it cleared with Williams, who, incredulous, stood nearby as Kraft did exactly as he'd threatened and tried to make the decision himself.) "Normal," in a sense, was all Glenn needed to hear; but with his own life in the balance—it was finally apparent that this was not about the fireflies—vital information was reaching him in fragments, through slips and inferences, if it was reaching him at all.

Then another fragment: this one from Schirra, in California. As the TIME TO RETROGRADE counter on Glenn's panel clicked down to 00:00:24, Schirra said, "John, leave your retropack on through your pass over Texas. Do you read?"

No explanation, just another command. Glenn replied: "Roger." Then Schirra counted down for the retrorockets to fire—one after the other, at five-second intervals, against the direction that the capsule was flying. The timing was crucial: a single second's delay would send Glenn five miles farther across the impact area than he was supposed to go. "Retros are firing," Glenn reported. "Are they ever! It feels like I'm going back toward Hawaii."

"Don't do that," Schirra laughed, "you want to go to the East Coast." Then, seconds later, as if Glenn needed a reminder: "Keep your retropack on until you pass Texas."

"That's affirmative."

Message delivered, Schirra seemed to relax, and he tried to get Glenn to do the same. "Pretty good-looking flight from what we've seen," Schirra told him.

"Roger," Glenn said, "everything went pretty well except for all this ASCS problem . . . I had a lot of trouble with it."

"Good enough for government work from down here."

"Yes, sir, it looks good, Wally. We'll see you back East."

By the time Glenn flew over Texas, a few minutes later, the story had changed. Once all three retrorockets had fired—and had, therefore, been emptied of fuel—Williams leaned over to Kraft and said, in a voice that signaled an end to all discussion, "We're coming in with the pack on." The decision was now conveyed to Glenn by the Texas CapCom, George C. Guthrie: "We are recommending that you leave the retropackage on through the entire reentry. Do you read?"

Recommending—this was politesse. Glenn understood that it was not a suggestion.

"This is *Friendship 7*. What is the reason for this? Do you have any reason? Over."

"Not at this time. This is the judgment of Cape Flight."

"Roger."

They talked for a moment, matter-of-factly, about the steps Glenn should take to override the automatic release of the retropack. "*Friendship 7*," Guthrie said, "Cape Flight will give you the reasons for this action when you are in view."

"Roger."

"Everything down here on the ground looks okay," the CapCom added feebly, but the weather, right now, was beside the point.

ON NORTH HARRISON STREET, they gathered again in front of the TVs. Annie Glenn, in an armchair, was flanked by her two children. Over the past few hours they had all endeavored to keep busy, though Dave remained riveted, listening for updates "so we could tell where my dad was." When Glenn had finished his first orbit, the women got up to serve the lunch that Annie and her mother had made the night before: chili, Jell-O with fruit. A neighbor brought out a chocolate layer cake she had baked weeks earlier and put in the freezer after the "big scrub" of January 27; A-O.K. JOHN, the icing read. Western Union, optimistically early, began to deliver telegrams of congratulations; the Glenns split the stack and read them. Five of Lyn's school friends arrived, excited to announce that they had just nominated her for ninth-grade class president. (Complicating matters, they had also nominated her boyfriend, whose football charm she wore around her neck. Lyn went on to win in a landslide.) The group went downstairs to the rec room. Soon two bellboys in red uniforms, sent to the house by the Marriott Motor Hotel, rang the bell, carrying twenty-five tubs of fried chicken and boxes of cheese rolls and apple pies. Annie opened the front door and stepped outside—her only appearance while her husband was in space—to offer some chicken and rolls to the restless and hungry reporters occupying her front lawn. "Help yourselves," she said.

During the four hours that followed the launch, David Glenn recalled years later, "we had sort of settled into it," but the reentry was "the second big sweat"—another moment when the dangers were at their most acute. The family and their guests stared fixedly at the TVs. With a few minutes left, the voice of Shorty Powers broke in. A signal, he said, had indicated a problem with the heat shield, though he was quick to note that "a check was made over Hawaii which confirmed that the system was operating properly and that the signal was probably erroneous." At the same time, he said, out of an excess of caution the retropack would be left attached. Dave, alert to any aberration in the flight plan, scowled at the news. On CBS, Walter Cronkite worried aloud about Powers's tone of voice but conceded that he might be projecting "our own tension" into it.

Across the country, people froze, went silent. Long-faced crowds, large and small, gathered wherever a television set was in sight: a hotel lobby; a hospital waiting room; a bar; the sidewalk outside an appliance store window; the Oval Office, where Kennedy stood watching the coverage with his aides. On Capitol Hill, in the otherwise empty chamber of the Senate, Majority Leader Mike Mansfield adjourned for the afternoon "in light of the circumstances" and went to see what would happen.

THE LAST VOICE Glenn would hear before reentry was Al Shepard's. Texas had promised that Cape Flight would explain itself, and this was left to Shepard. "Recommend you go to reentry attitude and retract the scope manually at this time," he began. The periscope had been meant to retract automatically when the retropack was jettisoned, but Mercury Control was overriding the chain of events. Glenn grabbed the handle and began winding the scope inside. The door beneath him closed behind it. "While you're doing that," Shepard continued, "we are not sure whether or not your landing bag has deployed. We feel it is possible to reenter with the retropackage on. We see no difficulty at this time in that type of reentry. Over."

There was nothing more to be said—not now, at any rate, with two minutes left before the capsule shot through the atmosphere and communications were lost. Glenn had pieced most of this together, anyway—but the hours of indirection, the obtuse questions about banging noises when his life was at risk, incensed him.

"Roger, understand" was all Glenn said.

And then both men changed the subject—filling what remained of the time with technical talk, the virtues of manual control versus fly-by-wire, and, from Shepard, a report on the weather in the recovery area. "Excellent," he told Glenn. "Three-foot waves, only one-tenth cloud coverage, ten miles visibility—"

"You're going out," Glenn said.

"We recommend that you—" Shepard began, but then the crackling on the line became a roar of static and John Glenn went alone into the heat.

THE BLUNT END hit the atmosphere with such force that it sent forth a shock wave, a pressure wave so intense that it shattered the molecules in its path, enveloping the capsule in a shroud of plasma—superheated, ionized gas, electrically charged, blocking all transmissions. The resin in the ablative heat shield began, as designed, to vaporize, carrying away some but hardly all of the heat. Flying backward, Glenn knew that any problem with the heat shield was going to reveal itself, at first, as heat along his spine; he also knew that if he felt heat along his spine it would all be over quickly. He braced himself to burn. His nerve endings prickled; they seemed to tell him it was starting.

Outside the window the plasma glowed orange, more and more intensely. As the capsule plunged and gravity returned, Glenn heard a hissing sound, as if fragments were scraping the skin of the spacecraft; then he heard a thump. He felt he knew what this meant. "This is *Friendship 7*," he said to no one, the tape recorder his witness. "I think the pack just let go." Seconds later one of its straps, freed from its base, swung across the window. It held there for a moment, then burned off and fell away. Suddenly a barrage, a cannonade, of flaming chunks of metal tumbled past him, banging against the capsule like a handful of rocks thrown at a tin can. Glenn watched them fly, leaving smoke trails behind them. Something, clearly, was coming apart. He hoped he'd been wrong about the retropack, because if these were not pieces of the pack, they were pieces of the heat shield. "I will be afraid," he had confessed to his children, "when the booster is getting ready to fire" on the launchpad, but this, now, was the first moment that he truly felt fear—fear of a contingency that would not yield to grit or goodness or skill or self-assurance.

"Hello, Cape, *Friendship 7*," he kept saying. "Hello, Cape, *Friendship 7*. Do you receive?"

SHEPARD, TOO, called into the void: "*Seven,* this is Cape. How do you read? Over."

The communications blackout was expected. Shepard himself had experienced it on his return from space. Glenn's silence was chilling regardless. Radar tracked an object moving through the atmosphere, speeding down the slant *Friendship 7* was to take, but the dot on the screen said nothing about the shape that object was in. The dead air lasted two minutes, three minutes, four minutes. A pall of fear hung over the room. Bill Douglas stood at his console, clasped his hands and prayed. Through it all Shepard sat immobile, leaning into his microphone, repeating those few words. ("Keep talking, Al," someone behind him said.) "*Seven,* this is Cape. How do you read? Over."

The line crackled to life. "Loud and clear. How me?"

"There he is!" a controller shouted. Shepard shot his arm into the air and gave the room a thumbs-up. "Roger," he said to Glenn, "reading you loud and clear. How are you doing?"

"Oh, pretty good," Glenn replied, without emotion. It was a bit hard to push the words out, harder still to lift his hand and reach the controls: he was passing through peak g nearly eight times the force of gravity. Shepard told him he would likely splash down within a mile of one of the recovery ships. "What's your general condition?" he asked. "You feeling pretty well?"

"My condition is good, but that was a real fireball, boy!" Glenn said. He had passed through peak g, and there was vigor in his voice. "I had great chunks of that retropack breaking off all the way through." No one needed to tell him it had not been the heat shield.

He was twelve miles high and falling at supersonic speed. Seconds later, ten miles from the water, the capsule began to swing back and forth, oscillating so violently that Glenn lost control. "It just plain overpowered me," he said a short time later. His contrail, following him downward, twisted like a corkscrew as he fought the capsule almost to a draw; it was nearly stable when the fuel cut out. Both tanks went dry. The control stick controlled nothing now. Immediately the spacecraft began rocking again. With 35,000 feet to go, concerned that he was going to hit the Atlantic nose-down, he reached for the emergency switch to fire his drogue parachute early, but the small chute shot out on its own. Watching through his smoke-sullied, resin-streaked window, he saw the chute open and saw that it held. At 10,000 feet, the main parachute opened, blooming outward, orange and white. "Beautiful chute," he declared. He rolled out his periscope and opened a valve that let the outside air in.

The last thing he did was hit the landing bag switch. The light glowed green, the pins holding the heat shield released, and the bag filled with air to cushion his impact.

"Here we go," he said.

CHAPTER 17

The Big Lift

FRIENDSHIP 7 SLAMMED INTO the ocean surface far more forcefully than Glenn had expected. "Impact," he said, as the capsule's momentum carried it, for a few seconds, almost completely underwater. It popped up and righted itself. The time was 2:43 p.m., four hours and fifty-five minutes since Scott Carpenter had wished him "Godspeed" at the start of the eighty-one-thousand-mile journey that ended here, 150 nautical miles due east of Grand Turk Island, the largest of the Turks and Caicos. The fresh air, gulped in through snorkel valves, was hot and humid. Soaked in sweat, Glenn was so uncomfortable that he considered climbing out through the neck of the capsule—a route he preferred to the hatch, which had put Grissom in such peril. But this would require a good deal of work: he would have to move the instrument panel, then detach the bulkhead and the parachute canister. He decided to wait it out, remaining as still as possible; any movement, he knew, would just make him hotter.

Lookouts on the destroyer *Noa* had spotted the capsule just before splashdown. The ship sped to meet it, followed by helicopters from the carrier *Randolph*, farther away. The *Noa* got there first, in about twenty minutes. As the ship approached, Glenn aimed his periscope at it and could read the numbers on the hull; he could see the sailors in their white T-shirts crowding the deck and holding up a stenciled sign: NOA's MEN WELCOME GLENN. A small crane hoisted the spacecraft from the water, letting it dangle for a moment while the landing bag drained. As the waves rolled, the capsule swung on the line and banged into the side of the ship—"probably the most solid jolt of the whole trip," Glenn reported later. But in short order, twenty-one minutes after splashing down, *Friendship 7* was on the deck. Now Glenn blew the hatch. When he hit the detonator, the firing pin recoiled and cut his knuckles. In the course of 159 combat missions, years of test flying, and now three orbits around the Earth, this was the only injury he had ever received.

The crewmen moved in to extract him from the capsule. "It was hot in there," he said. Someone ran to get him a glass of iced tea. Another sailor painted Glenn's footprints in white where he had stepped onto the deck—just as the French, in 1927, had marked the spot at Le Bourget Field where the *Spirit of St. Louis* touched down and Charles Lindbergh emerged, immortal, the first man to have flown across the Atlantic.

THEY REACTED AT FIRST with relief, not elation. At Grand Central Terminal, beneath the CBS screen, people shouted, "Thank God." At the Statler Hilton Hotel in Washington, the crowd by the TV in the lobby broke out not in cheers but in applause. The tension took a few minutes to release its grip. But then it did, almost all at once. In Trenton, whistles, sirens, and cannon fire sounded. Jazz bands began playing on Cocoa Beach. In Manhattan, office workers tossed shreds of paper and ribbons of adding-machine tape out skyscraper windows—the first impromptu ticker-tape celebration since the end of World War II. In the Mercury Control Center, there were tears—of joy, pride, an almost desperate gratitude.

On the South Grounds of the White House, a clutch of reporters and photographers stood in the cold of the late afternoon, waiting for the president. Jackie Kennedy, wearing a pink wool suit that matched the mood of the hour, burst out of the residence and ran the length of the colonnade. She let herself into Evelyn Lincoln's office. Minutes later, around half past three, the French doors of the Oval Office opened and JFK stepped outside. He had been scribbling notes—illegible, as usual, to any eyes but his own or Lincoln's—across a typewritten memo that happened to have been sitting on his desk. Then, thinking better of it, he left these behind and spoke off-the-cuff. His tone was not triumphal. This was Glenn's day, and reporters sensed that Kennedy did not want to steal the spotlight.

"I know that I express the great happiness and thanksgiving of all of us that Colonel Glenn has completed his trip," he began, as the first lady watched from the doorway. "A few days ago, Colonel Glenn came to the White House and visited me. And he was—as are the other astronauts—the kind of American of whom we are most proud." Kennedy praised the team at Cape Canaveral and closed, after a little more than a minute, with a renewed call to purpose. "We have a long way to go in the space race. We started late. But this," he said, "is the new ocean, and I believe the United States must sail on it and be in a position second to none."

Returning to the Oval Office, he spent a few minutes talking with Hugh Sidey (upbraiding him, actually, about a story in *Time* stating, wrongly, that JFK had modeled a suit for a fashion magazine). A military aide rushed in: Glenn was on the line. Kennedy walked to his desk and picked up the receiver. "Colonel Glenn?" he said.

"This is Colonel Glenn."

"Listen, Colonel, we are really proud of you, and I must say you did a wonderful job. . . . We are glad you got down in very good shape. I have just been watching your father and mother on television, and they seemed very happy."

"It was a wonderful trip," Glenn said, "almost unbelievable, thinking back on it right now. But it was really tremendous."

Kennedy explained that he would come down to the Cape on Friday, in three days' time. "We will be looking forward to seeing you there."

"Fine," Glenn said. "I will certainly look forward to it."

"How was the trip?" the president asked, unaware that the connection had been lost. He repeated the question. Hearing only silence, he smiled, put down the receiver, and returned to berating Sidey.

Glenn, on the *Noa,* settled into a padded chair in a quiet spot on the deck. He had shed his space suit by now, taken a shower, and put on light blue coveralls adorned with zippered pockets and the NASA insignia. He propped up his feet—outfitted in high-top Converse All Stars with bright white soles—and held a microphone tethered to a reel-to-reel recorder. He had been told to conduct a "self-debriefing," to capture his impressions while they were still fresh. "It's almost impossible to try and describe the sensations that I had during the flight," he began. "The thing that stands out more particularly than anything right now is the reentry during the fireball," he said, and he went on, for the first time, to tell the story that, half a century later, Americans would still ask to hear.

WITH GLENN BACK ON EARTH, his household was suddenly at loose ends. Annie and her guests stood around smiling, saying little, until the minister offered to lead a prayer of thanksgiving. Outside was pandemonium. The press had been there since daybreak, churning the front lawn to mud, demanding comment from anyone who came or went, keeping watch on the drawn curtains in the picture window.

Glenn's safe return meant that they might finally get what they were waiting for: an interview with Annie. But first she had phone calls to answer: one from JFK, who invited the family to fly to Florida with him, then, at long last, one from Glenn himself, on a crackling radio-phone from the ship. Lyn and Dave huddled nearby to listen. "Honey," he shouted, "how is everything out there?" Annie wept with relief. Quickly, she brightened and listened to Glenn for a moment. "Yes, I *am* happy!" she said. "By the sound of the things we heard you'd said, you must be the most excited man in the world."

It was time to face the press. However happy the news today, this was never something Annie relished. Her debilitating stutter made live TV treacherous. Flanked by her children on the front porch, she smiled apprehensively as reporters shouted, "Louder!" and a man from CBS fit headphones over her ears so she could hear Walter Cronkite speak. Her voice was halting as she made a brief statement: "This is the most wonderful day for my family, and we're quite—we're so proud of our—of their father, of the Mercury team, of everyone that's made this such a successful day." Now the networks, overplaying their hand, tried to engineer a family reunion: they patched Glenn's parents in from Ohio. Nudged along by reporters, the family exchanged awkward greetings and bantered ("I think Grandpa's a big ham," Lyn said), but most of their remarks were lost amid static and cross talk.

The networks moved on to New Concord, where Glenn's parents entered the Muskingum College gym to a standing ovation. State troopers led them through the crowd. The Glenns had watched their son's flight on the TVs in their living room; after liftoff, they invited a few highway patrolmen in for coffee and sent a fresh pot to the reporters outside. Now, dressed for the occasion—Herschel in a bow tie and boutonniere; Clara with a large corsage on her lapel—they fielded questions. Had they been afraid? a reporter asked. Not for a minute, Clara insisted.

"I'm so very, very thankful to God for his safety and for the success of the mission." Herschel—grinning widely, laughing at his own jokes—told stories about living for weeks with camera crews encamped on his front step. "They start introducing theirselves, you know—well, I can't remember one name, let alone a dozen! I said, 'Well, come on in. As far as I'm concerned, you're all 'Joneses'!" The Glenns were asked whether their son might become the first man on the moon. "Well, I think that's a long way off yet, myself," Herschel said, "but I wouldn't be a bit surprised when it comes time to go that he'll want to make the first trip."

NASA, too, held a press conference, perhaps the first in months that did not announce a delay. The whole lineup was euphoric. "I am very, very happy," Gilruth said, "very, very proud of John Glenn and the entire Mercury team." (*Very* was as close as Gilruth got to jubilation.) Glenn, he added, had done "a thoroughly impressive job" in the face of great pressure. Williams gave credit where due: "I have watched many a test pilot at work, and I have never seen a better job anywhere than John Glenn did today." He acknowledged the irony plain to all: that Glenn's success in overcoming the problems with the autopilot had made the case for the pilot's role more powerfully than a flawless mission would have. As for the heat shield, Williams pinned the blame on a "faulty micro switch"; but with the capsule, like Glenn, still on the *Noa,* that investigation had not begun.

As evening approached, a helicopter was sent to bring Glenn to the aircraft carrier. He went back to his capsule, tied in place to the deck, and collected the pouch that held his camera and film. The crewmen gathered to bid him farewell. They helped him into the straps of the hoist and cheered as he was pulled up into the chopper. The last they saw of him were the bright white soles of his sneakers. Airborne again, Glenn looked to the horizon and, for the fourth time that day, watched the sun set.

On the *Randolph* he was given a bite to eat, a debriefing, a chest X-ray, and an electrocardiogram; after that, he boarded a Navy S2F jet for the one-hour flight to Grand Turk, where he would spend the next two days on an auxiliary U.S. Air Force base in relative seclusion, reviewing the mission with Mercury officials and resting up for the public events ahead. Glenn's schedule had just begun to take shape, but it was already filling up: NASA and the Kennedy administration were planning a long (perhaps month-long) victory lap, starting with Friday's ceremony at the Cape. In the days before his flight, Glenn had recorded the possibilities on a notepad:

> Arr. ADW [Andrews Air Force Base] . . . Helo [helicopter] . . .
> 1000 arr White House
> Annie & kids front [at Andrews]
> Motorcade to Capitol

Next on the list: an address to a joint session of Congress, an honor not extended to Shepard. The speech had been Lyndon Johnson's suggestion. And there was one event Glenn could cross off his list:

> If abort—Gus-type conf[erence] at Cape

As the Navy pilot prepared to take off for Grand Turk, he offered Glenn the controls. Glenn declined. It had been too long, he said, since he had flown an S2F. Instead he acted as co-pilot: he ran through the checklist before they strapped in, settled into their seats, and were flung by the catapult into the moonlit sky. The g-forces—maybe 2 or 3 g's—were nothing to Glenn. It was hard to recall that there had been a time, nearly two decades ago, when a catapult had thrilled him. "That really is a boot in the tail," he wrote in his diary in 1944, after being shot off a ship for the very first time. "You sit there, get everything all ready . . . , rev up the engine to about full power, reach over and give the signal, and about one second later you are sitting out in space flying and not too sure about how you got there."

And now, having flown faster and farther than any other American, Glenn would again have a feeling of dislocation. During the next two days on Grand Turk, he caught glimpses of newspaper headlines and got an intimation of what awaited him back home, but it was bigger and more intense than he understood: less a celebration than a catharsis.

FOR LACK OF A PARADE—these were still days away—they converged on the post office. "The people just came from nowhere," marveled one of the workers at Benjamin Franklin Station, a few blocks east of the White House. Within minutes of Glenn's landing, TV reporters announced that the Post Office Department and the Bureau of Engraving and Printing—clandestinely, as if planning an invasion—had produced a four-cent stamp commemorating the flight, and that it would go on sale immediately across the country. The stamp, an image of a capsule glowing gold against a deep blue, starry sky, read, U.S. MAN IN SPACE and PROJECT MERCURY. (The Post Office had wisely rejected a design picturing the capsule during reentry, enveloped in flames.) It was only a stamp; but in the emotion of the day, it touched off a mania. At the main post office in Manhattan, the crowds grew so large that every window had to be dedicated to selling the stamps in blocks, in sheets, on first-day covers. Nationally, ten million were sold by the end of the day. In Columbus, people lined up overnight.

The occasion was marked in myriad ways. Schools, roads, babies were named for Glenn. (There were variations on the theme: a boy in Utah was given the name "Orbit.") At the National Theatre in Washington, before a performance of *Bye Bye Birdie,* someone in the audience began to sing "The Star-Spangled Banner." The whole crowd joined in. On Cocoa Beach, the marquee at the Holiday Inn lit up: OUR PRAYERS WERE ANSWERED. The manager, Henri Landwirth, finally got to take his month-old, spacecraft-sized cake out of its refrigerated van. Gilruth cut the first slice at a lamplit poolside ceremony. (The cake was stale but was said to be delicious.) President Kennedy, hosting a previously scheduled dinner at the White House that night in honor of Vice President Johnson, Speaker of the House John McCormack, and Chief Justice Earl Warren, raised his glass and said, "It's been a great day for an orbit. I've been in orbit, too."

In a special report that night on NBC, the anchor, Frank McGee, captured the feeling:

> A great sigh rose across the land as millions of Americans experienced the end of an intensely moving experience. . . . We had, after all, anticipated so long this country's first effort to place a man in orbit; we had encountered so many maddening delays; we had feared, each of us, for the safety of the chosen astronaut, John Glenn. . . . We have by this time, it seems, reached something bordering on hypnosis.

Melodrama, but deeply felt: it seemed as if a fever dream had broken. During the six months since the United States had last managed to put a man in space, the Cold War had intensified. All the while, Americans waited to see if their space program would finally prove more formidable than the weather, or if they had already lost the race to the moon, and for survival. Those were the stakes as they were perceived. At the White House dinner, Chief Justice Warren reflected on the "many discordant voices speaking about the failures, the weaknesses, the lack of progress of our country." Glenn's flight, he said, had restored the nation's self-belief. BIG LIFT FOR U.S., the *New York Times* agreed. HE'S BACK! cheered the *Miami News,* and what that meant was that America was back.

The United States was still behind the Soviets. And no one knew what they might do next. With this in mind, some critics of the program dismissed MA-6 as insignificant. "The nation would be deluding itself to believe that a breakthrough occurred," scoffed William Hines, science writer for the *Evening Star.* Glenn, he argued, had "demonstrated two truths which really did not need to be proved: that man can survive in space and that brave men are willing to try." But Glenn had demonstrated another truth. He had proved, as Bill Douglas pointed out, that "man is a heck of a lot better than a black box"—that in the end, the pilot proved more reliable than the equipment (which needed more fine-tuning). In this sense, as Bob Voas later said, Glenn delivered "a real rebuke to Chuck Yeager and those guys at Edwards" who had mocked the astronauts as "Spam in a can." This augured well for Apollo. NASA now said with confidence that astronauts would be able to perform the complex tasks of a lunar mission: docking two vehicles in orbit, navigating to the moon, landing with precision, and flying home. The path to the moon was still full of danger, but Glenn had swept away the passivity and inefficacy that had afflicted the space program from the start.

And the world took note. Across Western Europe, the news brought a shared sense of pride and relief. "A spell has been broken," proclaimed the *Frankfurter Allgemeine Zeitung.* The free world "need no longer stare as if hypnotized at Soviet space successes with pricks of doubt in their hearts as to whether there is not some deep deficiency in the democratic order." America's openness, its invitation (though grudging) for the world to watch it take this risk, to see it stumble and then succeed, now seemed a masterstroke: it sharpened the contrast with what *Paris Jour* called "the enigmatic and disquieting missile launchings in the Siberian no man's land." (The USSR launched its cosmonauts from Kazakhstan, not Siberia, a fact the Soviets worked to obscure. *Paris Jour*'s error underscored, in effect, its point.)

At the Vatican on February 21, Pope John XXIII expressed his wish to Robert and Ethel Kennedy—on their way to West Berlin—that Glenn's achievement "always and only" be used in service of peace. Yet its military significance was clear—not least to European diplomats, who welcomed the flight as an overdue demonstration of U.S. missile technology. In this regard it was well-timed. The reason Bobby Kennedy had been sent to Berlin was to reaffirm America's commitment in the face of Soviet provocation. Russian MiGs had been harassing Allied planes along the air corridor linking Berlin with West Germany—a harbinger, observers believed, of further escalation. But for now, at least, Glenn had given the United States its swagger back. On February 22, the day Bobby stood at the Berlin Wall and pledged that an attack on that city "would be the same thing as an attack on Chicago," Ted Sorensen boasted (off the record) to Hugh Sidey that U.S. pilots had been told not to turn back, not to be forced down, and were winning their battle of wills with the Russians. "Then there are the men in Vietnam, the Special Forces boys," Sidey wrote his editor, relaying the White House view. "Though we aren't doing so well there militarily, our guys are performing with daring. All this adds up to a good old display of real American guts, the stuff some people were beginning to wonder about."

In the Soviet Union, the initial response was a studied indifference. The morning after the flight, JFK's daily intelligence briefing explained that news outlets there had delivered prompt, dry reports on the launch, making sure to note exactly how many times it had been postponed.

Pravda, not known for subtlety, put an image of Titov, not Glenn, on its front page. Though the party organ found praise for Glenn's "great courage and self-control," it added that "Americans do not forget that the road to the cosmos was paved by Soviet people, Soviet scientists, engineers, technicians and workers"—all standard fare. It was therefore surprising when Khrushchev struck a more conciliatory note: in a letter to Kennedy, he welcomed Glenn "to the family of astronauts." More significantly, he opened the door to what he had rejected in Vienna: cooperation in space. "If our countries pooled their efforts—scientific, technical, and material—to master the universe," he wrote, it "would be joyfully acclaimed by all peoples who would like to see scientific achievements benefit man and not be used for 'cold war' purposes and the arms race."

"I regard it as most encouraging, this proposal," Kennedy said at a press conference on February 21, noting that he himself had suggested the idea in his first State of the Union address. Now that the Kremlin was showing interest, he declared that the United States would move quickly to develop peaceful, bilateral projects to discuss with the Russians. Still, he was careful not to appear too ardent. No statement by Khrushchev, whatever its tone, could be taken at face value. Kennedy cautioned that he had seen no evidence to indicate that the Soviets were truly inclined to cooperate in space. "But we, I might say, now have more chips on the table than we did some time ago. So perhaps," he added, allowing himself a hint of satisfaction, "the prospects are improving." What about a joint mission to the moon? a reporter asked. Ideas like that would have to wait, Kennedy replied, "until we see whether the rain follows the warm winds." There might be something to gain in either event. As Mac Bundy wrote Webb that day, Kennedy "knows that there are lots of problems in this kind of cooperation. . . . At the same time, there is real political advantage for us if we can make it clear that we are forthcoming and energetic in plans for peaceful cooperation. . . . It is even conceivable that progress on this front would have an automatic dampening effect on the Berlin crisis."

But the space race, for all its Cold War implications, was not only about the outward projection of power. It was a mirror: a reflection of the American character. As Sidey wired his editor at *Time,* "A nation that had nearly lost its capacity to idealize" now beheld "a deed and a personality perfectly matched": matter-of-fact in the face of mortal danger; composed and self-assured, yet warm, humble, honest, natural. This was Glenn's greatness, as commentators perceived it, and it was how many Americans wished their nation to be seen. The man had met his moment. In an age of unsettling, often terrifying technological change, Sidey concluded, "America needed this kind of hero . . . and got it."

LYNDON JOHNSON TALKED all through breakfast and then kept talking. To Glenn's surprise, the vice president had flown all the way to Grand Turk to escort him to Patrick Air Force Base, just south of Cape Canaveral. After two days on the island—days of debriefings and biomedical tests, and time on the water with Scott Carpenter and others—Glenn was about to return to the public eye. At Patrick, he would be reunited with his family; together with LBJ they would travel by motorcade to the Cape and welcome President Kennedy, who would present NASA's Distinguished Service Medal to Glenn and Bob Gilruth. Three days hence, on February 26, Glenn would return to Washington and deliver his address to Congress—an event orchestrated largely by Johnson, which he, at breakfast, was not loath to point out.

It was less a conversation than a lecture—warm, solicitous, relentless. "You are a real can-do man," Johnson said. "In my country"—the Texas Hill Country—"we'd say you're pretty tall cotton." He was building Glenn up, priming him for the role ("hero-diplomat," as *Life* put it) that Johnson believed he could play. "You produced a proposal two presidents attempted to produce without success," he said about Khrushchev's overture. "Your mission is to bring peace." Johnson urged Glenn to make that the theme of his address. And on that subject, Johnson added paternally, "Keep your speech simple and humble, the shorter the better"—injunctions that LBJ himself, in his own public speeches, rarely followed.

Around 8:35 a.m., the Glenn family—Annie and the children; Glenn's parents, Clara and Herschel—were driven onto the tarmac at Patrick. They had arrived in Florida the night before, passengers on the presidential plane. For all but Annie, the trip on Air Force One had been their first trip on a jet airplane. (They were typical of the times: the week Glenn orbited the Earth, a study showed that 78 percent of Americans had never traveled by air.) Before the plane took off, Kennedy had come out of his cabin to make sure the family was comfortable. They were given sandwiches and snacks; Dave got a *PT-109* tie clasp—a memento of JFK's wartime heroics—which he promptly put on. "Mighty good of you," Herschel told the president before they landed. "We never thought anything like this would happen to us." Kennedy went to stay at his "Winter White House" in Palm Beach; the Glenns traveled on to the Holiday Inn at Cocoa Beach, where Henri Landwirth greeted them from the balcony. He had saved them each a piece of cake.

The vice president's plane, a small Lockheed JetStar—"Air Force One-Half," he called it—landed just a few minutes after the Glenns had arrived at the base. Bending down to clear the door, Johnson stepped off first, followed by Glenn—who went straight to Annie, threw his arms around her, swung her back and forth. Tears in his eyes, he embraced his children and his parents. "It is a high privilege," Johnson said into a microphone, "to welcome one of the great pioneers of all history." Then, under cloudless skies, they all climbed into white convertibles for the ride to the Cape. In crimson red, including her pillbox hat, Annie was a bright counterpoint to Glenn and Johnson, in dark suits. On her lapel was a pin Glenn had just given her: a globe of gold flecked by rubies and encircled by an orbital band. He had flown with it in space. "He had it on him somewhere," Annie laughed to reporters.

"Security," Sidey wrote that day, "went to hell in a hurry." As the motorcade crept forward, the sea of onlookers parted, but only slightly. Like waves they receded, then surged; the cars moved with difficulty to exit the base. And outside it, another reporter noted, "there was bedlam." The route to the Cape, passing through Cocoa Beach, was lined with as many as a hundred thousand people: people along the sidewalks and in the street, forcing the cars to keep stopping; people pulling at Glenn; people waving flags from balconies, from ladders, from deck chairs on rooftops. Some held up babies for Glenn to see; some held up dogs. Children wore space helmets; school bands marched and majorettes flung batons; others simply stood and wept. Glenn called it a parade, but it was more manic than that; it was delirium. It went on like this for an hour, until the cars had made their way to the gate at Cape Canaveral. Glenn, fishing around his suit pocket for comic effect, pulled out his ID and showed it to the guard.

On the Skid Strip—the runway where cargo planes came and went, carrying the components of spacecraft—hundreds of notables stood in the sun, waiting for Kennedy to arrive. Wernher von Braun turned to Shepard. "I want to congratulate you for those last few minutes, what you did," von Braun said, referring to Shepard's steady calm in his communications with Glenn before reentry. "There are very few people who know what it was like." Shepard nodded: few did. Suddenly, the whine of Air Force One filled the air; the plane touched down and taxied to a stop; and then, at the sight of Kennedy coming down the steps and walking past the honor guard with Glenn, the spectators—among them senators, generals, NASA officials—abandoned all decorum and rushed the tarmac. “The scramble was incredible,” Sidey reported. “There were no police lines; the thin contingent of Secret Service agents and base cops couldn’t stem the tide.” As the crowd converged around them, Kennedy grabbed the hand of Clara Glenn. “Here,” he said, “you get in this car.” It was his car—one in a procession of nine that was about to tour the base. Kennedy guided her to the front seat, then got in the back, next to Glenn and General Leighton Davis, commander of the Missile Test Center at Patrick.

The motorcade pulled away, heading first toward the control center. Herschel Glenn had been left behind, lost in the crowd. Eventually he found his way to Hangar S, along which a stage was set up for the medal ceremony. On either side were two capsules: the one on the left was pristine and new; its matte-black shingles looked somehow bright in the sun. On the right was Glenn’s. His father stood and considered it. The *Friendship 7* logo had been scorched during reentry; the American flag, painted near the window, looked as if it had been flown in battle. A sheen of aluminum alloy—a melted remnant of the retropack—was spread across several shingles. “Boy,” Herschel said. “That thing really got hot.” He turned to a reporter. “I got separated from my wife,” he remarked with a shrug. He thought he knew how it had happened. “One of those senators took her. You know those senators and their weakness for women!”

By now the group had reached the Mercury Control Center, where Chris Kraft steered Kennedy among the consoles, pointing at maps and charts. The next stop was Pad 14. At the base of the gantry, behind a rope line, were hundreds of workers responsible for the launch. Glenn, greeting them, thrust his arms into the air with both thumbs up. "In view of the fact that you sent Colonel Glenn around the world," Kennedy said, "he wanted to come back and thank you for it." "Best launch crew in the business!" Glenn boomed, and the crowd cheered. The event had the feel of a campaign rally in which either man could be the candidate. Someone passed Glenn a green hard hat—a gift for the president. Above the brim it read, J. F. KENNEDY, PRESIDENT, U.S.A. and 1ST U.S. MANNED ORBITAL FLIGHT 2–20–62; it also bore the image, crudely if conscientiously rendered, of a rocket in flight. Glenn, in his exuberance, reached up to place the hat on the president's head. ("It seemed like the thing to do at the time," he said later.) And Kennedy let it stay there, if only for four seconds. He had little use for hats of any kind; but then, he had been cornered, live on television.

Finally, now, they arrived at Hangar S for the medal ceremony. In his opening remarks, Webb spoke of "turbulent forces" at loose in the world, praising Glenn for "the will to stand on the firing line, indeed to die on the firing line." Kennedy struck a more clement note. "Two weeks ago," he said, "when Colonel Glenn came by the White House, I asked him how he enjoyed the public attention, and he said that he wished that they were paying more attention to the scientific part of this voyage rather than to his wife's hair." Glenn, laughing loudly, turned to Annie and touched her hair. "My own feeling," Kennedy went on, "is that both are equally important. . . . We are proud of this trip because of its scientific achievement and we are also proud of it because of the men and women that are involved in it. Our boosters may not be as large as some others, but the men and women are."

After bestowing the medal on Gilruth, who was so overcome he could scarcely speak, Kennedy called on Glenn to step forward. The astronaut stood at attention, like a soldier. "Seventeen years ago,"

Kennedy said, "a group of Marines put the American flag on Mount Suribachi"—that iconic image of Iwo Jima—"so it is very appropriate that today we decorate Colonel Glenn of the United States Marine Corps, and also realize that in the not too distant future a Marine or a Naval man or an Air Force man will put the American flag on the moon." He pinned the medal to Glenn's lapel and asked him to say a few words. Glenn spoke without notes; his message was simple. All he wanted to say was that "it was a real team effort all the way," a team of thousands, "a crosscut of Americana" that ranged from the other six astronauts to NASA workers, private industry, the armed services, the federal government. "I was sort of the figurehead," Glenn concluded, and the gold medal he was proud to wear "represents all of our efforts, not just mine."

A few minutes later, when the ceremony was done, the president stood in front of *Friendship 7,* tapping the hull with his hand, peering inside. "Could you just give us the word here?" he asked Glenn. The astronaut's family stood behind him; so did LBJ, Gilruth, and concentric rings of Secret Service agents, photographers, senators, and other eavesdroppers, spreading outward across the concrete as Glenn, in response to Kennedy's questions ("What does this do?"), explained how the controls worked, how the retrorockets attached, how the resin in the heat shield boiled away into steam. "This is all burned under there now," Glenn said. Kennedy folded the sunglasses he had been fiddling with and reached under the capsule to feel it. "Tell him about your most anxious moment," Johnson said, leaning in—a prompt to tell the story of the reentry. "I thought possibly the heat shield was tearing up," Glenn told Kennedy. "If this had happened, why, of course the whole thing would have just disintegrated, it would have burned up. . . . It was an interesting return, let's put it that way."

At that the president grinned, shook his head, then stepped away as Glenn continued the lesson for his family. It was the first time any of them had seen the capsule with their own eyes. "What a tin can!" Lyn exclaimed, making LBJ laugh. "How did you ever learn what they all mean?" Glenn's mother asked, looking at the switches and buttons. "Training," he answered.

"A lot of trouble, wasn't it?" she said.

THE CEREMONY CONFERRED a finality on the mission. As Kennedy said, Glenn's flight was history now, a feat to be remembered, as Lindbergh's was. But like the crossing of the Atlantic, Glenn's trip around the Earth was not an end in itself. It was a step, perhaps a leap, toward future flights. With these in mind there were anomalies to be explained, mysteries to be resolved, design flaws to be probed and fixed. There was, in sum, unfinished business; and some of it came up at a sprawling press conference that afternoon beneath a tent on the tarmac. Before a rapt audience, Glenn recounted highlights of his trip, although he acknowledged that "we had some problems." He talked at length about his difficulties with the automatic control system and the heat shield warning light. The latter, the engineers had confirmed overnight, was indeed the fault of a defective switch. But this, by now, was not news. What reporters wanted to know more urgently was, as one of them said (and others soon echoed), "When did you know that you had a possible emergency on your hands?"

"I can't pin a time to it, exactly," Glenn demurred, but the press (politely) kept hammering at it: Was it late in the first orbit? Early in the third? "I don't believe this was of any significance," Glenn protested, but it *was* of significance that for some length of time Mercury Control had believed his life was in peril and did not tell him so. Since the flight, reporters had been trying to reconstruct the chain of events while NASA, just as assiduously, worked to blur it. "He had to know," Walt Williams had told the *Washington Evening Star* on February 22. "He made an on-board check for us over Hawaii." Schirra chimed in: "John had the picture when we were talking to him. He didn't need a fill-in from me." The *Star,* obligingly, presented this spin as fact: GLENN AWARE OF DEATH PERIL ON RE-ENTRY. The story ran while Glenn was in seclusion on Grand Turk. Now he had a chance to correct the record, but the most he would allow was that he had looked "a little bit askance" at leaving the retropack on. Still, he said, "I knew that there would not be any change of this magnitude made unless it had certainly been considered from every possible angle by many, many experts on the ground."

It was a pointed comment. The press took it at face value, this vote of confidence in the control center, but NASA officials knew it masked Glenn's anger. During the debriefings, Scott Carpenter and Kris Stoever later wrote, "Everyone saw John's eyes narrow as he appraised Kraft across NASA conference tables." Kraft was unrepentant: "He said that John had a lot of worries," Bob Voas recalled. "That was the worst thing you could have said about John." Kraft's view, which was hardening, was that fewer people should have a say in decisions. Others, like Kranz, came to regret cutting Glenn out of the loop. Glenn himself, having declined to air his complaint at the press conference, was still mad enough later that year to put it in print: in *We Seven,* a collection of essays by the astronauts, he stated sharply that mission control should never "keep the pilot in the dark, especially if you believe he might be in real trouble. . . . He can hardly be fully prepared if he is not being kept fully informed. On future space flights, when the spacecraft and its crew will get thousands of miles from earth, some of the apron strings will have to be cut."

The "death peril," for all the drama it engendered, had a simple cause: the bad switch. Ironically, the main problem with the autopilot was also traced to a switch. But in the latter case, faulty electronics were not to blame. The pilot was.

At the press conference Glenn speculated that “a combination of things” had been at issue, citing “thruster trouble.” He was partly right about that: the yaw thrusters had malfunctioned. But he was about to learn what else the engineers had discovered. Later that day at Hangar S, Robert Schepp and Jerry Roberts, experts on the autopilot, pulled the astronaut aside and asked if he had a minute to review some simulations they’d just run. They stepped together into a vacant room. Schepp and Roberts had drawn straws to determine who would have to tell Glenn what they’d found. Schepp had picked the short one.

He showed Glenn the simulation results. The engineers had replicated, exactly, the skating effect Glenn had experienced in orbit—the problem with the ASCS. “Wow,” he said, “you hit that on the money. That’s exactly what I saw. What in heaven’s name could cause it to do that?” Schepp had the answer. “The switch—” he began, but Glenn cut him off; he got it right away. “Crap!” he burst out. “I forgot that was there!” It had been part of his training, but only recently. The switch had been a late addition to the design—a concession to the astronauts, who had insisted on the ability to fly facing forward part of the time. When they did, they were supposed to flip the switch to reset their orientation. But as Glenn said: he’d forgotten. So while his capsule flew forward, its autopilot, in effect, believed it was still flying backward and sent it farther and farther out of alignment.

“The hero was human,” Roberts reflected. Glenn accepted that fact with grace—though he felt no compulsion, then or in later years, to share the story more widely than that room. The press, for its part, lost interest in these technical questions; and NASA, having resolved the matter, assessed no blame, issued no statements. Even its confidential

"Mission Critique" treated Glenn with kid gloves: it concluded, gently, that "astronaut training will be modified . . . to avoid this trouble."

THE LAST TIME there had been a parade like this along Pennsylvania Avenue, Glenn had been at home, writing an anguished letter to Gilruth. That had been Inauguration Day, a mere thirteen months ago. While Annie and Lyn and Dave had stood in the cold, watching the marching bands and the floats (FROM THE LONE STAR TO THE SPACE STARS, proclaimed the one from Texas) and the missiles on flatbed trucks, Glenn had been drafting his argument that Gilruth should reverse his decision that Shepard would fly first. It was, on its face, self-serving, probably ill-advised. But from the vantage point of February 26, 1962, as Glenn and his family rode in an open limousine along that same route in the rain and a quarter of a million people stood there cheering, the key contentions in his letter had proved correct.

He had written Gilruth that the astronauts were more than pilots. They were role models, flesh-and-blood exemplars of American ideals. This was not just public relations, as it was dismissively called by many at NASA, including some of the seven themselves; it was, as Glenn contended, central to the purpose and future of the program. Not long before that, Gilruth, considering the qualities an astronaut should have, had told Voas that "the only requirement is that they be able to fly this vehicle successfully." That was still the view when Glenn was assigned his flight. Only now, stunned by the response to *Friendship 7,* did NASA's leaders truly understand how much America wanted and needed from its astronauts. Not all of them welcomed it, but they understood it.

Though nearly a week had passed since the flight, the intensity of the crowds was undiminished. If anything, the adulation, the devotional interest in Glenn, had built on itself. The Glenns had flown to Washington that morning on Air Force One. On the plane, Jackie introduced Caroline to the man who, she said, had flown in a spaceship around the Earth. The four-year-old curtsied, then looked around, crestfallen, and asked, "But where's the monkey?" With JFK's blessing, Glenn added the story to the speech he would give to Congress that day. Capitol Hill was the terminus of Glenn's parade, but only for now: it would resume a few days later in New York and, two days after that, in New Concord.

Just after one o'clock, his thin hair damp from the ride in the rain, Glenn stepped up to the rostrum in the House and did as Johnson had advised: he kept it simple and humble and short, only seventeen minutes. That included the time during which he said nothing at all, just stood there and smiled while the applause cascaded from the galleries and swept across the floor. Sitting behind him, LBJ wore what *Life* described as the "look of a father listening to his son give the commencement address." But the mood in the chamber was one of reverence. More eloquent speeches had been given here, but few, surely, had moved as many officials unself-consciously to tears. "Amen!" some shouted as Glenn spoke. "I am certainly glad to see that pride in our country and its accomplishments is not a thing of the past," he said. As he had at the Cape, he shared the credit—hunting around to find Walt Williams in the gallery, sweeping his arm to bring his fellow astronauts to their feet, acknowledging "the real rock in my family, my wife, Annie."

"*Friendship 7* is just a beginning," he continued, "a successful experiment." He looked ahead to more ambitious flights. Still, he knew that critics remained—critics who would continue to press for quicker returns on the taxpayers' investment. Glenn asked for patience. "Exploration and the pursuit of knowledge have always paid dividends in the long run," he said, "usually far greater than anything expected at the outset." He closed with a prayer: "As our knowledge of the universe in which we live increases, may God grant us the wisdom and guidance to use it wisely." And with that, Glenn was carried out on a wave of more *amens* and what *Life* called "almost hysterical applause." *Time* detected "an all-but audible click in the minds of politicians and political connoisseurs: he was a campaign manager's dream."

NEW YORK, ON MARCH 1, outdid the rest, outdid even its outsized sense of itself. In 1927, three million people had come out to cheer Lindbergh and throw ticker tape; today, despite the bitter cold, Glenn drew four million—only half of whom, city officials estimated, actually caught sight of the ivory-colored Cadillac that Glenn, Annie, and LBJ rode in, or the cars carrying the other astronauts and their wives. To get a better view, one man scaled the Triborough Bridge; others climbed trees or inched out onto ledges. Even compared with Cocoa Beach, the celebration in New York seemed unhinged, almost violent: people swarmed; they wept with raw emotion. "Wild mob scenes," UPI reported. "Screaming humanity broke through police lines and isolated Glenn's car." When the Cadillac slipped free and rejoined the procession, a press bus got left behind; its driver, leaning on his horn, gave chase for forty blocks, scattering police officers and spectators along the way. The tour had reached its apogee. It was time to go home.

Two days later, he was there. The morning was cold and bright; and as the motorcade traveled the twelve miles from Zanesville, where the Glenns' flight had just landed, to New Concord, the family shared the road with other families, thousands of them, who were driving in from towns across central and eastern Ohio. The parade began on Liberty and turned first onto Main, passing Glenn Plumbing. Several blocks away, a crowd of sixteen hundred—nearly equal in number to the population of the town—waited for Glenn in the Physical Education Building at Muskingum College. The gym was being named for him. Across the quad was Brown Chapel, where Annie had played the organ for her senior recital and Glenn, just beforehand, had sat outside in his car, listening to a news bulletin from Pearl Harbor.

Now, as the college president announced that the P.E. Building would bear the astronaut's name, Glenn looked at the floor and shook his head slowly—as if, of all the honors he had received, this was the one he just couldn't believe. When it was his turn to speak, Glenn flashed a grin. "When you people leave," he said, "clean up my gym."

EPILOGUE

Escape Velocity

"IS THE MOON WORTH John Glenn," James Reston asked in the *New York Times* just after the flight of *Friendship 7,* "when we need him so badly here on earth?" A friend from New Concord posed a better question. In a letter to Annie, she paraphrased an old song: "How are you going to keep him down on the earth after he's seen the stars?" Glenn's love—or need—for flight seemed to him predestined. A Scottish relative had long ago given Clara Glenn a ring inscribed, in Latin, with the family motto: ALTA PETE. "Aim at high things." It bore the image of a bird in flight—but the bird had no feet. "Once it started flying," Glenn explained years later, "it had to go higher."

"This stuff," Glenn had said at the press conference on February 23, "you can become addicted to it rather rapidly." Webb leaned into the microphone. "Any man who flies as well as John Glenn," he said, "will fly again."

"I'm with you!" Glenn burst out, shooting his arm into the air.

Some were leery of his chances. "I personally doubt he will ever orbit again," Bill Shelton of *Time* wired his editor. "He is becoming and will become too valuable—both to the Administration and to the country—as a symbol the nation desperately needs . . . I think he'll be pulled out of the program, screaming and kicking, to become an ambassador of space." NASA, however, decided not to send its star astronaut on a goodwill tour, as the Soviets had done with Gagarin and Titov. "Here I am, clap for me, give me the flowers and the medal" was how Glenn described it. Better, he said, to stay on the job and "show the world that we were serious about this thing." USIA sent *Friendship 7* on a tour of its own—the "fourth orbit," they called it, and by cargo plane and trailer hitch the capsule made its way across twenty-three countries. In Cairo people waited for hours in 112-degree heat to see it; in Bombay the line stretched twelve miles, half a million people long.

Glenn, meanwhile, went back to work. He attended meetings on Gemini, the next phase of the manned space program, and contributed to the cockpit design for the Apollo Lunar Module. For Gordon Cooper's flight, he was the capsule communicator on a tracking ship near Japan. But there were few mission-critical tasks for Glenn. Headquarters "essentially kidnapped" him, Bob Voas recalled, for PR work. He was sent to Capitol Hill to argue for the space budget. He was parceled out to the TV networks: as a public affairs officer reported to Webb, NASA granted ABC's request to use Glenn as "the attractive force to insure a large . . . audience" for a show on science. The president himself had ideas about Glenn's utility. Kennedy's secretary wrote Webb that JFK "was wondering about the possibility of sending [Glenn] to some scientific meetings." The State Department, too, got into the game. In the spring of 1962 Glenn was asked to play host to Titov, who had come to Washington to appear at a conference. Trailed by news cameras, Glenn shepherded the cosmonaut around town; that night they grilled steaks under Glenn's carport in Arlington. And throughout this period, every month or so, Glenn would ask Gilruth about getting back into the flight rotation and would be told, without fail, that it might be a while.

Glenn tried to stay on task. By autumn 1962, he had turned down invitations to twenty-six hundred events. Still, anywhere he showed up, he made headlines, creating the sense that he had become—as one of those headlines put it—a "gadabout." Here were the Glenns in Hyannis Port, sailing with JFK and water-skiing with Jackie; here they were at parties at Hickory Hill, Bobby and Ethel Kennedy's home. In Houston, the location of the new Manned Spacecraft Center, there was a good deal of sniping—much of it fueled, as Powers acknowledged, by "petty jealousy." But it went beyond that. "Kraft crapped his pants worrying Kennedy'd hand NASA to John," Gus Grissom told Rene Carpenter a few years later. "John was bigger than *Jesus*!"

That fall, on the eve of his six-orbit flight, Wally Schirra told CBS News that Glenn's "commitments have just about wiped him out of the space program. He hasn't been able to maintain the currency that he should have with the rest of us. . . . John's falling behind, in other words." The flap was front-page news. Gilruth dismissed the attack as groundless; Glenn called it "a tempest in a teapot." But Schirra's remarks wounded him. In a letter to his superiors, he meticulously accounted for each and every event he had attended since his flight. Nearly all of them, he said, were at NASA's behest. Still, the story roiled for weeks. Some newspapers scolded Schirra, but others, for the first time, went after Glenn. "Leave the role of the celebrity for others," chided the *Atlanta Journal*. "We have felt for some time that Astronaut Glenn's halo was slipping," intoned the *Shreveport Times*. "Glenn seems to have been used by the Kennedys."

In the summer of 1963, Bobby Kennedy invited Glenn to dinner. He had an agenda. He and the president, he said, wanted Glenn to run for Senate in Ohio in 1964. Glenn, they thought, could give JFK's reelection campaign a lift in a state he had lost in 1960. Glenn had never declared his party affiliation—publicly, he described himself as an independent, and he was eagerly courted by leaders of both parties—but he had always been a Democrat. He told his children that when he was growing up, and there was a Democratic Party meeting in New Concord, the only people in attendance were Annie's parents and his own. It was widely assumed that the Kennedys were grooming him for political office, and the rumors, now, had been proved correct. Bobby implied that the Kennedy machine would throw its full weight behind his candidacy.

Glenn asked for time to mull it over. He sought out Voas, who had quietly been urging him to think about a presidential run in 1968. Glenn had laughed that off. But this was different. "He was excited—jubilant—that they asked him to run for Senate," Voas recalled. Yet he turned the offer down. He felt a loyalty to NASA; but mostly, he still held out hope for a return to orbit—despite what had come to feel like "an unending moratorium." Rumors that JFK himself had ruled out future flights—on the grounds that Glenn was too valuable to put at risk—were probably just that, rumors; no directive was issued. None was needed. Gilruth told Glenn that flights would be allotted first to astronauts who might someday be eligible to walk on the moon. By the late 1960s, the likeliest time frame for a lunar landing, Glenn would be almost fifty years old. Warning against wishful thinking, Gilruth "suggested that it might be wise for me to get into one of the management areas," Glenn recalled. A desk job.

Mercury was done; Gemini and Apollo were harbingers, for Glenn, of his own obsolescence. He had been grounded and, he now understood, would remain so. Late in 1963 he drafted notes for a tough conversation with Gilruth. He had had his fill, he wrote, of being "window dressing," lent out to yet another congressman to "show his home-folks what a big influence he has in Washington." NASA, he believed, was being run on the basis of "political favors . . . rather than on merit." The situation had become intolerable. "I've been loyal all the way through," he wrote. "Now I must look out for myself."

In the early afternoon of November 22 Bob Voas was at NASA headquarters, attending a meeting; when that finished, he hailed a taxi to National Airport. That was when he heard, on the car radio, the news from Dallas: the president was dead. Arriving at the airport, not knowing what else to do, Voas went ahead with his trip to Cleveland—a trip to meet with Democratic officials to begin setting up a Glenn campaign.

Such talk would have to wait. The nation had to pass through its grief. For Glenn, John Kennedy's death was a personal loss. They had not been close, as he and Bobby were becoming, but there was an affinity between them—a mutual esteem and even awe. In the weeks that followed, Glenn reflected on Kennedy's life and his own—asking how, in the years he had left, he could contribute the most. He went back to Robert Kennedy and said he had changed his mind; he was ready to run.

But Bobby, too, had reconsidered. Months had passed; the landscape had violently shifted. It was too late now, he told Glenn. There was not enough time to build a winning campaign. Glenn thanked his friend for the advice. Then he set forth—with energy and purpose—to disregard it.

THE FLIGHT OF JOHN GLENN did not win the space race. What it did was shift the momentum. It gave the U.S. effort what physicists call escape velocity—the speed required to break free of gravity and sail into space. In 1958, John Foster Dulles had said that Sputnik might "go down in history as Mr. Khrushchev's boomerang," for it jolted America awake; but it was *Friendship 7* that sent the boomerang on its return to Moscow.

Before the decade was out, the Soviets would achieve other "spectaculars": the first space walk, the first woman in space. And the United

States would face reversals, including, most tragically, the death of the crew of *Apollo 1*—Gus Grissom, Roger Chaffee, and Ed White—in a fire during a preflight test of their capsule in 1967. But by then, for the Americans, the space race had become chiefly a race against time—Kennedy's deadline—not against the Russians. Once the United States had built an arsenal of boosters that could lift heavier spacecraft, brute force no longer gave the Soviets an advantage. For years, Soviet engineers had been able to give Khrushchev what he demanded: "first" after "first," propaganda that suggested progress. But the artifice was unsound, weakened by mismanagement. Kennedy's moon shot called Khrushchev's bluff. For all the latter's crowing about domination in space, he was ambivalent about racing America to the moon. "He wanted to be ahead . . . , but for free," his son Sergei said. In October 1964, when Khrushchev was ousted in a coup, he left a program in disarray. Nearly five more years would pass before Neil Armstrong and Buzz Aldrin walked on the moon, but the race, in effect, had already been decided.

Not long after Glenn's flight, the United States and the Soviet Union agreed to share data from weather satellites and jointly map the Earth's magnetic field—a pretty paltry accord, given the hopes that had preceded it. But JFK never abandoned his idea that Americans and Russians might venture together to the moon. Two months before his death, in a speech at the U.N., he made what turned out to be his closing argument. "Why," Kennedy asked, "should man's first flight to the moon be a matter of national competition?" He knew the reasons, but never ceased to question them.

ON FEBRUARY 20, 1997, John Glenn went to New Concord to give a speech. It was the thirty-fifth anniversary of his orbital flight. He had spent twenty-two of those years in the U.S. Senate. And now, he told his hometown crowd, he had decided not to seek reelection in 1998. At the age of seventy-five, he said, "my health remains excellent," and it was "time to serve my country in other ways."

His path to office had been more tortuous than he had expected. On the campaign trail in Ohio, only a month after launching his candidacy, he slipped on a bathroom rug and banged his head on the tub, severely damaging his inner ear. In a cruel irony, his symptoms—vertigo, nausea—were exactly what some at NASA had worried he might experience while weightless. From his hospital bed, in his pajamas, Glenn announced that he was ending his bid for the Senate. The months ahead were bleak and long. Every morning, his hands gripping chairs and tables and pressing against walls for support, Glenn made his way from the bedroom to the living-room couch. It took him nearly ten minutes. By fall, he had recovered enough to accept a position on the board of Royal Crown Cola; the campaign, however brief, had left him deeply in debt.

He tried again in 1970. This time, he stayed in long enough to lose. The glow of *Friendship 7* was still an asset, but it also obscured the fuller picture Glenn hoped to present—a picture that, in truth, was thinly drawn. He had a pedestal, not a platform, and voters decided that it was not enough. Howard Metzenbaum, a businessman and local pol, edged him out in the primary before losing to the Republican, Robert Taft Jr. In 1974, Glenn faced Metzenbaum again. As before, he ran mainly on his name, his fame, but he did so now with fierce efficiency. He beat his rival handily. In November he took every county in Ohio, all eighty-eight. Before election night was through, TV commentators were calling him a presidential prospect.

In the Senate he stood apart—by virtue of his achievement and, it seemed, by choice. As a legislator, no less than as an astronaut, Glenn was serious-minded, detail-oriented, disciplined. He approached problems as an engineer might: they were contraptions to be taken apart, examined, repaired. Unlike many senators, he had little interest in hearing himself talk; he did not seek to have his picture taken; he resisted taking credit, even when he deserved it. While others, in the evenings, hit the steak houses and bars, talking shop, Glenn went home to Annie. "He was never a slap-on-the-back, hail-fellow-well-met" type, Lyn Glenn said later; and it cost him. So did what the *New York Times* called his "prickly sense of integrity." He had the respect of most senators but the affection of few. These were politicians, after all, and what they saw in Glenn was a distaste for politics: the horse trading, the deal cutting, the cultivation of key groups and conditional allies. On issues of national security—arms reduction, nuclear nonproliferation—others looked to him and followed. But he remained, by and large, a loner. When he finally ran for president, in 1984, all of this was manifest. Early on, Ronald Reagan's team saw Glenn as their strongest challenger, but his campaign never got off the ground; he failed to win a single primary.

One night in the 1980s when the Senate wrapped up its business late—some contentious issue had kept the senators bickering—Lyn joined her father for the ride home. They drove down the slope of Capitol Hill, and as the car approached the National Air and Space Museum, Glenn pulled over. The building was closed, but the lights were on. Through the window they could see *Friendship 7,* tipped upward on its stand, illuminated. They sat in silence for a while. Finally, Lyn asked, "Do you come here often, Dad?"

"Oh, just when I want to." He shrugged. There was nothing, he said, looking at his capsule, that could ever compare with a shared mission—with working alongside others who, in pursuit of that mission, were willing to put their lives on the line.

He had one more in him. He was not called to it, as he had been in December 1941; rather, he campaigned for it, with a persistence that surprised no one who knew him. In 1995, reviewing materials for a hearing on NASA's shrinking budget, he stumbled onto a reason—or, some said later, a pretext—to be sent back into orbit. In a book he spotted a list of symptoms that astronauts experienced after extended stays in zero gravity: osteoporosis, disrupted sleep patterns, other echoes of the effects of aging. Glenn had the notion that NASA could learn more about this by sending an older person—say, a former astronaut in his mid-seventies—up on the space shuttle and putting him through a battery of tests. "You're serious about this, aren't you?" Dan Goldin, the NASA administrator, asked him. There was little doubt. Over the next two years, Glenn lined up scientists who backed his idea; took—and passed—every physical exam the younger candidates did; and pitched his plan to President Bill Clinton. In January 1998, Goldin announced what had come to seem inevitable: John Glenn would fly again. He was assigned to the *Discovery* on an October flight—a nine-day mission.

The scientific agenda was sound; its results, someday, might have practical effect; but this, it was clear, was not really the point. "The nation needs heroes," Goldin said frankly. NASA hoped that bringing back the star—and evoking the spirit—of the early days of space exploration might give the agency a boost. NASA "picked an old pro to do a sales job," said Wally Schirra. Critics called it a joy ride. Even Glenn's family, at first, was opposed. "I was just so angry at him," his son, Dave, recalled, "that he would make us go through this again." Lyn wrote him a farewell letter—in case things went wrong. Glenn, of course, understood the risks. In the early 1980s, every time the shuttle flew, he stopped what he was doing and watched the launch on TV. He expected—and said so privately—that someday one would explode.

When the *Challenger* blew up in 1986, he reacted with sorrow but not surprise.

Yet there was no denying him now. "He was so excited about it," Dave said. "Like a little kid." So was the press. Cronkite came out of retirement to cover the event. *Time* put Glenn back on its cover, against a field of stars. This time, his pressure suit was orange, not silver; his face, today, showed the lines of age, though in some true and vital sense he looked more like himself than he ever had in the Senate chamber. "A trip back in time," the *New York Times* declared. When the launch date came, the crowds did, too, filling the beaches and causeways where they had stood in 1962. "Godspeed, John Glenn," Scott Carpenter said once more. This time, Glenn heard him.

When the rocket boosters on the *Discovery* fired—with twenty times the power of John Glenn's Atlas—the roar, a shock wave of sound, hit the reviewing stands with such force that Glenn's son burst out in tears. The shuttle rolled; it shed its boosters; it flew through the keyhole, where Glenn had passed before. Once in orbit, he slipped free of his restraints. He floated up to the flight deck and caught sight through the window of the Earth below, its blues as brilliant as memory held. "Zero G," he said, "and I feel fine"—his familiar words an echo from a high and distant place, a transmission breaking through, reaching home.

ACKNOWLEDGMENTS

It took more than eighteen thousand people, working in countless capacities, to get *Friendship 7* off the ground and then bring it home. This book, of course, was a smaller-scale, lower-risk operation, but the principle holds.

In the years since John Glenn's death in 2016, at age ninety-five, he has not been forgotten: his capsule and space suit are still big draws at the Smithsonian Air and Space Museum; he is front and center in a television reboot of Tom Wolfe's *The Right Stuff*. But the man himself is not well understood. He is lauded as a hero, but of the blandest sort. Wolfe's portrayal of Glenn as Mercury's moralist—the priggish "Presbyterian Pilot" with a "pipeline to the dear Lord"—stuck, and echoes through other accounts. That flat caricature sells Glenn and his significance short. It contains elements of truth but fails to explain how he managed to achieve what he did and why it had the electric effect it did—why Glenn, and not any other astronaut, was the one to lift the nation's hopes, expand its sense of possibility, and embody John Kennedy's New Frontier.

In creating, I hope, a fuller portrait of Glenn, I was fortunate to hear, firsthand, the stories of some who were close to him, worked with him, or contributed, in indispensable ways, to the success of his mission. I am enormously grateful to Lyn and David Glenn for speaking to me with such candor—about the strength and struggles of their mother, the pride they had in their father and his achievements, and the feeling of watching him ride a missile and wondering, as David put it, "if he's going to turn into a ball of fire."

One of the greatest pleasures of this project was the chance it gave me to talk at length with Robert Voas. Bob deserves a special mention—many special mentions—in any history of Project Mercury. He helped select the seven astronauts, trained them for spaceflight, and designed a number of the experiments they would conduct in space (Glenn, in fact, carried a device known as a "Voasmeter"). Bob's stories deepened my understanding on crucial questions; his comments on the manuscript enriched it greatly. It was also an honor to get to know the late Jerry Roberts, one of the McDonnell engineers who, in 1960, had the good fortune of being sent to Cape Canaveral and told to stay for the duration. More than half a century later, Jerry was able to remember the look on a man's face in the blockhouse or the location of a switch on the capsule's instrument panel—stories he punctuated with "hells bells!" and his roar of a laugh. Jerry passed away in 2020, a year after the death of his friend and colleague Bob Schepp—another brilliant engineer and storyteller to whom this book is indebted.

That debt extends to Manfred "Dutch" von Ehrenfried, whose tenure in mission control began on February 20, 1962, and who shared his memories of the debate over how—and whether—Glenn's life could be saved if his heat shield had come loose. I'm also thankful to Don Morway, who recalled the rivalries among the astronauts during their training, and the late Arnold Aldrich, who recounted his experience at the tracking site in California during the flight of MA-6. Glenn's post-NASA life was vividly described by Greg Schneiders and Mike

McCurry, key members of his political team, and by Scott Parazynksi, one of the space shuttle astronauts who flew with Glenn in 1998 and found him "as happy as a human being could be."

I am profoundly grateful to Kris Stoever, who, in 2017, responded warmly to a cold call and has, in the years since, eloquently answered every subsequent query (there have been many). With her father, Scott Carpenter, Kris wrote the best of the Mercury memoirs, *For Spacious Skies*, and she was kind to share with me the surviving volume of the journal that her parents kept together. In reviewing my manuscript, Kris applied the insights of a historian and the precision of a professional editor—both of which, in fact, she is. Kris challenged my assumptions, pressing me to question the received wisdom about the Mercury Seven. And her reports of the interviews she conducted with her mother, the remarkable Rene Carpenter, on my behalf, were a delight and a gift. Rene, at ninety, was as keen a judge of character and as sparkling a wit as she ever was before, which is really saying something.

No scholar has contributed more to our understanding of the space race than John Logsdon, the peerless dean of U.S. space studies. Professor Logsdon offered guidance from the start (it was Glenn, he said in our first meeting, who "made credible Kennedy's commitment to a moon landing") and suggested important changes to the manuscript. My friends Bill Causey and George Leopold have both written essential books on the space race—Bill, on the NASA engineer John Houbolt; George, on Gus Grissom—and their expertise and encouragement have been invaluable at every stage of the process. Many thanks to Diane McWhorter for generously offering research she had unearthed while writing a forthcoming biography of von Braun. Danny Parker, at work on a much-needed reappraisal of Scott Carpenter, passed along a fascinating find about the Kona Kai confrontation among the astronauts.

And Frank Van Riper shared wisdom he gained while working on his excellent 1983 biography of Glenn.

The collection at the John Glenn Archives at Ohio State was indispensable; the files there are bursting with materials that had never before been cited or quoted, and that shed a new and more nuanced light on Glenn, the space program, and the era. (It didn't hurt that Glenn saved just about everything, from high school papers to notes in the margins of flight plans.) Halle Mares and Carly Dearborn swiftly responded to every request and made my time in Columbus hugely productive. Bill Barry, NASA's Chief Historian, and Robyn Rodgers, its Chief Archivist, helped me navigate a complex network of collections across multiple facilities. At the NASA Historical Reference Collection in Washington, Colin Fries helped locate even the most obscure documents, and others at NASA—David DeFelice, Connie Moore, and Mary Wilkerson—provided help in a range of ways. I'm also thankful for an illuminating talk with Margaret Weitekamp, chair of the Space History Department at the National Air and Space Museum and author of *Right Stuff, Wrong Sex: America's First Women in Space Program.*

I owe special thanks to Mike Hill, for whom research is not just a vocation but an art. Mike's counsel, his contributions, and his passion for the subject enliven every chapter; his friendship has made writing books less of a solitary venture. Much gratitude goes to my good friend David Greenberg for meaningfully improving my manuscript, not least its account of President Kennedy and the politics of space. My colleagues Riley Roberts and Ellie Schaack also provided thoughtful comments, and Claire Blumenthal, Brooke Davies, and Amy Muller tracked down sources and nailed down citations at the John F. Kennedy Presidential Library, the Houghton Library at Harvard, and elsewhere. Thanks, too, to Peter Baker, Matt Dallek, Ben Feist, Emily Ludolph, Eleanor O'Rangers, J. L. Pickering, Regina Robinson-

Easter, Adam Sackowitz, Jon Siefken, Myra Taub Specht, Ron Specht, Jay Strell, Bonnie Thompson, and my partners at West Wing Writers—Jonas Kieffer, Vinca LaFleur, Jeff Nussbaum, Paul Orzulak, and Ben Yarrow—for all manner of support.

This is the third book I've had the privilege to write under the editorial guidance of Starling Lawrence, and I don't use the word "privilege" cavalierly. In 1994, when I had just begun work on *Mutual Contempt*, my book on Lyndon Johnson and Robert Kennedy, my first editor left W. W. Norton and I was—in one of the greatest strokes of luck in my professional life—effectively kicked upstairs to Star, the editor in chief. Every discussion with Star is a master class, whether its topic is narrative structure, a character sketch, or, sometimes, a single word. Others at Norton, especially Nneoma Amadi-obi, Louise Brockett, and Rachel Salzman, have been essential allies and advocates. And as hard as it is for me to believe it, I have been working with Rafe Sagalyn, my literary agent, for thirty years now—ever since I was a senior in college. Most people think of an agent as a dealmaker, and Rafe is certainly that; but even more, he has an unerring instinct for what makes a book worth writing or reading. Rafe is not only an agent but a trusted adviser and valued friend.

In writing this book, as in all things, I have been lucky to have such a loving and encouraging family. My parents, Susan and Barry Shesol, read each chapter the moment it was finished, like installments in an old-fashioned serial, and offered ideas and several shots in the arm; my brother Rob weighed in from the other side of the world. My in-laws, Bea and Stephen Epstein, as ever, provided their effervescent brand of support. (Bea also shared stories of building a life-size *Friendship 7* with her students at P.S. 198 in East Harlem in 1962; for the hull, they used cardboard flour barrels from the Silvercup Bakery.)

A book, especially as it nears completion, can become a kind of isolation chamber, but I've been lucky to be spared that feeling: at just about any hour, our home has been filled with the exuberant sounds of an exuberant family. No teenagers in America hoped to spend the better part of a year inside the house with their parents; but during the difficulties of 2020, Jonah's and Anna's joy and creativity and maturity carried them through, and lifted Rebecca and me, too, every day. As many have said, it's been a year (in our case two) for remembering what matters most, but I can't say it's ever escaped my mind: my wife, Rebecca Epstein, has always, for me, been the center of it all. Her across-the-board brilliance and her radiance have brought depth and, I hope, warmth to these pages, just as she does to every part of our lives together. Rebecca, with love, I dedicate this book to you, but so much more than that.

NOTES

ABBREVIATIONS USED IN THE NOTES

AAE	*Astronautical and Aeronautical Events*
AW	*Aviation Week*
HL	Houghton Library, Harvard University, Cambridge, MA
JFK	John F. Kennedy
JFKL	John F. Kennedy Presidential Library, Boston, MA
JGA	John H. Glenn Archives, Ohio State University, Columbus, OH
JHG	John H. Glenn Jr.
JSC	Johnson Space Center, Houston, TX
NASA	National Aeronautics and Space Administration
NASA HRC	NASA Headquarters Historical Reference Collection, Washington, DC
NSF	National Security Files, John F. Kennedy Presidential Library, Boston, MA
NYT	*New York Times*
OH	Oral history
POF	President's Office Files, John F. Kennedy Presidential Library, Boston, MA
PPP	Pre-Presidential Papers, John F. Kennedy Presidential Library, Boston, MA
TCS	Theodore C. Sorensen
TNO	Swenson, Grimwood, and Alexander, *This New Ocean*
WES	*Washington Evening Star*
WHCF	White House Central Files, John F. Kennedy Presidential Library, Boston, MA
WP	*Washington Post*

INTRODUCTION

1 **John F. Kennedy spent most of the morning:** Hugh Sidey to Robert Parker, "Waiting for Glenn," Feb. 21, 1962, Box 28, Folder 542, HL; Barbour, *Footprints on the Moon,* 48; *WP,* Feb. 21, 1962; *NYT,* Feb. 14 and 21, 1962.

2 **Congress was losing patience:** *NYT,* Apr. 15 and 16, 1961; *WP,* Apr. 13, 1961; *Hearing Before the Committee on Science and Astronautics,* U.S. House of Representatives, 87th Cong., 1st Sess., Apr. 12, 1961, 3, 6, 21; *Nation,* Apr. 29, 1961.

3 **And John Glenn was the most famous:** *NYT,* July 17, 1957.

3 **Americans were subdued:** *NYT,* Feb. 21, 1962; *WP,* Feb. 21, 1962.

3 **Kennedy had his own reasons:** Beschloss, *The Crisis Years,* 308; Schlesinger, *Robert Kennedy and His Times,* 705.

4 **Since the launch of Sputnik:** Gabriel A. Almond, "Public Opinion and the Development of Space Technology," *Public Opinion Quarterly* 24, no. 4 (Winter 1960): 559, 564, 566, 569–70; Erskine, "The Polls," 485–86.

4 **Kennedy held an index card:** "Voice Procedures," POF, JFKPOF-030-001, JFKL.

5 **Glenn was on his second circuit:** *Results of the First United States Manned Orbital Space Flight,* 164, 171; Kraft, *Flight,* 157–58; Kranz, *Failure,* 68–69.

CHAPTER 1: THE NEAREST TO HEAVEN I WILL EVER GET

7 **When the airplane . . . landed:** Powers OH, NASA HRC; Cooper diary, Apr. 9, 1959, accessed at https://www.rrauction.com/bidtracker_detail.cfm?IN=4064.

8 **The offices of the National Aeronautics and Space Administration:** Powers, ibid.; Haney OH, JSC; Glennan, *The Birth of NASA,* 20; John DeFerrari, "Dolley Madison's House on Lafayette Square," Oct. 31, 2010, http://www.streetsofwashington.com/2010/10/dolley-madisons-house-on-lafayette.html; Joel D. Treese, "The Dolley Madison House on Lafayette Square," May 25, 2016, https://www.whitehousehistory.org/dolley-madison-house-on-lafayette-square; "Press Conference, Mercury Astronaut Team," Apr. 9, 1959, p. 2, Box 67, Folder 31, JGA.

8 **They entered to applause:** "Press Conference," 2–3; Makemson, *Media, NASA, and America's Quest for the Moon,* 45.

9 **Shepard . . . seemed to consider himself:** Thompson, *Light This Candle,* 6, 37, 46, 112, 166, 130; Wolfe, *The Right Stuff,* 118.

9 **Walter Schirra, like Shepard:** Schirra with Billings, *Schirra's Space*, 28; French and Burgess, *Into That Silent Sea,* 43–44; author int., Stoever, July 5, 2018; Francis French, "'I Worked with NASA, Not for NASA,'" collectSPACE, Feb. 22, 2002, at http://www.collectspace.com/news/news-022202a.html.

9 **the third naval aviator, Scott Carpenter:** Carpenter and Stoever, *For Spacious Skies,* 188–89; *Life,* May 18, 1962, 36; Stoever, e-mail to author, May 7, 2019.

10 **The Air Force pilots:** Wolfe, *The Right Stuff,* 118; Leopold, *Calculated Risk,* 67, 96; Cooper diary, Apr. 9, 1959; Carpenter and Stoever, *For Spacious Skies,* 229; Cooper with Henderson, *Leap of Faith,* 15–17; Slayton with Cassutt, *Deke!,* 69, 72.

10 **"the lonesome Marine":** "Press Conference," 9; *New York Mirror,* July 17, 1957; *Los Angeles Evening Herald Express*, July 16, 1957; Betty Jane Sullivan to JHG, Sept. 26, 1957, Box 18, Folder 25, JGA; *Cong. Record,* 85th Cong., 1st Sess., vol. 103, no. 127, July 18, 1957, A5773.

11 **He had seen more combat:** *Time,* Apr. 20, 1959. For the astronauts' body language, see the video of the news conference at https://www.youtube.com/watch?v=d7WzC9BN2Dc.

11 **The first question from the press:** "Press Conference," 6. The NASA transcript is inaccurate in spots; I have made corrections according to the audio recording available at https://archive.org/details/FirstSevenAstronautsPressConference.

11 **"What I do . . . is pretty much my business":** "Press Conference," 7.

12 **The questions they might have liked:** Ibid., 8–10, 15.

12 **"I think you will find":** Ibid., 17.

13 **"That's a real tough one":** Ibid., 27.

13 **"The sky's no longer the limit":** *NYT,* Apr. 12, 1959.

14 **"I never saw anybody":** *Life,* Feb. 2, 1962.

14 **New Concord sits:** Kettlewell, *Our Town,* 13–14; JHG OH, JGA; Glenn and Taylor, *Memoir,* 6–8; program, "Labor Day Potato Show," Sept. 1957, Box 18, Folder 27, JGA; "18 Years Of—," p. 6, Box 2, Folder 1, JGA.

15 **It was also close-knit:** JHG OH, JGA; Kettlewell, *Our Town,* 58, 117; Annie Glenn OH, JGA; Glenn and Taylor, *Memoir,* 55, 88. When Bud was around four years old, his mother lost a baby in childbirth, and the Glenns adopted a girl, Jean, from an orphanage in Columbus. Jean was close to Bud in age, but not in other ways. She was "not quite all there," as neighbors put it, in the cruel parlance of the time, and as she got older she took a smaller and smaller part in family life. During her later years, Glenn himself provided for her—but rarely spoke of her (Van Riper, *Glenn,* 78–79; William Shelton to Robert Parker, "Astronaut John Glenn Cover—Bioperse—Take I—Early Days," Jan. 19, 1962, Box 28, Folder 535, HL).

15 **"You were either patriotic":** JHG OH, JGA; Glenn and Taylor, *Memoir,* 4–5.

15 **"Dad was my hero":** Glenn and Taylor, *Memoir,* 4, 33; JHG OH, JGA; "18 Years Of—"; *Official Book of the Fair* (Chicago: A Century of Progress, 1932), 5, 42–43, accessed at https://archive.org/details/officialbookoffa00cent/page/n0. The fair opened in 1933; Glenn visited during its second summer, in 1934 (*Memoir,* 33).

16 **Glenn had been fascinated:** "Autobiography," Box 1, Folder 17, JGA; "The John Glenn Story," 1963, transcript at https://www.archives.gov/files/social-media/transcripts/transcript-john-glenn-story-45022.pdf; JHG OH, JGA; Glenn and Taylor, *Memoir,* 15–16; *Zanesville Sunday Times Recorder,* Jan. 21, 1962; AP, Mar. 23, 2018.

16 **One summer day:** Glenn and Taylor, *Memoir,* 13–15; JHG OH, JGA.

17 **His family muddled through:** Glenn and Taylor, *Memoir,* 29–31, 56; brochures in Box 1, Folders 20 and 35, JGA; "18 Years Of—."

17 **In January 1941, Glenn was walking:** Glenn and Taylor, *Memoir,* 60–61; JHG OH, JGA; Theresa L. Kraus, "CAA Helps America Prepare for World War II," n.d., at https://www.faa.gov/about/history/milestones/media/The_CAA_Helps_America_Prepare_for_World_WarII.pdf.

18 **At the dinner table:** Glenn and Taylor, *Memoir,* 60–61; *St. Louis Post-Dispatch,* Apr. 12, 1959; "America's Wings: Taylorcraft" (Taylorcraft Aviation Corp., 1941), Box 2, Folder 4, JGA.

18 **Glenn sat in the college chapel:** JHG OH, JGA; Glenn and Taylor, *Memoir,* 67–69.

19 **Months passed; no orders came:** Glenn and Taylor, *Memoir,* 69–71; JHG OH, JGA.

19 **Nearly two years went by:** Glenn and Taylor, *Memoir,* 84–86, 94–95; JHG OH, JGA; JHG diary, n.d. [after Mar. 10, 1944], Box 13, Folder 33, JGA.

20 **But the plane could be perilous:** JHG OH, JGA; Berg, *Lindbergh,* 447–50; Glenn and Taylor, *Memoir,* 103–05.

20 **By the time Glenn arrived:** Herman, *Douglas MacArthur,* 480; Glenn and Taylor, *Memoir,* 119–20.

20 **He quickly showed why:** Glenn and Taylor, *Memoir,* 114, 121, 124–26, 134–35; JHG OH, JGA; JHG diary, n.d. [after Mar. 10, 1944], and n.d. [after Dec. 3, 1944], Box 13, Folder 33, JGA.

21 **This last question brought Lindbergh:** Berg, *Lindbergh,* 455; JHG OH, JGA; JHG diary, n.d. [after Mar. 10, 1944].

21 **"a part of, and the brains of":** JHG diary, n.d. [after Mar. 10, 1944].

21 **At El Centro, the left elevator:** Glenn and Taylor, *Memoir,* 114–16, 121–24, 132; JHG diary, n.d. [after Mar. 10, 1944], n.d. [after Dec. 3, 1944]; JHG OH, JGA.

22 **"You need not worry":** JHG to Rev. C. E. Houk, Dec. 7, 1942, Box 3, Folder 10, JGA; *Memoir,* 136.

CHAPTER 2: SUPERSONIC

23 **Glenn's consuming worry:** Glenn and Taylor, *Memoir,* 162–67; JHG OH, JGA.

24 **Glenn got his orders:** Glenn and Taylor, *Memoir,* 167–68; JHG, "Private (Confidential) Statement," Feb. 18, 1953, Box 16, Folder 44, JGA; Annie Glenn OH, JGA.

24 **"America had to be bled":** Eisenhower, address in Detroit, MI, Oct. 24, 1952, accessed at https://www.eisenhower.archives.gov/research/online_documents/korean_war/I_Shall_Go_To_Korea_1952_10_24.pdf; Patterson, *Grand Expectations,* 210–11, 225, 232–34, 260; *NYT,* Feb. 8 and July 12, 1953.

24 **In February 1953, Glenn arrived:** JHG OH, JGA; Glenn and Taylor, *Memoir,* 171, 174–75.

25 **"The man is crazy":** JHG OH, JGA; Glenn and Taylor, *Memoir,* 180.

25 **During the late spring of 1953:** Goldman, *The Crucial Decade—and After,* 245; Glenn and Taylor, *Memoir,* 186–87.

26 **Its pilots flew F-86 Sabres:** Smithsonian National Air and Space Museum, "North American F86-A Sabre," https://airandspace.si.edu/collection-objects/north-american-f-86a-sabre; Glenn and Taylor, *Memoir,* 186–88; JHG OH, JGA.

26 **LYN, ANNIE, and DAVE:** Glenn and Taylor, *Memoir,* 189–90; *NYT,* May 19, 1953; JHG to David Glenn, June 21, 1953, Box 118, Folder 26, JGA; JHG to Annie Glenn, Sept. 28, 1953, Box 118, Folder 27, JGA.

26 **On June 16, 1953, Glenn was flying wing:** JHG, "Statement," June 17, 1953, Box 16, Folder 52, JGA; Glenn and Taylor, *Memoir,* 190–91; JHG OH, JGA.

27 **"I orbited in the area":** "Statement"; Glenn and Taylor, *Memoir,* 191.

27 **He was gliding now:** JHG OH, JGA.

28 **The debriefing that night:** Ibid.

28 **"Never had anyone":** JHG to Annie Glenn, Aug. 2, 1953, Box 118, Folder 27, JGA; JHG to David Glenn, June 21, 1953.

28 **"I am singing a slightly different tune":** JHG to Annie, David, and Lyn Glenn, July 12, 1953, Box 118, Folder 26, JGA (emphasis in original); JHG manuscript, n.d. [1954], Box 16.1, Folder 5, JGA.

29 **"Got another one today":** JHG to Annie, David, and Lyn Glenn, July 19, 1953, Box 118, Folder 26, JGA; Glenn and Taylor, *Memoir,* 194–96; JHG manuscript.

29 **"Nailed another one!":** JHG to Annie Glenn, July 22, 1953, Box 118, Folder 26, JGA.

30 **Glenn's first reaction:** JHG manuscript; Glenn and Taylor, *Memoir,* 196–97; JHG OH, JGA.

30 **Back in 1945:** Glenn and Taylor, *Memoir,* 142–43, 203.

30 **It no longer made sense:** JHG OH, JGA; *Life,* Feb. 2, 1962; Williams quoted in Richard Sisk, "Battle Buddies: John Glenn and Baseball's Ted Williams," Dec. 9, 2016, at https://www.military.com/daily-news/2016/12/09/battle-buddies-john-glenn-baseballs-ted-williams.html, and Brent Larkin, "John Glenn: The Twilight Years of an American Hero," Mar. 13, 2015, at https://www.cleveland.com/opinion/index.ssf/2015/03/john_glenns_rock_is_still_anni.html.

31 **Yet he knew he was never going to get to the top:** JHG OH, JGA; Jones to Parker, "Astronaut Cover: Ice Cream and Burgundy," Jan. 24, 1962, Box 28, Folder 536, HL; Glenn and Taylor, *Memoir,* 203–08, 211–12.

31 **In November 1956, Glenn transferred:** JHG OH, JGA; Glenn and Taylor, *Memoir,* 220–21; Jones to Parker, "Astronaut Cover: Ice Cream and Burgundy."

32 **"a perfect capstone":** Glenn and Taylor, *Memoir,* 221–22; JHG OH, JGA.

32 **Project Bullet, he called the plan:** "Narrative Account of 'Project Bullet,'" appendix to "Nomination for Award of the Harmon International Trophy; Case of Maj. John H. Glenn, Jr., USMC," Commandant of the Marine Corps to the Chief of Naval Operations, Apr. 1, 1958, Box 18, Folder 31, JGA; *NYT,* July 17, 1957; Glenn and Taylor, *Memoir,* 226; JHG, address at the National Exchange Club Convention, Atlantic City, NJ, Sept. 7, 1957, Box 18, Folder 32, JGA.

33 **Just before 12:30 p.m.:** "Jet Record," Universal-International News, accessed at https://www.youtube.com/watch?v=JcA96_ookN0; *NYT,* July 17, 1957.

33 **"it was a slow news day":** JHG OH, JGA; *NYT,* July 17, 1957; on 1950s sitcoms, see Halberstam, *The Fifties,* 508–10.

34 **He was conventional:** Halberstam, *The Fifties,* 521–23; *NYT,* Aug. 16, 1959; Wilson, *The Man in the Gray Flannel Suit,* 13; *NYT,* Mar. 3 and June 12, 1957.

34 **The invitations flooded in:** "Outline of Instruction, Part III," Box 16, Folder 34, JGA; JHG, address at the National Exchange Club Convention.

35 **"Future—to Moon":** JHG, address at the National Exchange Club Convention (emphasis in original).

35 **between appearances he took Dave shopping:** JHG OH, JGA; Lansing Lam-

ont to Parker, "Glenn Shot Cover—Glenn Bio—'Annie and Uncle Johnny,'" Jan. 18, 1962, Box 28, Folder 535, HL; Glenn and Taylor, *Memoir,* 229–31.

35 **The show's producers:** *Name That Tune,* 1957, VHS, Audiovisual Collection, Box 128, Item 11, JGA.

36 **On October 8, Glenn and Eddie:** "Eddie and the Major" script, Oct. 8, 1957, Box 18, Folder 16, JGA; *Cleveland Plain Dealer,* Oct. 5, 1957; *San Francisco Chronicle,* Oct. 5, 1957; PBS, "Sputnik's Impact on America," *Nova,* Nov. 6, 2007, accessed at https://www.pbs.org/wgbh/nova/article/sputnik-impact-on-america/.

36 **"Major," George DeWitt asked:** *Name That Tune,* 1957, VHS; *NYT,* Oct. 8, 1957.

37 **"reaching the practical age limit":** *NYT,* July 17, 1957.

CHAPTER 3: RED MOONLIGHT

38 **Dwight Eisenhower stood:** Eisenhower, *Public Papers of the Presidents of the United States,* 1957: 724, 731; *NYT,* Oct. 5 and Nov. 10, 1957; McDougall, . . . *the Heavens and the Earth,* 146, 148.

39 **one correspondent after another rose:** Eisenhower, *Public Papers,* 1957: 722, 728, 730. In November 1956, the NSC warned Eisenhower that the Soviet Union might well become the first to put a satellite into orbit, and that the United States, in that event, would face a loss of national prestige. But Eisenhower was dubious of prestige as a measure of power (Mieczkowski, *Eisenhower's Sputnik Moment,* 47).

39 **PRESIDENT CALM IN RED MOONLIGHT:** *Pittsburg (KS) Sun,* Oct. 10, 1957; Mieczkowski, *Eisenhower's Sputnik Moment,* 67–68; *Fortune,* Jan. 1958; Walker, *The Cold War,* 114; *NYT,* Oct. 18, 1957.

40 **Even Sputnik's weight was worrisome:** *NYT,* Oct. 6 and 16, 1957; Mieczkowski, *Eisenhower's Sputnik Moment,* 13.

40 **space exploration stirred no excitement:** *Time,* Dec. 8, 1952; Mieczkowski, *Eisenhower's Sputnik Moment,* 187; NSC Report (NSC 5520), May 20, 1955, accessed at https://history.state.gov/historicaldocuments/frus1955-57v11/d340; *Fortune,* Jan. 1958.

40 **Vanguard was a trial run:** Mieczkowski, *Eisenhower's Sputnik Moment,* 43–44, 77–79; Neufeld, *Von Braun,* 281; McDougall, . . . *the Heavens and the Earth,* 108–11, 123.

41 **The best way to avoid an attack:** Mieczkowski, *Eisenhower's Sputnik Moment,* 41–43, 46–47; Herken, *Cardinal Choices,* 89, 96; McDougall, . . . *the Heavens and the Earth,* 123–24; *NYT,* Apr. 14, 1959; Green and Lomask, *Vanguard,* v–vi.

41 **experts knew otherwise:** Herken, *Cardinal Choices,* 98–100; Siddiqi, *The Red Rockets' Glare,* 345, 352; Taubman, *Khrushchev,* 378–79; Wiesner OH, NASA HRC.

42 **Nikita Khrushchev . . . had failed to anticipate:** Siddiqi, *The Red Rockets' Glare,* 357, 360–62; Burrows, *This New Ocean,* 195; *NYT,* Oct. 13, 1957.

42 **The shock of Sputnik:** *NYT,* Oct. 6, 13, 1957; USIA, "World Opinion and the

Soviet Satellite: A Preliminary Evaluation," Oct. 17, 1957, accessed at https://history.nasa.gov/monograph10/doc1.pdf.

43 **sent the White House a confidential report:** USIA, "World Opinion."

43 **the United States had itself to blame:** USIA, "World Opinion"; Siddiqi, *The Red Rockets' Glare,* 360–61; *Life,* Nov. 18, 1957.

43 **He ordered a second launch:** Siddiqi, *The Red Rockets' Glare,* 360; Alex Wellerstein, "Remembering Laika, Space Dog and Soviet Hero," *New Yorker,* Nov. 3, 2017, accessed at https://www.newyorker.com/tech/annals-of-technology/remembering-laika-space-dog-and-soviet-hero; *NYT,* Nov. 5, 1957.

44 **"we are in for a few more shocks":** *Life,* Nov. 18, 1957; *NYT,* Nov. 8, 10, 1957.

44 **On November 7, Eisenhower delivered:** *NYT,* Nov. 5, 1957; Eisenhower, *Public Papers,* 1957: 791–93, 796, 798.

45 **the president had provided no plan:** *NYT,* Oct. 8, 1957.

45 **the newswires carried a photograph:** *NYT,* Nov. 5, 6, 1957.

45 **He had reached for the reins:** Mieczkowski, *Eisenhower's Sputnik Moment,* 137–38; McDougall, . . . *the Heavens and the Earth,* 141, 148–49; Caro, *Master of the Senate,* 1024–25.

46 **he spoke at the annual Rose Festival:** Mieczkowski, *Eisenhower's Sputnik Moment,* 139; *NYT,* Oct. 19, 1957.

46 **The hearings opened:** Mieczkowski, *Eisenhower's Sputnik Moment,* 140; Caro, *Master of the Senate,* 1023–24; *Life,* Jan. 20, 1958.

47 **Vanguard, the three-pound satellite:** Green and Lomask, *Vanguard,* 209; Caro, *Master of the Senate,* 1028.

47 **"How long," he lamented:** Caro, *Master of the Senate,* 1028–29; Mieczkowski, *Eisenhower's Sputnik Moment,* 141; *NYT,* Jan. 24, 1958.

47 **This last recommendation:** Mieczkowski, *Eisenhower's Sputnik Moment,* 167–68, 170; Sherwin Badger to Ben Williamson, "Space Lede—II," Feb. 6, 1958, Box 10, Folder 187, HL.

48 **The idea of a civilian space organization:** Herken, *Cardinal Choices,* 106–07; Mieczkowski, *Eisenhower's Sputnik Moment,* 134, 169–71; *TNO,* 76, 83, 97–98; Eisenhower, *Public Papers,* 1958: 269–73, 573.

49 **Earlier that year, on January 31:** Green and Lomask, *Vanguard,* 217–18; *NYT,* Feb. 5, 1958.

49 **The Air Force had been staking its claim:** *TNO,* 34–36, 69–72, 99–100; "America's First Spaceman," NPR, July 25, 2003, accessed at https://www.npr.org/templates/story/story.php?storyId=1356467.

50 **the laboratories and wind tunnels:** *TNO,* 94–96; Faget, Memorandum for Files, Mar. 7, 1958, NASA HRC; Berger, *The Air Force in Space, Fiscal Year 1962,* USAF Historical Division Liaison Office, June 1966, 1–2, 4.

50 **the president had to step in:** *TNO,* 101; L. A. Minnich Jr., "Legislative Leadership Meeting, Supplementary Notes," Feb. 4, 1958, accessed at https://history.nasa.gov/monograph10/doc3.pdf; Amy Shira Teitel, "The Space Plane That Wasn't," *Popular Science,* June 12, 2015, accessed at https://www.popsci.com/space-plane-wasnt-everything-you-never-needed-know-about-dyna-soar/#page-6.

51 **Seizing the mantle:** *TNO,* 72, 90, 106; Glennan, *The Birth of NASA,* 13; Hansen, *Spaceflight Revolution,* 37.

51 **"as the rising of the sun":** Von Braun, "Crossing the Last Frontier," *Collier's,* Mar. 22, 1952; *Time,* Dec. 8, 1952; Neufeld, *Von Braun,* 259, 267; *TNO,* 69, 99; S. Paul Johnston to J. R. Killian Jr., "Preliminary Observations on the Organization for the Exploitation of Outer Space," Feb. 21, 1958, accessed at https://history.nasa.gov/monograph10/doc4.pdf.

52 **Even the scientists:** Green and Lomask, *Vanguard,* 120–23; Special Committee on Space Technology, "Recommendations to the NASA Regarding a National Civil Space Program," Oct. 28, 1958, accessed at https://history.nasa.gov/monograph10/doc9.pdf.

52 **Would a man in a weightless state go mad:** "Picking the Men in Space," *Collier's,* Feb. 28, 1953; *TNO,* 36–37, 40–41, 49–50, 109.

53 **On December 17 . . . two thousand leaders:** *WP,* Dec. 18, 1958; *WES,* Dec. 18, 1958; *NYT,* Dec. 19, 1958. While the *Star* buried the mention of Mercury, the *Post* did put the story on its front page—though its focus was not Mercury but NASA's announcement that it would begin a four-to-six-year process of developing a booster with 1.5 million pounds of thrust, five times the power of existing U.S. rockets.

CHAPTER 4: ZERO G

55 **The press, a bit too eagerly:** *NYT,* Aug. 9, 1958; *TNO,* 41; Glenn and Taylor, *Memoir,* 236–37.

55 **Glenn did one run:** Glenn and Taylor, *Memoir,* 235–39.

56 **But BuAer offered Glenn one advantage:** JHG OH, NASA HRC; JHG OH, JGA; *TNO,* 36, 118–19; Glenn and Taylor, *Memoir,* 236–38.

56 **Servicemen had a term:** Michael Kramer, "John Glenn: The Right Stuff?," *New York,* Jan. 31, 1983.

57 **the Glenns settled into a ranch-style house:** Diane Sawyer to Robert Parker, "Annie Glenn—Waiting for Wednesday," Jan. 20, 1962, Box 28, Folder 536, HL; *NYT,* Oct. 5, 1958; Glenn and Taylor, *Memoir,* 240–41.

57 **NASA had considered opening:** NASA, "Invitation to Apply for Position of Research Astronaut-Candidate," Dec. 22, 1958, accessed at https://history.nasa.gov/40thmerc7/invite.pdf; Donlan OH, NASA HRC; Atkinson and Shafritz, *The Real Stuff,* 35–36; Burgess, *Friendship 7,* 66; Weitekamp, *Right Stuff, Wrong Sex,* 44; Allen O. Gamble, "Personal Recollections of the Selection of the First Seven Astronauts," paper presented to the Men's Club of the Bethesda United Methodist Church, Bethesda, MD, Mar. 10, 1971, provided to the author by Kris Stoever.

58 **The Navy's representative:** Author int., Voas, Feb. 14, 2020; Van Riper, *Glenn,* 123; Gamble, "Personal Recollections."

58 **Thus it was probably unnecessary:** JHG OH, JGA; Glenn and Taylor, *Memoir,* 239, 242–43, 247; Kramer, "John Glenn"; Van Riper, *Glenn,* 123–25.

59 **When he got home:** Glenn and Taylor, *Memoir,* 243–45; JHG, draft speech, n.d. [Sept. 1957], Box 18, Folder 32, JGA.

59 **"It just sounded":** *Logan (OH) Daily News,* Apr. 18, 1959.

60 **The selection process was a marathon:** Atkinson and Shafritz, *The Real Stuff,* 41–42; NASA, "Sample Questions from Project Mercury Tests," May 12, 1959, "Astronaut Qualifications" file, NASA HRC; Donlan OH, NASA HRC; Donlan OH, JSC.

60 **Donlan showed Glenn some drawings:** Donlan OH, NASA HRC; Donlan OH, JSC.

61 **On February 13, Glenn received a letter:** Donlan to JHG, Feb. 13, 1959, and "Instructions to Men Not Going with the Group to Wright," n.d. [Feb. 1959], Box 64, Folder 39, JGA.

61 **The tone was set:** "Instructions for Enema," n.d. [Feb. 1959], Box 64, Folder 39, JGA; Wolfe, *The Right Stuff,* 70–73; Glenn and Taylor, *Memoir,* 249; JHG to "Troops," Mar. 3, 1959, Box 47, Folder 11, JGA.

61 **"Psycho tests galore":** "Subject #1," n.d. [Mar. 1959], Box 64, Folder 39, JGA.

61 **Glenn was Subject #1:** Glenn and Taylor, *Memoir,* 249–51; Wolfe, *The Right Stuff,* 74–75.

62 **There was a chair:** JHG OH, JGA; JHG, notes from isolation chamber, Mar. 1959, p. 1, Box 69, Folder 39, JGA.

63 **Glenn began to write some poetry:** Notes from isolation chamber, 9–18.

63 **Glenn flew to St. Louis:** *TNO,* 148; Grimwood, *Chronology,* 45.

64 **After the meeting:** *TNO,* 163; Atkinson and Shafritz, *The Real Stuff,* 44; Voas OH, JSC; Burgess, *Friendship 7,* 68.

64 **On April 6, the phone rang:** Glenn and Taylor, *Memoir,* 253.

64 **The rapture of the press:** Schirra OH, JSC; *NYT,* Apr. 11, 1959; *WP,* Apr. 11, 1959; Wainwright, *The Great American Magazine,* 252, 263, 267.

65 **"We've joined a whole new world!":** Schirra OH, JSC; *Richmond Times-Dispatch,* Apr. 10, 1959.

65 **Keith Glennan . . . declared himself pleased:** Caidin, *The Astronauts,* 89; the AP story, n.d. [Apr. 1959] is in Annie's scrapbook, Box 9.1, Folder 2, JGA; *Richmond Times-Dispatch,* Apr. 10, 1959; *St. Louis Post-Dispatch,* Apr. 12, 1959; Betty Smith to the Glenns, Apr. 13, 1959, Box 19, Folder 5, JGA.

66 **But Annie's good cheer was a pose:** Shelton to Parker, "Astronaut John Glenn Cover—Bioperse—Take I—Early Days," Jan. 19, 1962, Box 28, Folder 535, HL; author int., Lyn Glenn, May 14, 2020.

66 **Annie became so anxious:** *Life,* Sept. 21, 1959; *St. Louis Post-Dispatch,* Apr. 12, 1959; Jane Hosey to Annie Glenn, Apr. 28, 1959, Box 19, Folder 5, JGA.

67 **NASA gave the astronauts two weeks:** "First Month's Program for Astronauts," n.d. [Apr. 1959], Box 69, Folder 45, JGA; *Life,* Mar. 3, 1961; Glenn and Taylor, *Memoir,* 269–70; Carpenter and Stoever, *For Spacious Skies,* 209–10; Lamont to Parker, "Glenn Shot Cover—Glenn Bio—'Annie and Uncle Johnny,'" Jan. 18, 1962, Box 28, Folder 535, HL.

67 **"Monday and Tuesday":** "First Month's Program"; Glenn and Taylor, *Memoir,* 270; Schirra OH, JSC.

68 **Glenn took notes:** JHG notes, n.d. [Apr. 1959], Box 69, Folder 45, JGA; JHG

to GPO, Apr. 30, 1959, Box 64, Folder 39, JGA; Schirra with Billings, *Schirra's Space*, 59.

68 **He was happier on the centrifuge:** *TNO*, 237–39; Cooper diary, June 3, 1959. Stockdale later became an admiral and one of the most highly decorated officers in the history of the U.S. Navy; in 1992, he was a candidate for vice president as Ross Perot's running mate.

68 **The astronauts also made the rounds:** *TNO*, 237–41; Glenn and Taylor, *Memoir*, 284–86, 295; author int., Aldrich, Dec. 20, 2018.

69 **big problems remained:** *TNO*, 141–42, 168; Grimwood, *Chronology*, 45, 53–54; Maxime A. Faget, "The Spacecraft's Safety System," Part 10, *First U.S. Man in Orbit*, Feb. 1962, pp. 1, 3, accessed at https://osdn.net/projects/sfnet_mscorbaddon/downloads/Research/Mercury%20MA-6/MA6_Reports.pdf/; *WP*, July 15, 1979; Gilruth OH, NASA HRC; Gilruth, "Memoir: From Wallops Island to Mercury, 1945–1958," manuscript, 38, Gilruth Papers, Box 4, Folder 24, Virginia Tech.

69 **The press, with NASA's encouragement:** *TNO*, 25, 122–23; author int., Roberts, Dec. 17, 2018.

70 **Most missiles had their champions:** *TNO*, 21, 90; Faget OH, JSC; Matthews OH, JSC; Kraft OH, NASA HRC; April 14 test described at http://www.astronautix.com/a/atlasd.html.

70 **On the night of May 18:** Glenn and Taylor, *Memoir*, 274–75; the explosion is seen in "Atlas Flight Test Review," 1959, at https://www.youtube.com/watch?v=MYvj8XWI9T0.

71 **"That's our ride?":** Cooper with Henderson, *Leap of Faith*, 3; Thompson, *Light This Candle*, 182–83; Glenn and Taylor, *Memoir*, 275; French and Burgess, *Into That Silent Sea*, 56–57.

71 **the seven had a frank talk:** Author int., Schneiders, Feb. 12, 2019.

CHAPTER 5: A SEVEN-SIDED COIN

72 **It was something short:** Cooper diary, entry for May 28, 1959, courtesy of Jon Siefken, Oct. 12, 2018; Slayton to Ken Hechler, Mar. 17, 1993, and Hechler memoranda, Mar. 15 and 26, 1993, NASA HRC, File 011017; *WES*, May 28, 1959.

72 **The seven made the rounds:** Cooper diary; Schirra OH, JSC; "Meeting with the Astronauts (Project Mercury—Man-in-Space Program), Hearing Before the Committee on Science and Astronautics, U.S. House of Representatives," 86 Cong., 1st Sess., May 28, 1959.

73 **A question about the men's motivation:** "Meeting," 4, 6, 10–11.

73 **new "space heroes":** *Life*, June 9, 1959; *NYT*, May 29, 1959; *WES*, May 28, 1959.

74 **"Able Baker perfect":** *Life*, June 9, 1959; *NYT*, May 29, 30, 31, 1959.

74 **"monkey-shit talk":** Slayton with Cassutt, *Deke!*, 82; Wolfe, *The Right Stuff*, 147–49; *Life*, June 15, 1959.

75 **"It made them madder":** Author telephone int., Roberts, Dec. 17, 2018.

75 **A passive role:** Amy Shira Teitel, "A History of the Dyna-Soar," Oct. 5, 2011, accessed at https://vintagespace.wordpress.com/2011/10/05/a-history-of-the-dyna-soar/; Voas OH, JSC; author int., Voas, Apr. 23, 2020; Hansen, *Spaceflight Revolution,* 56–57; French and Burgess, *Into That Silent Sea,* 159; *TNO,* 71, 87; Joachim P. Kuettner, "The Launching of a Manned Missile," paper prepared for delivery at the American Rocket Society meeting, June 8, 1959, in Box 3, Folder 24, Kraft Papers, Virginia Tech; *Training of Astronauts,* 15; author int., Roberts, Dec. 17, 2018.

76 **A question remained:** Voas OH, JSC; *TNO,* 174, 217; *Training of Astronauts,* 16.

77 **The volume of complex information:** Author int., Voas, Feb. 14 and Apr. 30, 2020; Voas OH, JSC.

78 **During the summer of 1959:** Neufeld, *Von Braun,* 337–38.

78 **In late 1959, they went to NASA's design team:** Glenn and Taylor, *Memoir,* 285–87.

79 **"to just sit there":** Caidin, *Man into Space,* 143–44; *WES,* July 8, 1959; Wolfe, *The Right Stuff,* 120. The subsonic trainer, received as an insult, was intended as insurance: NASA was not eager to let its carefully selected, highly trained astronauts risk their lives needlessly in high-performance aircraft (Voas OH, JSC).

79 **On July 9, George Miller:** *WES,* July 9, 1959; "Meeting," 14–16; Wolfe, *The Right Stuff,* 121; Voas OH, JSC.

80 **The astronauts felt:** "Meeting," 16; Voas, "Project Mercury: The Astronaut Training Program," paper for presentation at the Symposium on Psychophysiological Aspects of Space Flight, San Antonio, TX, May 26–27, 1960, p. 3, provided to the author by Voas; JHG notes, Aug. 1960, Box 67, Folder 10, JGA.

80 **It was all for one:** Slayton OH, NASA HRC; "Summary of Compliance," Aug. 16–18, 1960, Box 67, Folder 10, JGA; JHG OH, NASA HRC; Cooper OH, JSC; *NYT,* July 9, 1959.

81 **The national parlor game:** *NYT,* Apr. 11, 1959; *WP,* Apr. 10, 1959; "Meeting," 12; *Life,* Sept. 14, 1959; Carpenter et al., *We Seven,* 21.

81 **And Glenn was the man to beat:** "Isabel" to JHG, May 13, 1959, Paul B. Montagne to JHG, Nov. 28, 1959, Bob Soule to JHG, Apr. 10, 1959, all in Box 19, Folder 5, JGA; *Life,* June 18, 1956; author int., Stoever, July 5, 2018.

82 **"In the advance betting":** Wainwright, *The Great American Magazine,* 274–75; Haney OH, NASA HRC; William Shelton, [first name missing] Burkhart, and Ed Reingold to Parker, "Space Shot—Take Three," May 5, 1961, Box 25, Folder 487, HL; *WP,* Feb. 20, 1975.

82 **Glenn's original sin:** Schirra OH, JSC; Slayton with Cassutt, *Deke!,* 74; Cooper OH, JSC: 4; Cooper diary; Cooper with Henderson, *Leap of Faith,* 17.

82 **Glenn . . . had been pouring it on:** Schirra OH, JSC; Cooper with Henderson, *Leap of Faith,* 21; Schirra with Billings, *Schirra's Space,* 63.

83 **In an April 24 memo:** Powers to the Astronauts, "Public Relations," Apr. 24, 1959, Box 68, Folder 28, JGA; Powers OH, NASA HRC; Slayton OH, NASA HRC; Wainwright, *The Great American Magazine,* 252; Glenn OH, NASA HRC.

83 **It was Glenn, in fact:** JHG OH, JGA; JHG OH, NASA HRC; Slayton OH, NASA HRC; JHG to Lawrence Spivak, May 26, 1959, accessed at https://www.rrauction.com/preview_itemdetail.cfm?IN=4012; Williams, "Go!," 9.

85 **"Extracurricular activities":** Leopold, *Calculated Risk,* 106; Van Riper, *Glenn,* 149–50.

85 **"We were always looking":** Slayton with Cassutt, *Deke!,* 88; *Life,* May 16, 1960; Wolfe, *The Right Stuff,* 130–33; Stoever int. with Rene Carpenter, Mar. 9, 2020, and Stoever e-mail with author.

85 **He didn't flirt:** Author int., Stoever, July 5, 2018; Carpenter and Stoever, *For Spacious Skies,* 209; Lamont to Parker, "Glenn Shot Cover," Jan. 18, 1962, Box 28, Folder 535, HL.

86 **a prude and a scold:** Van Riper, *Glenn,* 151–52.

86 **Glenn's concerns:** Thompson, *Light This Candle,* 205–07.

87 **"We have a split":** Williams, "Go!," 10.

87 **"Follower" put it too strongly:** French and Burgess, *Into That Silent Sea,* 130; Carpenter OH, JSC: 3; Carpenter and Stoever, *For Spacious Skies,* 232; author int., Stoever, June 1 and July 5, 2018; Wolfe, *The Right Stuff,* 138–39; Wainwright, *The Great American Magazine,* 271–72.

88 **Shepard's relationships:** Williams, "Go!," 8–10; *Life,* Mar. 3, 1961; Shepard OH, JSC.

88 **they were the standouts:** Thompson, *Light This Candle,* 179–81; Haney OH, NASA HRC; "Astronauts Press Conference, Air Force Ballistic Missile Division, Inglewood, CA, Sept. 16, 1959," Box 68, Folder 28, JGA; Wolfe, *The Right Stuff,* 140.

89 **On September 12 the astronauts:** Powers to Colonel Lindell, "West Coast Trip," Aug. 31, 1959, Box 69, Folder 45, JGA; Schirra OH, JSC; *San Diego Union,* Sept. 20, 1959; *San Diego Evening Tribune,* Sept. 24, 1959; Slayton with Cassutt, *Deke!,* 88; Annie Glenn scrapbook, Box 9.1, Folder 2, JGA.

89 **Convair was still bickering:** *San Diego Union,* July 13, 1958, and Sept. 20, 1959; Caidin, *The Astronauts,* 122; *San Diego Evening Tribune,* Sept. 24, 1959.

89 **In the off-hours:** *San Diego Union,* Sept. 20 and 21, 1959; *San Diego Evening Tribune,* Sept. 24, 1959; "Officer Bars Interview with 6 Astronauts," San Diego newspaper and date [Sept. 1959] unavailable, in Annie Glenn's scrapbook, Box 9.1, Folder 2, JGA.

90 **After midnight:** Thompson, *Light This Candle,* 226; Van Riper, *Glenn,* 152; *Life,* Sept. 21, 1959.

90 **Shepard was worried:** "Single, Free, and 43; A Woman Still," unpublished article by Don A. Schanche, n.d. [1971], 3–4, and Schanche notes of an interview with Rene Carpenter, July 2, 1971, 16, both in Schanche Papers, Box 12, Folder 1, Hargrett Rare Book and Manuscript Library, University of Georgia (courtesy of Danny Parker); Stoever email to author, Nov. 13, 2020. Carpenter, ironically, was rooming with Schanche as a favor to Shepard. Carpenter had been assigned a room with a double bed; Shepard, whose room had twin beds, convinced Car-

penter to trade—on the grounds that Shepard would make better use of the double bed (Wolfe, *The Right Stuff*, 134–35; Parker email to author, Nov. 16, 2020).

91 **Glenn asked Powers to come by:** Van Riper, *Glenn*, 152–53; Thompson, *Light This Candle*, 226; Wolfe, *The Right Stuff*, 135; Glenn and Taylor, *Memoir*, 293–94.

92 **The others sat sullenly:** Thompson, *Light This Candle*, 227; Wolfe, *The Right Stuff*, 137; Van Riper, *Glenn*, 153; author int., Stoever, July 5, 2018.

CHAPTER 6: SECOND IN SPACE

93 **Khrushchev boarded a plane:** *NYT*, Sept. 13 and 15, 1959; Chertok, *Rockets and People*, 419.

93 **The spacecraft, Lunik II:** *NYT*, Sept. 14, 1959.

94 **On the morning of September 15:** Eisenhower, *Public Papers*, 1959: 655–56.

94 **Eisenhower got in a few digs:** Eisenhower, *Public Papers*, 1959: 671–72; *NYT*, Sept. 14 and 25, 1959.

94 **The next blow:** Chertok, *Rockets and People*, 419; *WES*, Oct. 4, 1959; *NYT*, Sept. 13, 14, 25, and Oct. 4, 5, 7, 1959; *TNO*, 213.

95 **Americans were left:** *NYT*, Oct. 5, 7, 8, and 11, 1959.

95 **The picture was not entirely bleak:** *NYT*, Oct. 4 and 5, 1959.

96 **And it was undercut . . . by Nixon:** *WES*, Oct. 4, 1959; *NYT*, Oct. 4 and 7, 1959.

96 **In 1940, at the beginning:** JFK, *Why England Slept*, xxiii, 217.

97 **On November 13, the candidate:** *Los Angeles Times*, Nov. 14, 1959; *NYT*, Nov. 14, 1959; *WP*, Nov. 14, 1959; JFK, *Strategy of Peace*, 193. "Eight gray years," as JFK pointed out, had been Franklin Roosevelt's phrase in 1928, after two terms under Republican presidents (*Strategy of Peace*, 198).

97 **"the unchallenged leaders":** JFK, *Strategy of Peace*, 194–95, 198.

97 **In the late 1950s, as a member:** "Harvard Board of Overseers, 1957," POF, JFKPOF-135–024, JFKL; JFK, "Remarks at the Annual Freedom Award to the Hungarian Freedom Fighters," Oct. 23, 1957, PPP, Sen. Files, JFKSEN-0898-020, JFKL.

98 **Kennedy shared the president's instinct:** McDougall, . . . *the Heavens and the Earth*, 301; TCS OH, JFKL; JFK, *Strategy of Peace*, 7.

98 **He had framed the challenge:** *NYT*, July 14, 1960; Nitze to JFK, Nov. 23, 1959, and JFK to Archibald Cox, Sept. 2, 1960, NASA HRC, File 012492.

99 **"impossible . . . to foresee":** Nitze to JFK; Donald H. Menzel to JFK, Sept. 7, 1960, NASA HRC, File 012492.

100 **In a blur of events:** On Kennedy's selection of LBJ, see Shesol, *Mutual Contempt*, 41–57. "Position Paper on Space Research," Sept. 7, 1960, NASA HRC, File 012492.

100 **"a revolution of automation":** JFK, "The New Frontier," acceptance speech, Democratic National Convention, July 15, 1960, PPP, Sen. Files, JFKSEN-0910-015, JFKL; JFK, "Remarks at Lawrence Stadium, Wichita, KS," Oct. 22, 1960, PPP, Sen. Files, JFKSEN-0913-059, JFKL; "Briefing Paper on Space,"

[1960], PPP, Box 993A, JFKL; Rostow to Cox, Aug. 11, 1960, and Menzel to JFK, Apr. 5, 1960, NASA HRC, File 012492. JFK mentioned the space issue twice as often as Nixon did during the campaign (McDougall, . . . *the Heavens and the Earth,* 225).

101 **Keith Glennan had been a studio manager:** Glennan, *The Birth of NASA,* xi; *Meet the Press,* vol. 4, no. 15, Apr. 10, 1960, JGA.

101 **No matter how Glennan put it:** Glennan, *The Birth of NASA,* 31, 66–67, 187.

102 **Mercury, meanwhile, proceeded:** *TNO,* 256; JHG notes, n.d. [1959], Box 69, Folder 45, and JHG notes, Jan. 4, 1960, Box 67, Folder 28, JGA.

102 **In early May, NASA conducted:** Grimwood, *Chronology,* 100, 103.

103 **desert survival training:** Grimwood, *Chronology,* 105; Schirra with Billings, *Schirra's Space,* 67; "Desert Survival Training Program, Project Mercury, 11 July 1960, Stead AFB, Nevada," JHG notes, Box 69, Folder 35, JGA. The astronauts had also been undergoing water-survival training since May 1959 (Carpenter and Stoever, *For Spacious Skies,* 204).

103 **Air Force helicopters:** JHG OH, JGA; Glenn and Taylor, *Memoir,* 297–99; JHG notes, July 15, 1960, Box 69, Folder 35, JGA; Thompson, *Light This Candle,* 225; "Astronaut Comments on Desert Survival Training," July 1960, Box 69, Folder 35, JGA.

104 **Glenn recorded stray observations:** JHG notes, July 15, 1960, Box 69, Folder 35, JGA.

104 **"We have been 'stewing'":** Glennan, *The Birth of NASA,* 252, 256; *NYT,* Oct. 22, 1960.

105 **After further discussion:** Glennan, *The Birth of NASA,* 256; *NYT,* Oct. 31, 1960.

106 **In the campaign's closing weeks:** Mieczkowski, *Eisenhower's Sputnik Moment,* 245; *Final Report of the Committee on Commerce,* 601; JFK, Remarks at Municipal Auditorium, Canton, Ohio, Sept. 27, 1960, accessed at https://www.jfklibrary.org/asset-viewer/archives/JFKSEN/0912/JFKSEN-0912-006.

106 **Nixon's strained insistence:** *Final Report of the Committee on Commerce,* 268–69, 277; *NYT,* July 25, 1959.

106 **NASA readied another test:** Grimwood, *Chronology,* 117; Glennan, *The Birth of NASA,* 259n; *NYT,* Nov. 9, 1960.

CHAPTER 7: SUSPENDED ANIMATION

108 **"We are in a state":** Glennan, *The Birth of NASA,* 286; Seamans, *Aiming at Targets,* 74–76; Slayton with Cassutt, *Deke!,* 89; *AW,* Mar. 6, 1961.

108 **Amid this ambiguity:** *NYT,* Nov. 22, 1960.

109 **When the countdown reached zero:** *TNO,* 293–96; Kranz, *Failure,* 29–32; *Life,* Dec. 5, 1960; Neufeld, *Von Braun,* 354–55.

109 **To Gene Kranz:** Kranz, *Failure,* 32; Glennan, *The Birth of NASA,* 280; *NYT,* Dec. 3, 1960; *AW,* Jan. 23, 1961; *NYT,* Dec. 16, 1960; Jerry Main to Parker, "Project Mercury—I," Nov. 23, 1960, Box 23, Folder, 450, HL.

110 **"We trained harder":** Glenn and Taylor, *Memoir,* 307–08; Dryden OH, JFKL.

110 **On December 16:** President's Science Advisory Committee, "Report of the Ad Hoc Panel on Man-in-Space," Dec. 16, 1960, NASA HRC.

110 **Three days later:** *NYT,* Dec. 20, 1960; *TNO,* 297; Glennan, *The Birth of NASA,* 291.

111 **The next day, December 20:** Glennan, *The Birth of NASA,* 292–93; Gilruth OH, NASA HRC; *AW,* Jan. 30, 1961; McDougall, *. . . the Heavens and the Earth,* 225.

111 **"There is kind of a hush":** Glennan, *The Birth of NASA,* 272, 300; Seamans, *Aiming at Targets,* 75.

112 **Kennedy was not . . . indifferent:** *NYT,* Jan. 20, 1961; *The Report of the President's Commission on National Goals,* 2–4, 15–16, 19.

112 **Which brought the conversation back:** *NYT,* Jan. 15, 1961; Schlesinger, *A Thousand Days,* 317; McDougall, *. . . the Heavens and the Earth,* 235; Mieczkowski, *Eisenhower's Sputnik Moment,* 244–46; Schlesinger, *A Thousand Days,* 315–17; *NYT,* Jan. 17, 1961.

113 **new Zenith television:** *Life,* Sept. 21, 1959; Logsdon, *John F. Kennedy and the Race to the Moon,* 12; Erskine, "The Polls," 485–86; Almond, "Public Opinion and the Development of Space Technology," 559, 561, 569–71; *Final Report of the Committee on Commerce,* 601.

113 **No one understood this better:** McDougall, *. . . the Heavens and the Earth,* 235–36, 253, 285, 288; Glennan, *The Birth of NASA,* 239; Taubman, *Khrushchev,* 491; *AW,* Apr. 3, 1961.

113 **"Launching Sputniks":** McDougall, *. . . the Heavens and the Earth,* 251; Kempe, *Berlin 1961,* 4, 130–32.

114 **"Space Problems":** BeLieu to LBJ, Dec. 22, 1960, NASA HRC, File 12288; Sidey, *John F. Kennedy,* 96–98.

114 **All these problems:** BeLieu to LBJ, Dec. 17 and 22, 1960; Logsdon, *Race to the Moon,* 28–31; *Space Digest,* Jan. 1961.

115 **That effort was already underway:** TCS to Wiesner, Dec. 9, 1960, Wiesner Papers, Box 10, MIT; Logsdon, *Race to the Moon,* 31–32; Wiesner, "Kennedy," in Rosenblith, *Jerry Wiesner,* 268, 270, 278; TCS, "A View from the White House," in Rosenblith, *Jerry Wiesner,* 42; *NYT,* Sept. 3, 1961; *New Yorker,* Jan. 19 and 26, 1963; Edward M. Kennedy, "Foreword," in Rosenblith, *Jerry Wiesner,* xiii.

115 **Wiesner was not an expert on space:** *New Yorker,* Jan. 26, 1963, 39–40; Alex Wellerstein, "America at the Atomic Crossroads," *New Yorker,* July 25, 2016, accessed at https://www.newyorker.com/tech/annals-of-technology/america-at-the-atomic-crossroads; Herken, *Cardinal Choices,* 128–29; Gavin, *War and Peace in the Space Age,* 248; Wiesner OH, NASA HRC.

116 **"We are interested":** TCS to Wiesner, Dec. 9, 1960, Wiesner Papers, Box 10, MIT.

116 **A month later, on January 11:** Logsdon, *Race to the Moon,* 32; *AW,* Jan. 16, 1961; Wiesner to TCS, Dec. 19, 1960, Wiesner Papers, MIT; *NYT,* Jan. 12,

1961; "Report to the President-Elect of the Ad Hoc Committee on Space," Jan. 10, 1961, pp. 1, 3–4, Wiesner Papers, MIT.

117 **National prestige was one possible reason:** "Report," 2, 19–21; Wiesner draft, n.d. [Jan. 1961], Wiesner Papers, MIT.

117 **Here its tone shifted:** "Report," 20–23.

118 **"Highly informative":** Logsdon, *Race to the Moon,* 34; *NYT,* Jan. 12, 1961.

118 **"What a slap":** Glennan, *The Birth of NASA,* 304; *Miami Herald,* Jan. 12, 1961; Seamans, *Aiming at Targets,* 74; "Report," 7.

118 **Before the holidays:** Abe Silverstein to Glennan, "Astronaut Selection Procedure for Initial Mercury-Redstone Flights," Dec. 14, 1960, NASA HRC; Williams, "Go!," 12.

119 **"It took a moment":** Glenn and Taylor, *Memoir,* 309–10; Slayton with Cassutt, *Deke!,* 93; French and Burgess, *Into That Silent Sea,* 137.

120 **"long tiresome debate":** Williams, "Go!," 11; *Life,* Sept. 14, 1959.

120 **Gilruth finally summoned his board:** *NYT,* July 9, 1959; Thibodaux OH, "Space Stories"; Kleinknecht OH, JSC; Thompson, *Light This Candle,* 231–32; Williams, "Go!," 14; Andrew Chaikin, "Bob Gilruth, the Quiet Force Behind Apollo," *Air & Space Magazine,* Feb. 2016, accessed at https://www.airspacemag.com/history-of-flight/quiet-force-behind-apollo-180957788/.

121 **Much the same could be said:** Johnston OH, JSC; Williams, "Go!," 13–14.

121 **But this last quality:** Williams, "Go!," 14–15; Kraft, *Flight,* 55, 86–87.

121 **Snow paralyzed the capital:** *NYT,* Jan. 20, 1961; Shepard OH, JSC; French and Burgess, *Into That Silent Sea,* 53.

122 **The office was empty:** French and Burgess, *Into That Silent Sea,* 53.

122 **Gilruth entered and shut the door:** Cooper with Henderson, *Leap of Faith,* 26; Glenn and Taylor, *Memoir,* 310; Shepard OH, JSC.

122 **"I didn't think":** Wainwright, *The Great American Magazine,* 273; Van Riper, *Glenn,* 156.

123 **"I kept thinking that Bob Gilruth":** Glenn and Taylor, *Memoir,* 311; Shepard et al., *Moon Shot,* 64; French and Burgess, *Into That Silent* Sea, 137; author int., Lyn Glenn.

124 **"Together," the new president declared:** JFK, *Public Papers,* 1961: 2.

124 **Keith Glennan listened:** Glennan, *The Birth of NASA,* 308–10.

CHAPTER 8: THE PROBLEMS OF MEN ON EARTH

125 **"Americans . . . ," John Kennedy said:** JFK, *Strategy of Peace,* 9.

125 **Khrushchev still had no fixed notion:** Taubman, *Khrushchev,* 484–85.

126 **"He had nightmares":** Taubman, *Khrushchev,* 482; Gaddis, *The Cold War,* 112.

126 **By 1961, 2.7 million East Germans:** Gaddis, *The Cold War,* 114–15; Kempe, *Berlin 1961,* 4, 6, 8–9, 41, 75, 130; Taubman, *Khrushchev,* 483, 485–86.

127 **Kennedy, too, wanted to meet:** Schlesinger, *A Thousand Days,* 298–99, 301–02, 304; Kempe, *Berlin 1961,* 53–55, 76–79; Taubman, *Khrushchev,* 487–88.

127 **The Kremlin did not know:** Walker, *The Cold War,* 146–47; Kempe, *Berlin 1961,*

81–82; Taubman, *Khrushchev,* 487–89; JFK, *Public Papers,* 1961: 2, 22–23; *Life,* Feb. 10, 1961.

128 **A Pathé newsreel:** Pathé, *Man and Monkey Train for Space Flight,* Jan. 1961, accessed at https://www.britishpathe.com/video/man-and-monkey-train-for-space-flight/query/man+and+monkey; JFK, *Public Papers,* 15; Dryden OH, JFKL.

129 **If fresh attention was being paid:** Richard Neustadt, "Memorandum on Organizing the Space Council," Feb. 27, 1961, TCS Papers, Box 120, JFKL; Logsdon, *Race to the Moon,* 57–58; *AW,* Jan. 16 and Feb. 13, 1961.

129 **Still, he had been unable:** Lambright, *Powering Apollo,* 82–83; Logsdon, *Race to the Moon,* 39–41.

129 **Webb and Kerr were mutually indebted:** Lambright, *Powering Apollo,* 5–7, 11, 15–19, 26–28; *NYT,* Mar. 29, 1992; Bizony, *The Man Who Ran the Moon,* 9–13.

130 **Square in frame:** Lambright, *Powering Apollo,* 5; Haney OH, JSC; Logsdon, *Race to the Moon,* 41; memorandum, Lambright conversation with Wiesner, Nov. 15, 1990, 1–2, NASA HRC, File 7106.

130 **The White House operator:** Lambright, *Powering Apollo,* 82–84; Lambright memorandum; *AW,* Feb. 6, 1961.

131 **flight of the chimpanzee:** *TNO,* 312–14; "Information Guide for Animal Launchings in Project Mercury," July 23, 1959, NASA HRC, File 18674; *Life,* Feb. 10, 1961; Pathé newsreel, Jan. 1961; Slayton with Cassutt, *Deke!,* 82. "Ham" was an acronym for his training facility: Holloman Aerospace Medical Center (Wolfe, *The Right Stuff,* 175).

131 **Just before noon:** *TNO,* 314–16; Gilruth, *Memoir,* 49, Gilruth Papers, Virginia Tech; Thompson OH, JSC; Thompson OH, NASA HRC; *AW,* Feb. 6, 1961; *Life,* Feb. 10, 1961.

132 **But the flight of Ham:** *TNO,* 318, 342; *Life,* Feb. 10, 1961; Gilruth diary, Feb. 13, 1961, Box 4, File 13, Gilruth Papers, Virginia Tech.

132 **"drove Al crazy":** Glenn and Taylor, *Memoir,* 312.

132 **"he is bitterly disappointed":** Williams, "Go!," 15–16; Wainwright, *The Great American Magazine,* 273; *WES,* Nov. 30, 1961; Van Riper, *Glenn,* 157; *NYT,* Nov. 13, 1983.

133 **It was no use assuring Glenn:** Wainwright, *The Great American Magazine,* 273; *NYT,* Nov. 13, 1983; JHG to Annie, Sept. 22 and 28, 1953, Box 118, Folder 27, JGA; Slayton with Cassutt, *Deke!,* 94.

133 **It was clear, in fact, to everyone:** Williams, "Go!," 16 (emphasis in original); Wolfe, *The Right Stuff,* 180–81; French and Burgess, *Into That Silent Sea,* 55.

134 **As the weeks passed:** Slayton with Cassutt, *Deke!,* 94; Wolfe, *The Right Stuff,* 180; *Columbus Dispatch,* Mar. 25, 1961; Gilruth diary, Mar. 1–5, 1961, Box 4, File 13, Gilruth Papers, Virginia Tech.

135 **"Venus," observed the *New York Times*:** *NYT,* Feb. 13, 1961; *WES,* Feb. 14, 1961; *AW,* Feb. 20, 1961.

135 **Kennedy was told the news:** *NYT,* Feb. 13, 1961; Overton Brooks, Memorandum for White House Conference, Feb. 13, 1961, POF, Box 82, JFKL; *WES,* Feb. 14, 1961; *NYT,* Feb. 16, 1961; *Life,* Apr. 7, 1961.

135 **That evening Kennedy held a press conference:** *Life,* Feb. 3, 1961; Dallek, *An Unfinished Life,* 335; JFK, *Public Papers,* 1961: 95.

136 **"We must expect":** Wiesner to JFK, Feb. 20, 1961, POF, Box 307, JFKL; Wiesner to JFK, Feb. 23, 1961, POF, Box 67, JFKL; Don D. Cadle to David Bell, "Meeting with Mr. James E. Webb," Feb. 15, 1961, NASA HRC, File 012457; *TNO,* 379.

137 **In the vacuum:** *AW,* Jan. 16 and 23, Mar. 6 and 13, 1961; Logsdon, *Race to the Moon,* 21–22, 47; "Report of the Air Force Space Study Committee" [Gardner Report], Mar. 20, 1961, POF, Box 307, JFKL.

137 **The battle for control:** Lambright, *Powering Apollo,* 89–91; *NYT,* Dec. 16, 1960; Logsdon, *Race to the Moon,* 45.

137 **On the morning of February 21:** *TNO,* 318–22; *AW,* Feb. 27, 1961; *NYT,* Feb. 22, 1961.

138 **Gilruth felt compelled:** Gilruth diary, Feb. 15–16, 1961.

138 **"In the Redstone program":** *NYT,* Feb. 22, 1961; *Life,* Mar. 3, 1961; Gilruth diary, Feb. 21, 1961.

139 **"It's an understatement":** *NYT,* Feb. 22, 1961; Thompson, *Light This Candle,* 235; Schirra with Billings, *Schirra's Space,* 72; Slayton with Cassutt, *Deke!,* 94; Reingold to Parker, "Glenn Cover, Take One," Jan. 19, 1962, Box 28, Folder 535, HL. "The astronauts themselves watched all these conjectures with amusement," NASA's official history of Mercury cheerfully claimed (*TNO,* 349).

139 **Webb understood:** Logsdon, *Race to the Moon,* 62.

139 **David Bell, Kennedy's budget director:** Logsdon, *Race to the Moon,* 62, 67; Lambright, *Powering Apollo,* 91–92.

140 **Webb took his case to Kennedy:** Seamans, *Aiming at Targets,* 82–83; Logsdon, *Race to the Moon,* 63–64; Robert Sherrod, memo, n.d., and Webb, "Administrator's Presentation to the President," Mar. 21, 1961, NASA HRC, File 012504.

140 **The president listened closely:** Seamans, *Aiming at Targets,* 82–83; Willis H. Shapley to Andreas Reichstein, Mar. 14, 1989, NASA HRC, File 012506—JFK WH Correspondence, 1961; *TNO,* 349–50; Haney OH, JSC; Makemson, *Media, NASA, and America's Quest for the Moon,* 51.

140 **After the meeting, Bell quickly sent:** David E. Bell to JFK, "NASA Budget Problem," n.d. [Mar. 22, 1961], POF, Box 82, JFKL (emphasis in original).

141 **Bell, LBJ, and Wiesner were present:** Logsdon, *Race to the Moon,* 65–66; Lambright, *Powering Apollo,* 92–93; Seamans, *Aiming at Targets,* 83; Administrator's Presentation; *NYT,* Mar. 29, 1961.

141 **NASA's ambitions:** Logsdon with Launius, *Exploring the Unknown,* vol. 7, 458–59, 471.

CHAPTER 9: THE FIRST MAN

142 **John Kennedy entered the ballpark:** *WP,* Apr. 11, 1961.

142 **At the end of the second inning:** Sidey, *John F. Kennedy,* 92–93; *WP,* Apr. 13, 1961; Andrew Hatcher, "Memorandum for the President," Apr. 10, 1961, WHCF, Box 654, JFKL.

143 **Other news outlets:** *WP,* Apr. 11, 1961.

143 **Back in Washington:** *Hearings Before the Committee on Science and Astronautics,* Apr. 11, 1961, 289, 292, 295, 297, 316, 320, 337, 340; Logsdon, *Race to the Moon,* 49.

144 **By late afternoon the CIA:** Sidey, *John F. Kennedy,* 93; Murrow memo quoted in Logsdon, *Race to the Moon,* 69–70, citing Murrow to Bundy, "Recommended U.S. Reaction to Soviet Manned Space Shot Failure," Apr. 3, 1961, NSF, Box 307, JFKL. ("Tell him I agree," Bundy wrote in the margin of the memo. Logsdon, *Race to the Moon,* 256n.)

145 **Salinger put his feet up:** Sidey, *John F. Kennedy,* 93.

145 **A week earlier, on April 5:** French and Burgess, *Into That Silent Sea,* 109; Chertok, *Rockets and People,* 71; Siddiqi, *Challenge to Apollo,* 266–67, 276; Cadbury, *Space Race,* 175–76; *NYT,* Apr. 6, 1986.

146 **The choice was Gagarin:** French and Burgess, *Into That Silent Sea,* 1–5, 8–11; Chertok, *Rockets and People,* 60–61, 69.

146 **At 5:30 a.m., Moscow time:** Siddiqi, *Challenge to Apollo,* 273–74, 276–77; Cadbury, *Space Race,* 233–36; Taubman, *Khrushchev,* 491.

147 **"Off we go!":** Gerovitch, *Soviet Space Mythologies,* 82; Siddiqi, *Challenge to Apollo,* 261–62, 276, 278, 281–82; Chertok, *Rockets and People,* 83. Unlike Project Mercury, which planned for, and preferred, a water landing, the Soviet program was "totally unprepared to rescue a cosmonaut from the water," as Nikolai Kamanin wrote in his diary (Gerovitch, *Soviet Space Mythologies,* 85).

147 **Radio Moscow:** *NYT,* Apr. 13, 1961.

147 **The United States had tracked:** "Main" to Cate, "Space Cover (Science)," Apr. 13, 1961, Box 25, Folder 481, HL; Sidey, *John F. Kennedy,* 94; Logsdon, *Race to the Moon,* 70.

148 **Phones were ringing:** *WP,* Apr. 13, 1961; Powers OH, NASA HRC. A few days later, NASA ordered Powers to apologize to the reporter. Powers refused. He telegraphed a colleague: "It is my intention to write a letter to the president of the UPI advising him that I think his agency owes me and the American people an apology for irresponsible reporting" (Powers OH, NASA HRC; Makemson, *Media, NASA, and America's Quest for the Moon,* 60).

148 **President Kennedy was already awake:** Sidey, *John F. Kennedy,* 94–95; JFK, *Public Papers,* 1961: 257; Bundy to Murrow, Apr. 11, 1961 [postscript on or after Apr. 12], POF, Box 307, JFKL.

149 **When Kennedy arrived at the Oval Office:** JFK Appointment Book, Apr. 12, 1961, JFK-MPF-PAB, JFKL; TCS, *Kennedy,* 525; TCS, *Counselor,* 334.

149 **On the president's desk:** Logsdon, *Race to the Moon,* 49–51; "Report of the Ad Hoc Mercury Panel," Apr. 12, 1961, TCS Papers, Box 38, JFKL. The poem is "High Flight," by John Gillespie Magee.

149 **The panel was impressed:** "Report," 1–2, 13–14; Logsdon, *Race to the Moon,* 49–51; Dryden OH, JFKL; Gilruth diary, Mar. 1–5, 1961; "Panel Findings Related to the Medical Aspects of Mercury," 24–25, POF, Box 086a, JFKL. Gilruth believed that the panel's biomedical experts were spectacularly misin-

formed: he and his colleagues were shocked by the comments of one expert who based his claims on information that was "completely untrue" (Gilruth diary, Mar. 5, 1961).

150 ***Izvestia*'s description:** *NYT,* Apr. 13, 1961.

150 **"The Russian accomplishment":** *NYT,* Apr. 13, 1961. Glenn's most oft-quoted response to Gagarin's flight appears to have been invented by Tom Wolfe in *The Right Stuff*: "Well, they just beat the pants off us, that's all, and there's no use kidding ourselves about that." Wolfe introduces the quote with the words "[Glenn] as much as said." That, paired with the fact that the line cannot be found in contemporaneous sources, suggests that Wolfe was granting himself poetic license (Wolfe, *The Right Stuff,* 189).

150 **the feeling at Langley:** *NYT,* Apr. 13, 1961; Faget OH, JSC.

150 **The astronauts . . . were furious:** Shelton to Cate, "Requested Man in Space, Take 1," Apr. 13, 1961, Box 25, Folder 481, HL; *NYT,* Apr. 15, 1961.

151 **Shepard in particular:** French and Burgess, *Into That Silent Sea,* 57; Thompson, *Light This Candle,* 239; *WP,* Apr. 13, 1961; *Life,* Apr. 21, 1961; *NYT,* Apr. 13, 1961.

151 **At 10:00 a.m., Overton Brooks began:** *Hearing Before the Committee on Science and Astronautics,* Apr. 12, 1961, 3, 6, 21.

152 **All the old fears:** "Draft of a Report on MIDAS by DDR&E Ad Hoc Group," Nov. 1, 1961, accessed at https://nsarchive2.gwu.edu/NSAEBB/NSAEBB235/04.pdf; *NYT,* Apr. 13 and 16, 1961; *WP,* Apr. 13, 1961.

152 **Berlin, not space:** *NYT,* Apr. 13, 1961; JFK, *Public Papers,* 1961: 265.

152 **It was a day of exultation:** *NYT,* Apr. 13, 1961.

153 **If the White House was looking:** *NYT,* Apr. 13 and 16, 1961.

153 **Observers anticipated:** *NYT,* Apr. 13, 14, and 16, 1961.

153 **Just before 4:00 p.m.:** JFK, *Public Papers,* 1961: 258–59.

153 **"Could you give us":** Ibid., 259 (emphasis added).

154 **This failed to put the matter to rest:** Ibid., 262–63.

154 **Kennedy professed confidence:** Ibid., 261–62.

155 **The press conference was panned:** *WP,* Apr. 13, 1961; *NYT,* Apr. 13, 1961; ABC News Special, Apr. 12, 1961, accessed at https://www.youtube.com/watch?v=ZYucEQFYNf0.

156 **"Ten days ago":** ABC News Special.

156 **"sparing no effort":** *Hearing Before the Committee on Science and Astronautics,* Apr. 13, 1961: 3, 7, 12; Lambright, *Powering Apollo,* 93–94; *NYT,* Apr. 16, 1961.

156 **was "in a mood":** Webb to Kenneth O'Donnell, Apr. 21, 1961, WHCF, Box 652, JFKL; *Hearing Before the Committee on Science and Astronautics,* 13, 15.

157 **a scene of jubilation:** *Vecherniaia Moskva,* "The Capital Meets Its Hero," Apr. 14, 1961, accessed at http://soviethistory.msu.edu/1961-2/first-cosmonaut/first-cosmonauts-texts/the-capital-meets-its-hero/; *NYT,* Apr. 15, 1961.

157 **And now the man in question:** *Life,* Apr. 21, 1961; Chertok, *Rockets and People,* 83; *WP,* Apr. 11, 2011; Taubman, *Khrushchev,* 491–92. Khrushchev was unaware that a series of small errors had plagued the flight of *Vostok* virtually

from start to finish; as a matter of course, the managers of the Soviet space program hid their missteps from the Soviet leadership (Gerovitch, *Soviet Space Mythologies,* 88–89, 92, 97).

158 **impatience was growing:** *NYT,* Apr. 16, 1961; Logsdon, *Race to the Moon,* 75; Sidey to Robert Sherrod, Jan. 6, 1971, NASA HRC, File 012504; Sidey to Salinger, Apr. 14, 1961, NASA HRC, File 012506.

158 **Webb and Dryden arrived early:** TCS, *Counselor,* 335; Wiesner to JFK, Apr. 14, 1961, TCS Papers, Box 38, JFKL; Logsdon, *Race to the Moon,* 75–76; Welsh to LBJ, "Discussion of Space Program Friday Evening, April 14," Apr. 18, 1961, NASA HRC, File 012506.

158 **The sun had set:** Sidey, *John F. Kennedy,* 100–01; *Life,* Apr. 21, 1961; Sidey to Sherrod.

159 **"What can we do?":** Sidey, *John F. Kennedy,* 101–02; Logsdon, *Race to the Moon,* 77; Sidey to Sherrod. Kennedy's trust in Sidey was not misplaced. The article in *Life* did show the White House in a state of indecision, but Sidey kept Kennedy's exasperation—and his edgier comments ("I don't care if it's the janitor")—in confidence until after Kennedy's death (Sidey to Hugh Moffett and Robert Parker, Apr. 15, 1961, attachment by Robert Sherrod to Eugene M. Emme, Jan. 18, 1971, NASA HRC, File 012504; Dwayne A. Day, "Pay No Attention to the Man with the Notebook," *Space Review,* Dec. 5, 2005, accessed at http://www.thespacereview.com/article/511/1).

159 **On April 14, a brigade:** Schlesinger, *A Thousand Days,* 267, 269.

160 **His overconfidence was shared:** TCS, *Kennedy,* 295, 297, 299, 303; Schlesinger, *A Thousand Days,* 274–77; Dallek, *An Unfinished Life,* 358–61, 366; Goodwin, *Remembering America,* 170–73, 180; Sidey, *John F. Kennedy,* 105–06.

160 **The shock and dismay:** Schlesinger, *A Thousand Days,* 291; *NYT,* Mar. 30 and Apr. 16, 1961.

161 **From that point forward:** Schlesinger, *A Thousand Days,* 296–97; Dallek, *An Unfinished Life,* 368–72; Dallek, *Camelot's Court,* 153–56; Edward C. Welsh OH, NASA HRC; Donald F. Hornig to TCS, Apr. 18, 1961, TCS Papers, Box 38, JFKL; unknown writer to Cate, "Space Cover—I," Apr. 13, 1961, Box 25, Folder 481, HL.

161 **Kennedy walked the grounds:** Logsdon, *Race to the Moon,* 79–80.

162 **Sorensen drafted the memo:** JFK to LBJ, Apr. 20, 1961, NASA HRC, File 12288.

162 **The next morning:** Logsdon, *Race to the Moon,* 82; JFK, *Public Papers,* 1961: 309–10; press conference footage accessed at https://www.youtube.com/watch?v=AYx6MG6NkjU.

CHAPTER 10: GO OR NO GO

164 **the First Team:** Carpenter et al., *We Seven,* 175–76; Glenn and Taylor, *Memoir,* 296. The crew quarters can be seen at https://www.nasa.gov/image-feature/project-mercury-crew-quarters-in-hangar-s.

164 **The flight of Mercury-Redstone 3:** *TNO,* 343, 345–46; Wolfe, *The Right Stuff,* 185.

165 **an air of unreality:** JHG OH, JGA; Glenn and Taylor, *Memoir,* 314–15; Thompson, *Light This Candle,* 241.

165 **The press—still in the dark:** *TNO,* 349; *Norfolk Ledger-Star,* Mar. 25, 1961; *Columbus Dispatch,* Mar. 25, 1961.

166 **Glenn . . . kept up the charade:** Annie Glenn scrapbooks, Box 9.1, Folder 2, and Box 10, Folder 2, JGA; Wolfe, *The Right Stuff,* 187.

166 **Project Mercury proceeded:** Welsh to LBJ, Apr. 26, 1961, NASA HRC, File 012506; Faget OH, JSC; Faget and Purser OH, "Space Stories"; Gilruth OH, NASA HRC.

167 **"a man is fairly useless":** *Nation,* Apr. 29, 1961.

167 **And while the Soviets:** *NYT,* May 2, 1961; Makemson, *Media, NASA, and America's Quest for the Moon,* 66.

167 **"Many persons involved":** Wiesner to Bundy, Mar. 9, 1961, WHCF, Box 176, JFKL; Makemson, *Media, NASA, and America's Quest for the Moon,* 65.

168 **Kennedy's popularity rating:** Sidey, *John F. Kennedy,* 130–31; *NYT,* Apr. 21, 1961; *Life,* May 5, 1961.

169 **the role he had expected as vice president:** See Shesol, *Mutual Contempt,* esp. 61–87; Neufeld, *Von Braun,* 361; Logsdon, *Race to the Moon,* 86; "Outline," n.d. [Apr. 24, 1961], NASC Files, Box 3, JFKL.

169 **first there was a recitation of grievances:** "Outline."

169 **This meant the moon:** Logsdon, *Race to the Moon,* 87, 90–91; Lambright, *Powering Apollo,* 95; Milton W. Rosen to Webb, Apr. 19, 1961, NASA HRC, File 012504.

170 **Johnson moved swiftly:** Welsh to Eugene M. Emme, n.d. [1976], NASA HRC, File 012506; LBJ to JFK, "Evaluation of Space Program," Apr. 28, 1961, NASA HRC.

170 **"more resources and more effort":** "Evaluation of Space Program"; *Aeronautical and Astronautical Events of 1961,* 17; Grimwood, *Chronology,* 132; Leopold, *Calculated Risk,* 101; *AW,* May 1, 1961.

171 **As Kennedy weighed the costs:** Welsh to Emme; Logsdon, *Race to the Moon,* 95–96.

171 **the press had built a garrison:** *NYT,* May 2, 1961.

172 **Webb released a public statement:** *NYT,* Apr. 30 and May 2, 1961; *AAE of 1961,* 19.

172 **on May 1:** Dallek, *An Unfinished Life,* 394; *TNO,* 350; JFK Appointment Book, Apr. 1961, JFK-MPF-PAB, JFKL; "Vice President's Ad Hoc Meeting," May 3, 1961, in Logsdon, *Exploring the Unknown,* vol. 1, 436; Haney OH, JSC.

173 **Amid the motels:** *NYT,* Apr. 30 and May 1, 1961; *WES,* May 1, 1961; Clara Glenn and John Glenn Sr. to JHG, Apr. 30, 1961; *Zanesville Times Recorder,* May 2, 1961.

174 **At 1:30 a.m. on May 2:** *TNO,* 350; Thompson, *Light This Candle,* 340; *WES,* May 2, 1961.

174 **No one outside:** Thompson, *Light This Candle,* 340–41; *WES,* May 2, 1961; *WP,* Feb. 20, 1975. The correspondent was Julian Scheer of the *Charlotte News;* in 1962, Scheer joined NASA's public affairs team.

175 **Foul weather:** French and Burgess, *Into That Silent Sea,* 59; Thompson, *Light This Candle,* 241–42; Glenn and Taylor, *Memoir,* 315.

175 **At 1:10 a.m. on May 5:** NASA Space Task Group, "Postlaunch Report for Mercury-Redstone No. 3 (MR-3)," June 16, 1961, Box 4, Folder 28, Kraft Papers, Virginia Tech; *TNO,* 350–51; Thompson, *Light This Candle,* 243.

175 **a *Playboy* centerfold:** Glenn saved the "no handball" sign and, years later, put it in his archives (Box 67, Folder 8). The centerfold has been lost to history.

176 **At 5:15 a.m. Shepard strode:** "Postlaunch Report"; French and Burgess, *Into That Silent Sea,* 62; Thompson, *Light This Candle,* 245–46.

176 **The National Security Council meeting:** JFK Appointment Book, May 5, 1961, JFK-MPF-PAB, JFKL; handwritten note and "Record of Actions by the National Security Council," NSC Meetings, 1961: No. 483, May 5, 1961, NSF, JFKL; Evelyn Lincoln to JFK, May 5, 1961, accessed at https://www.christies.com/lotfinder/Lot/kennedy-john-f-1917-1963-lincoln-evelyn-6145985-details.aspx; Sidey, *John F. Kennedy,* 131; Logsdon, *Race to the Moon,* 96.

177 **At 10:34 the Redstone fired:** "Postlaunch Report"; *Proceedings of a Conference on Results of the First U.S. Manned Suborbital Space Flight,* June 6, 1961, accessed at https://msquair.files.wordpress.com/2011/05/results-of-the-first-manned-suborbital-space-flight.pdf; Wolfe, *The Right Stuff,* 204–05; *TNO,* 352–57; French and Burgess, *Into That Silent Sea,* 65–69; Thompson, *Light This Candle,* 252–57; TCS, *Counselor,* 338; Sidey, *John F. Kennedy,* 131–32.

177 **Aboard the recovery ship:** *TNO,* 357, 361; *WES,* May 5, 1961.

178 **"He made it":** *NYT,* May 6, 1961; Wolfe, *The Right Stuff,* 136; Thompson, *Light This Candle,* 6–10.

178 **NASA's concessions:** *NYT,* May 6, 1961; "NBC Special News Report: *Freedom 7,*" May 5, 1961, accessed at www.youtube.com/watch?v=Jc0ccOmcPno; Logsdon, *Race to the Moon,* 96; "Address by Edward R. Murrow Before the National Press Club," May 24, 1961, POF, Series 7, Box 091, JFKL.

179 **The Soviets . . . took the posture:** *NYT,* May 6, 1961; *AW,* Apr. 3 and May 15, 1961; "Meeting with the Astronauts," 8; *TNO,* 100; *Life,* May 12, 1961; Slayton with Cassutt, *Deke!,* 101; *WES,* May 6, 1961.

179 **"we have a long way to go":** JFK, *Public Papers,* 1961: 355, 358.

179 **"The Vice President just called":** Webb to O. B. Lloyd, May 5, 1961, NASA HRC, File 12288; Seamans, *Aiming at Targets,* 88; Lambright, *Powering Apollo,* 97–98; LBJ to JFK, May 8, 1961, TCS Papers, Box 38, JFKL. On McNamara's motivations—chiefly his concern about the financial health of the aerospace industry at a time when he was planning cutbacks—see McDougall, *. . . the Heavens and the Earth,* 320–21.

180 **Sweeping in scope:** Logsdon, *Exploring the Unknown,* vol. 1, 446–47.

180 **three Marine Corps helicopters:** Bisney and Pickering, *The Space-Age Presidency of John F. Kennedy,* 9–12; photo of greeting, accessed at https://www.jfklibrary.org/asset-viewer/archives/JFKWHP/1961/Month%2005/Day%2008/JFKWHP-1961-05-08-A?image_identifier=JFKWHP-AR6569-E; JFK, *Public Papers,* 1961: 366; film of the ceremony accessed at https://www.youtube.com/watch?v=4Qd-t3vFsxc.

181 **Behind them on the platform:** JFK, *Public Papers,* 1961: 366; NASA photo accessed at https://www.nasa.gov/content/president-john-f-kennedy-and-alan-shepard-in-washington-dc.

181 **There was still a parade:** *WES,* May 8, 1961; Bisney and Pickering, *Space-Age Presidency,* 12–14; French and Burgess, *Into That Silent Sea,* xviii; Shepard OH, JSC; *NYT,* May 28, 1961.

181 **Some of Kennedy's questions:** French and Burgess, *Into That Silent Sea,* xviii; Logsdon, *Race to the Moon,* 103–04; Gilruth OH, National Air and Space Museum, accessed at https://airandspace.si.edu/research/projects/oral-histories/TRANSCPT/GILRUTH4.HTM.

182 **Two days later:** JFK Appointment Book, May 1961, JFK-MPF-PAB, JFKL; Logsdon, *Race to the Moon,* 107; *WES,* May 10, 1961.

CHAPTER 11: HOLOCAUST OR HUMILIATION

183 **Kennedy had already given a speech:** *NYT,* May 26, 1961; JFK Appointment Book, May 1961, JFK-MPF-PAB, JFKL; JFK, *Public Papers,* 1961: 396–97.

183 **Wiesner had extracted:** Memorandum, Lambright conversation with Wiesner, Nov. 15, 1990, NASA HRC, File 7106; "Policy Guidance for Space Task Group Speakers," Gilruth to Staff, May 16, 1960, Box 68, Folder 28, JGA; JFK, *Public Papers,* 1961: 403–04.

184 **"I believe that this nation":** JFK, *Public Papers,* 1961: 404.

184 **He put the question to Congress:** *NYT,* May 26, 1961; JFK, *Public Papers,* 1961: 404–05; Reading copy, "Special Message to Congress on Urgent National Needs," May 25, 1961, p. 72, POF, Series 3, Box 34, JFKL; *Chasing the Moon,* directed by Robert Stone, episode 1, 55:16, aired July 10, 2019, on PBS, https://www.pbs.org/wgbh/americanexperience/films/chasing-moon/>.

184 **To Ted Sorensen:** *NYT,* May 26, 1961; TCS, *Kennedy,* 526.

184 **Kennedy had noticed:** *NYT,* May 26, 1961; *WP,* May 26, 1961; Burrows, *This New Ocean,* 332; *AW,* June 5, 1961.

185 **It was far from clear:** William Sims Bainbridge, "The Impact of Space Exploration on Public Opinions, Attitudes, and Beliefs," in Dick, ed., *Historical Studies in the Societal Impact of Spaceflight,* 12; Erskine, "The Quarter's Polls," 664. Also see Launius, "Public Opinion Polls and Perceptions of U.S. Human Spaceflight," 163–75.

185 **Network anchors and newspapers:** Stone, *Chasing the Moon,* episode 1, 56:00; *Los Angeles Times,* May 26, 1961; *Arizona Daily Star,* May 26, 1961; *WP,* May 26, 1961; *Bakersfield Californian,* May 26, 1961; *Ludington (MI) Daily News,* May 26, 1961; *Time,* June 2, 1961; Neufeld, *Von Braun,* 363–64; Kraft, *Flight,* 143; author int., Roberts, Dec. 17, 2018; Kranz, *Failure,* 56–57.

186 **The astronauts kept their own counsel:** Shepard OH, JSC; JHG OH, JGA; JHG calendar, May 1961, Box 19, Folder 2, JGA.

186 **An ill wind:** Kempe, *Berlin 1961,* 12–13, 53–55, 242, 245; *NYT,* July 18, 1961; Schlesinger, *A Thousand Days,* 303–05.

186 **"The Soviet-American conflict":** JFK, *Strategy of Peace,* 10–11; O'Donnell and Powers, *"Johnny, We Hardly Knew Ye,"* 305; Schlesinger, *A Thousand Days,* 348; *Time,* June 2, 1961.

187 **"Both nations," he said:** JFK, *Public Papers,* 1961: 26–27, 93–94.

187 **Some of this was for show:** Fursenko and Naftali, *"One Hell of a Gamble,"* 120; Stone and Andres, *Chasing the Moon,* 103; "Draft Proposals for US-USSR Space Cooperation," Apr. 13, 1961, POF, Series 09, Box 126, JFKL; Wiesner to JFK, May 16, 1961, and "Possible U.S.-Soviet Cooperative Space Projects," n.d. [May 12, 1961], POF, Series 09, Box 126, JFKL.

188 **Overtures were made:** Fursenko and Naftali, *"One Hell of a Gamble,"* 121–22.

188 **The start of the summit:** "President's Meeting with Khrushchev, Vienna, June 3–4, 1961, Background Paper, Scientific Cooperation," May 25, 1961, POF, Series 09, Box 126, JFKL; Kempe, *Berlin 1961,* 228–30; O'Donnell and Powers, *"Johnny, We Hardly Knew Ye,"* 316; Dallek, *An Unfinished Life,* 403–06.

188 **Khrushchev's belligerence:** Schlesinger, *A Thousand Days,* 368, 372–73; Kempe, *Berlin 1961,* 233–34, 244–46; Memorandum of Conversation, "Vienna Meeting Between the President and Chairman Khrushchev," June 3–4, 1961, POF, Series 09, Box 126, JFKL.

189 **Kennedy had raised the idea:** "Vienna Meeting"; Beschloss, *The Crisis Years,* 197.

189 **"I've got a terrible problem":** Fursenko and Naftali, *"One Hell of a Gamble,"* 125, 130–31; Beschloss, *The Crisis Years,* 221; Kempe, *Berlin 1961,* 235, 250, 257–58; Taubman, *Khrushchev,* 495.

190 **"We're obviously getting in trouble":** *AW,* June 26, 1961; *U.S. News & World Report,* July 3, 1961; *Hearings Before the Committee on Aeronautical and Space Sciences,* 11; *Hearings Before the Subcommittee of the Committee on Appropriations,* 649–53, 673.

190 **The flight assignment would seem:** According to NASA's official history of Project Mercury, Grissom was told in January "that he would *probably* be the pilot for Mercury-Redstone 4" (emphasis added; *TNO,* 365); Glenn and Taylor, *Memoir,* 319.

190 **In June, Gilruth reaffirmed that Grissom:** *TNO,* 368–69; Glenn and Taylor, *Memoir,* 321.

191 **Not until July 17:** *NYT,* July 15, 16, 17, and 18, 1961.

191 **Glenn celebrated:** *WES,* July 17 and 19, 1961; Glenn and Taylor, *Memoir,* 319; *NYT,* July 19, 1961.

192 **The crowds in Cocoa Beach:** *WES,* July 18 and 19, 1961; *TNO,* 383.

192 **As the new launch date . . . approached:** *WES,* July 16, 1961; Haney OH, NASA HRC; *NYT,* July 22, 1961; Leopold, *Calculated Risk,* 122–26; *TNO,* 370–73; French and Burgess, *Into That Silent Sea,* 79–81.

192 **the escape hatch blew:** Leopold, *Calculated Risk,* 127–33; *TNO,* 373, 375; Wolfe, *The Right Stuff,* 225–29; French and Burgess, *Into That Silent Sea,* 88; *NYT,* July 22, 1961.

192 **NASA sat Grissom in front of the press:** Leopold, *Calculated Risk,* 129–31, 134–40; *AW,* Aug. 7, 1961; Wolfe, *The Right Stuff,* 231–33.

193 **"I hope it does something":** Leopold, *Calculated Risk,* 142–43; Wolfe, *The Right Stuff,* 235–36.

193 **An atmosphere of crisis:** Kempe, *Berlin 1961,* 277–79; Taubman, *Khrushchev,* 150, 500.

194 **born of desperation:** Taubman, *Khrushchev,* 503–05; *NYT,* July 18, 1961; *WES,* July 16, 17, and 18, and Aug. 6, 1961.

194 **Since returning from Vienna:** JFK, *Public Papers,* 1961: 513–14, 518; TCS, *Kennedy,* 587–88; Schlesinger, *A Thousand Days,* 381–88; Dallek, *An Unfinished Life,* 418–21; Kempe, *Berlin 1961,* 280–83; *Time,* July 7, 1961.

195 **By late July, Kennedy had settled:** Sorensen, *Kennedy,* 588–89; JFK, *Public Papers,* 1961: 534–40.

195 **for now it quieted:** *AW,* July 31, 1961; Dallek, *An Unfinished Life,* 424; Taubman, *Khrushchev,* 501–03.

196 **On August 6:** *NYT,* Aug. 6 and 7, 1961; Chertok, *Rockets and People,* 187–88, 191.

196 ***Vostok II* lifted off:** *NYT,* Aug. 7, 1961; Siddiqi, *Challenge to Apollo,* 292–94; French and Burgess, *Into That Silent Sea,* 112–15; Chertok, *Rockets and People,* 63, 191–94, 201.

196 **RED SPACEMAN LANDS!:** *New York Journal-American,* Aug. 7, 1961; French and Burgess, *Into That Silent Sea,* 99–100; *NYT,* Aug. 7, 1961; Stone, *Chasing the Moon,* 58:10, 58:30. In April, a few commentators had claimed the Gagarin flight was a hoax, but after *Vostok II,* these fantasies gained a following. That fall, Arthur C. Clarke, the science-fiction writer, told a panel at the American Rocket Society meeting that the United States had "a minor industry . . . dedicated to prove that the Russian space achievements are faked." Hugh Dryden added that he had "spent many hours" on Capitol Hill, debunking such claims ("The London Letter No. 105," May 2, 1961, Box 25, Folder 486, HL; *WES,* Aug. 8 and 10, 1961; Grey and Grey, *Space Flight Report,* 168–69).

197 **In Moscow, Titov was feted:** *WES,* Aug. 9, 1961; *NYT,* Aug. 10 1961; Siddiqi, *Challenge to Apollo,* 360.

197 **At a Kremlin reception:** *WES,* Aug. 9, 1961; *NYT,* Aug. 10, 1961; *AW,* Aug. 14, 1961. "I felt myself the complete master of the ship," Titov insisted at a news conference that week, underscoring Khrushchev's message. "It obeyed my will, my hands." In response to an American reporter's question, he said that "the *Vostok II* ship is not adapted for carrying bombs"—"adapted" again implying that it would be easy to arm a spacecraft (*NYT,* Aug. 12, 1961).

198 **Khrushchev had other plans for Berlin:** *Time,* Aug. 25, 1961.

CHAPTER 12: TALKING OUR EXTINCTION TO DEATH

199 **The "space gap":** *TNO,* 377–38; *NYT,* Aug. 13, 1961.

199 **NASA retired Mercury-Redstone:** *TNO,* 377–78, 382–83; *NYT,* Aug. 18, 1961; Jerome B. Hammack and Jack C. Heberlig, NASA Space Task Group, "The Mercury-Redstone Program," Paper No. 2238-61, presented to the American Rocket Society, Space Flight Report to the Nation, Oct. 9–15, 1961, accessed at https://ntrs.nasa.gov/archive/nasa/casi.ntrs.nasa.gov/20150018552.pdf.

200 **Some of them . . . cheered:** *New York,* Jan. 31, 1983; Gus Grissom to Cecile Grissom, Oct. 7, 1961, accessed at https://www.rrauction.com/PastAuctionItem/3355717.

200 **Earlier that year:** Glenn and Taylor, *Memoir,* 321.

200 **"He's imprisoned":** Dallek, *An Unfinished Life,* 418; *NYT,* Aug. 15, 16, and 17, 1961; *Time,* Aug. 25, 1961. The cabinet member was Stewart Udall, secretary of the interior (Schlesinger, *A Thousand Days,* 391, 395).

201 **The wall was an abomination:** TCS, *Kennedy,* 593–94; Dallek, *An Unfinished Life,* 426–28; Sidey, *John F. Kennedy,* 199; *NYT,* Aug. 19 and 20, 1961.

201 **As Kennedy had anticipated:** Dallek, *An Unfinished Life,* 428; Schlesinger, *A Thousand Days,* 398; *NYT,* Aug. 18, 20, and 25, 1961; JFK, *Public Papers,* 1961: 568–69; *WES,* Aug. 25, 1961; TCS, *Kennedy,* 595; *NYT,* Sept. 3, 1961.

202 **Then, on August 30:** Sidey, *John F. Kennedy,* 200–01; TCS, *Kennedy,* 618–19; Sorensen OH, JFKL; Herken, *Cardinal Choices,* 134; *NYT,* Sept. 3 and 5, 1961; Schlesinger, *A Thousand Days,* 483.

202 **The first Soviet blast:** TCS, *Kennedy,* 620, 622; JFK, *Public Papers,* 1961: 589–90; Dallek, *An Unfinished Life,* 429.

203 **By the end of September:** *AW,* Nov. 6, 1961; *NYT,* Sept. 3, 1961; Taubman, *Khrushchev,* 504, 536.

203 **The tests were shadowboxing:** Marc Raskin to Bundy, July 25, 1961, Box 299, NSF, JFKL; *NYT,* Sept. 3, 1961; *AAE of 1961,* 42; *AW,* Sept. 4, 1961.

204 **In a poem:** Beschloss, *The Crisis Years,* 308; *Life,* Sept. 15, 1961; *NYT,* Sept. 3, 22, and 23, 1961.

204 **The mood was not panic:** *NYT,* Sept. 3, 1961; *Life,* Sept. 15, 1961.

205 **On October 4:** *WES,* Oct. 4, 1961.

205 **Among the astronauts:** *TNO,* 381–82, 389; *AAE of 1961,* 46; Voas to Associate Director, "Astronauts' Preparation for Orbital Flight," Sept. 25, 1961, Box 69, Folder 1, JGA. Voas shared the memo with Glenn in advance of a Sept. 26, 1961, meeting of STG personnel. Also, Carpenter diary, Sept. 27, 1961; author int., Voas, Feb. 14, 2020.

206 **On the evening of October 3:** Carpenter diary, Oct. 4, 1961.

206 **In the morning Gilruth gathered:** Carpenter diary. Glenn's own account, provided off the record to a *Time* reporter in February 1962, was consistent with Carpenter's. "We're not a bunch of giddy, back-slapping kids," Glenn added. "You're happy for the guy they pick—or you're sorry it's not you. So you congratulate the pilot, and that's all there is to it." The reporter found this "deceptively nonchalant" (Billings to Parker, "Glenn Observed and Observing," Jan. 26, 1961, Box 28, Folder 537, HL).

207 **"Well," said Grissom:** Carpenter diary, Oct. 4, 1961; Scott Carpenter entry, Rene Carpenter diary, Oct. 9, 1961; Carpenter and Stoever, *For Spacious Skies,* 219.

207 **"All of us are mad":** Gus Grissom to Cecile Grissom, Oct. 7, 1961; Leopold, *Calculated Risk,* 144. Grissom's surprise might have been warranted. Within the program, rumors had suggested not just that he and Shepard were being considered for the orbital flight but that they were the *only* astronauts being considered (David Nevin to Parker, Jan. 22, 1962, Box 26, Folder 536, HL).

207 **"I tried to take the news":** Glenn and Taylor, *Memoir,* 323–24; Lansing Lamont to Parker, "Glenn Cover—Glenn Bio—'The All-American' II," Jan. 17, 1961, Box 28, Folder 534, HL.

208 **"Until we have a man on the moon":** JFK, *Public Papers,* 1961: 662.

208 **There were no follow-up questions:** JFK, *Public Papers,* 1961: 656–62.

208 **Given all this:** *AW,* Sept. 11, 18, 25, and Oct. 2, 1961.

209 **Frustrations flared:** *NYT,* Oct. 10 and 14, 1961, and Dec. 9, 2008; Grey and Grey, *Space Flight Report,* 16, 127–28, 130, 144, 150, 182; *WES,* Oct. 13, 1961; *Space Digest,* Dec. 1961.

209 **this marked, for Johnson:** Grey and Grey, *Space Flight Report,* 182; Lambright, *Powering Apollo,* 102–03; Logsdon, *Race to the Moon,* 121–22; *AW,* Oct. 9, 1961.

210 **During a typical week:** JHG, 1961 Monthly Planning Book, week of Oct. 15–22, 1961, Box 19, Folder 2, JGA; *TNO,* 392–97, 402; Grimwood, *Chronology,* 151–52; *WP,* Nov. 2 and 13, 1961; Franklyn W. Phillips to Bundy, "Highlights of Current NASA Activities," Nov. 14, 1961, NSF, Box 282, JFKL.

211 **The chimp flight, MA-5:** *WP,* Nov. 26, 1961; *Guardian,* Dec. 16, 2013, accessed at https://www.theguardian.com/science/animal-magic/2013/dec/16/ham-chimpanzee-hero-or-victim; Lamb, *Inside the Space Race,* 170–72.

211 **And no one faulted Enos:** *TNO,* 402–07, 589n; Matthews OH, JSC.

212 **The safe return:** *WP,* Nov. 30 and Dec. 1, 1961; Reingold to Parker, "Glenn Cover, Take One," Jan. 19, 1962, Box 28, Folder 535, HL; *AW,* Dec. 4, 1961; *TNO,* 405, 407.

212 **what Gilruth said next:** *TNO,* 407.

212 **"I knew all along":** Eleanore E. to JHG, Nov. 29, 1961, Box 19, Folder 6, JGA; *Columbus Dispatch,* Nov. 30, 1961; *NYT,* Feb. 21, 1962; Shelton to Parker, "Astronaut John Glenn Cover," Jan. 19, 1962, Box 28, Folder 535, HL; author int., Morway.

213 **"the first major flight":** Voas OH, JSC.

CHAPTER 13: CONTINGENCIES

214 **On November 29:** Helmut A. Kuehnel, "MA-6-13 Flight Plan," Dec. 6, 1961, and JHG handwritten notes on "MA-6—Flight Plan Review," Nov. 29, 1961, Box 65, Folder 15, JGA; JHG notes, n.d. [Nov. 1961], Box 65, Folder 21, JGA.

215 **"We have had troubles":** JHG notes, "MA-6—Flight Plan Review"; JHG notes, n.d.; Reingold to Parker, "Glenn Cover—What Will We Learn?" Jan. 25, 1962, Box 28, Folder 536, HL; *Proceedings of a Conference on the Results of the First U.S. Manned Suborbital Space Flight,* 61–62.

216 **"as a result of this presentation":** "MA-6-13 Flight Plan."

216 **"One politico":** Carpenter diary, Oct. 31, 1961.

216 **The pressure to put the first American:** *TNO,* 409, 412; Reingold to Parker, "Glenn Cover, Take One," Jan. 19, 1962, Box 28, Folder 535, HL.

217 **In early December:** French and Burgess, *Into That Silent Sea,* 74; Glenn and Taylor, *Memoir,* 326–28; Shelton to Parker, "Man Orbit," Jan. 26, 1962, Box 28, Folder 537, HL.

217 **his regimen of running:** Glenn and Taylor, *Memoir,* 325; Carpenter and Stoever, *For Spacious Skies,* 239; Carpenter diary, Oct. 9 and an unspecified date, 1961; Billings to Parker, "Glenn Observed and Observing," Jan. 26, 1962, Box 28, Folder 537, HL.

218 **On December 6:** *TNO,* 409; *NYT,* Dec. 7, 1961.

218 **Over the course of the year:** LBJ to JFK, Jan. 1, 1962, WHCF, Box 114, JFKL; Lamont to Parker, "Glenn Shot Cover—Space Yearender Updater (Nation)," Jan. 16, 1962, and Lamont to Parker, "Glenn Shot Cover—Space Yearender Updater II (Nation)," Jan. 17, 1962, Box 28, Folder 534, HL; "U.S. Aeronautics and Space Activities 1961," Jan. 1962, NSF, Series 05, Box 307, JFKL; TCS OH, JFKL.

218 **During the past three months:** JHG, Monthly Planning Book, Dec. 1961, Box 19, Folder 2, JGA; *WES,* Dec. 23, 1961; Glenn and Taylor, *Memoir,* 330; *WES,* Jan. 26, 1962; Lamont to Parker, "Glenn Shot Cover—Glenn Bio—'Annie and Uncle Johnny' (Nation)," Jan. 18, 1962, Box 28, Folder 535, HL.

219 **"just another step":** "Glenn Shot Cover—Glenn Bio—'Annie and Uncle Johnny' (Nation)."

219 **Great Falls Park:** Glenn and Taylor, *Memoir,* 330–32; *WES,* Jan. 26, 1962; author int., Lyn Glenn; Van Riper, *Glenn,* 162.

220 **"It's been a tough first year":** *Time,* Jan. 5, 1962; JFK, *Public Papers,* 1961: 725–27.

220 **"wiser, more mature":** *Time,* Jan. 5, 1962; JFK, *Public Papers,* 1961: 726; *NYT,* Jan. 20, 1962; *WES,* Jan. 11 and 20, 1962; *Life,* Jan. 5, 1962.

220 **On January 11:** *AAE of 1961,* 77; Welsh to TCS, Dec. 14, 1961, WHCF, Box 652, JFKL; Dryden to TCS, Jan. 9, 1962, TCS Papers, Box 70, JFKL; JFK, *Public Papers,* 1962: 10–11; "Glenn Shot Cover—Space Yearender Updater II (Nation)."

221 **rumors made the rounds:** "Glenn Shot Cover"; *AAE of 1961,* 71.

221 **At the Cape, the year began:** *TNO,* 411, 413; Williams, "Go!," 1–3.

222 **The mood was brighter:** [first name unavailable] Jones to Parker, "The D Bird," Jan. 17. 1962, Box 28, Folder 534, and Reingold to Parker, "Glenn Cover, Take One," Jan. 19, 1962, Box 28, Folder 535, HL.

222 **Glenn would crash-land:** Voas, "Astronauts' Preparation for Orbital Flight," Sept. 25, 1961, Box 69, Folder 1, JGA; JHG OH Project, Session 18, 44–45; Marvin W. Robinson to Warren North, Jan. 12, 1962, Box 66, Folder 8 (emphasis in original), JGA.

223 **"Take me to your leader":** The phrase seems to derive from a 1953 *New Yorker* cartoon in which two aliens, stepping off their flying saucer, approach a horse in a field and say, "Kindly take us to your President!" (*New Yorker,* Mar. 21, 1953; *NYT,* Feb. 11, 2012).

223 **Gallows humor:** "Ode to the Astronauts," n.d. [1961–62], Box 64, Folder 42, JGA.

223 **Capsule No. 13:** Like some pilots, Glenn might also have seen the number 13 as auspicious: during World War II and the Korean War, one of the most popular good-luck charms was a 1921 Morgan silver dollar, which featured thirteen stars and a date (Glenn's birth year, no less) whose digits added up to thirteen. *TNO,* 412–13; Shelton to Parker, "Glenn Cover, Take Two," Feb. 21, 1962, Box 28, Folder 542, HL; Bill Wallrich, "Superstition and the Air Force," *Western Folklore* 19, no. 1 (Jan. 1960), 11–16; Glenn and Taylor, *Memoir,* 324.

224 **The children went:** Glenn and Taylor, *Memoir,* 324; JHG OH, JGA; "Names," n.d. [1962], Box 65, Folder 7, JGA; author int., Lyn Glenn. The "7" was a relic of the McDonnell numbering system. Shepard had been assigned Capsule No. 7 and had named it, therefore, *Freedom 7.* Yet the "7" was widely assumed to refer to the Mercury Seven, and thus became a tradition: *Liberty Bell 7, Friendship 7, Aurora 7,* and so on.

224 **Shepard and Grissom let it be known:** Cecelia Bibby, "Story Time: A Naked Lady," accessed at http://freepages.rootsweb.com/~cecebibby/genealogy/nasa-stories/naked-lady.htm; Lawrence McGlynn, "Breaking Through the Glass Gantry," Aug. 7, 2005, collectSPACE, accessed at http://www.collectspace.com/news/news-080705a.html.

225 **Bibby had never met:** McGlynn, "Breaking Through the Glass Gantry"; Bibby, "Story Time."

225 **The astronauts came to treat Bibby:** Ibid.

226 **Pranks did little:** Williams, "Go!," 5–7.

226 **For Glenn, the flashpoint:** Kraft, *Flight,* 154–55; JHG notes, Jan. 1962, Box 48, Folder 37, JGA (emphasis in original).

227 **"We don't look at this flight":** "Eyewitness: Flight Plan for Orbit," CBS News, Jan. 19, 1962, Box 98-331/31, Cronkite Papers, Briscoe Center; Makemson, *Media, NASA, and America's Quest for the Moon,* 75–76; JHG, James Reston int., Nov. 10, 1964, Box 98.1, Folder 22, JGA.

228 **Glenn wrote a letter:** JHG to "Troops," n.d. [Jan. 1962], Box 47.1, Folder 23, JGA.

229 **during his long stints:** On Glenn's recordings, see Van Riper, *Glenn,* 161.

229 **"If you hear this":** JHG notes, n.d. [Jan. 1962], Box 47.1, Folder 23, JGA. Emphasis in original; some (not all) abbreviations have been corrected for clarity.

231 **NASA, too, was at work:** Lloyd to Webb, Dryden, and Seamans, "MA-6 Contingencies," Jan. 16, 1962, Box 47.1, Folder 23, JGA.

232 **"There is a strong rumor":** Shelton to Parker, "Astronaut John Glenn Cover—Bioperse—Take II—Miscellaneous," Jan. 19, 1962, Box 28, Folder 535, HL.

CHAPTER 14: THE BIG SCRUB

233 **the life-support system:** *AAE of 1962,* 7; Williams, "Go!," 8–13; Carpenter et al., *We Seven,* 258; French and Burgess, *Into That Silent Sea,* 82; *WES,* Jan. 22, 1962; *NYT,* Jan. 23, 1962.

233 **Glenn drew a neat line:** JHG, 1962 Monthly Planning Book, Jan. 1962, Box 19, Folder 2, JGA; *TNO,* 413.

234 **under the direction of Jocelyn Gill:** JHG notes, n.d., ca. 1961, Box 63, Folder 8, JGA; Jocelyn R. Gill, Summary Minutes, Dec. 1, 1961 meeting of the Ad Hoc Committee on Astronomical Tasks for the Mercury Astronaut, Jan. 11, 1962, NASA HRC, File 6803; Nancy G. Roman and Jocelyn R. Gill, Memorandum for File, Jan. 23, 1962, NASA HRC, File 6803. Roman was NASA's chief of astronomy from 1959 to 1979; she is often described as the

"mother" of the Hubble Space Telescope, which launched in 1990 (*WP,* Dec. 28, 2018).

234 **Bob Voas . . . had proposed:** *TNO,* 414, 590n; Glenn and Taylor, *Memoir,* 332–33; French and Burgess, *Into That Silent Sea,* 81; Jennifer Levasseur, "Another Journey for John Glenn's Ansco Camera," posted May 27, 2011, at https://airandspace.si.edu/stories/editorial/another-journey-john-glenn%E2%80%99s-ansco-camera.

234 **Finally Glenn pleaded:** Glenn and Taylor, *Memoir,* 333–34; Levasseur, "Pictures by Proxy," 110–11, 123. Glenn's modified Ansco can be viewed at https://airandspace.si.edu/collection-objects/camera-35mm-glenn-friendship-7. He and others frequently referred to his camera as a Minolta Hi-Matic, which is in a sense correct: the camera was manufactured by Minolta and sold under that name in many locations; but the one that Glenn bought in Cocoa Beach had been repackaged and renamed by the Ansco Company, based in New York.

235 **Sightings of Glenn:** Wolfe, *The Right Stuff,* 245; Glenn and Taylor, *Memoir,* 325–26.

235 **An unruly battalion:** *TNO,* 420; *Columbus Dispatch,* Jan. 21, 1962; Makemson, *Media, NASA, and America's Quest for the Moon,* 73–74; *Chicago Daily Tribune,* Jan. 31, 1962; *Space Digest,* Mar. 1962.

236 **NASA was losing control:** Shelton to Parker, "Man-Orbit Update," Jan. 24, 1962, Box 28, Folder 536, HL; *WP,* Jan. 19 and 25, 1962; *NYT,* Jan. 27 and 28, 1962; *Time,* Feb. 2, 1962; *Chicago Daily Tribune,* Jan. 25 and 29, 1962; "Ranger 3," accessed at https://nssdc.gsfc.nasa.gov/nmc/spacecraft/display.action?id=1962-001A.

237 **News coverage took a darker turn:** *Chicago Daily Tribune,* Jan. 27, 1962; *NYT,* Jan. 21, 27, and Feb. 3, 1962.

237 **David Bell, the White House budget director:** *January 1962 Economic Report of the President,* 115–17.

238 **Glenn himself, amid final preparations:** Glenn and Taylor, *Memoir,* 335.

238 **At two o'clock:** French and Burgess, *Into That Silent Sea,* 83–84; Williams, "Go!," 14–17.

239 **Cloud cover shrouded:** *Chicago Daily Tribune,* Jan. 28, 1962; *WP,* Jan. 28, 1962; Glenn and Taylor, *Memoir,* 338; *Time,* Feb. 2, 1962; *Newsweek,* Feb. 5, 1962.

240 **Glenn's parents watched:** *Chicago Daily Tribune,* Jan. 27, 1962; Shelton to Parker, "Astronaut John Glenn Cover—Bioperse—Take II—Miscellaneous," Jan. 19, 1962, Box 28, Folder 535, HL.

240 **In Arlington, Annie Glenn:** JHG OH, JGA; Lamont to Parker, "Glenn Shot Cover—Glenn Bio—'Annie and Uncle Johnny' (Nation)," Jan. 18, 1962, Box 28, Folder 535, HL; *Chicago Daily Tribune,* Jan. 27, 1962; *Newsweek,* Feb. 5, 1962.

241 **After Glenn was inserted:** *Time,* Feb. 2, 1962; McGlynn, "Breaking Through the Glass Gantry"; Williams, "Go!," 17; *Chicago Daily Tribune,* Jan. 28, 1962.

241 **It was a "peculiar situation":** *Chicago Daily Tribune,* Jan. 27, 1962; *NYT,* Jan. 28, 1962; *AW,* Feb. 19, 1962.

242 **The countdown began again:** *TNO,* 420; Williams, "Go!," 19–20; French and Burgess, *Into That Silent Sea,* 87; *NYT,* Jan. 28, 1962; *Chicago Daily Tribune,* Jan. 27, 1962.

242 **In the Muskingum gym:** *Chicago Daily Tribune,* Jan. 27, 1962; *WP,* Jan. 28, 1962; *NYT,* Jan. 28, 1962.

243 **"Well, there'll be another day":** JHG OH, JGA; Van Riper, *Glenn,* 169; Wainwright, *The Great American Magazine,* 278–79.

243 **Annie Glenn—having endured:** Van Riper, *Glenn,* 168; author int., Lyn Glenn; Wolfe, *The Right Stuff,* 248–49; Thompson to JHG, Jan. 13, 1962, Box 19, Folder 8, JGA.

244 **Glenn was unyielding:** JHG OH, JGA; Wainwright, *The Great American Magazine,* 278–79. The most familiar treatment of the Johnson incident—in the movie version of *The Right Stuff*—is, in large part, an invention. LBJ—wearing, of course, a Stetson hat—sits in his limo and shouts at an aide: "What do you mean . . . '*no*'? . . . They better get somebody to tell her to play ball!" Glenn picks up the phone and tells Annie, "I don't want Johnson or any of the rest of them to set as much as one *toe* inside our house! You tell them that Astronaut John Glenn told you to say that." A shouting match with Webb follows, broken up by the other astronauts. Johnson is left sputtering, pounding the seat so hard his whole car shakes. Wolfe's account, in his book, is only slightly more restrained.

244 **For the press corps:** *NYT,* Jan. 28, 29, and 30, 1962; *WP,* Jan. 28, 1962.

245 **Gagarin, on a similar note:** Glenn and Taylor, *Memoir,* 337; Barbour, *Footprints on the Moon,* 46; *Chicago Daily Tribune,* Jan. 29, 1962; *Columbus Evening Dispatch,* Feb. 4, 1962; "Man-Orbit Update"; JHG OH, JGA.

245 **The Glenns had at least two more weeks:** *TNO,* 420; *Results of the First United States Manned Orbital Space Flight, February 20, 1962,* 67; *NYT,* Jan. 31, 1962; *Chicago Daily Tribune,* Jan. 31, 1962; *Space Digest,* Mar. 1962, 64; Makemson, *Media, NASA, and America's Quest for the Moon,* 74; French and Burgess, *Into That Silent Sea,* 90; *AW,* Feb. 26, 1962.

246 **his audience had thinned:** *NYT,* Feb. 17, 1962; *Newsweek,* Feb. 5, 1962; *TNO,* 420–21.

246 **Glenn, too, went home:** JHG, 1962 Monthly Planning Book, Jan. 1962; *WP,* Feb. 3, 1962; *WES,* Feb. 3, 1962; *NYT,* Feb. 4, 1962; Lamont to Parker, "Glenn Shot Cover—Glenn Bio—'The All-American,'" Jan. 17, 1962, Box 28, Folder 534, HL.

247 **Yet the longer the launch vehicle sat:** *AW,* Feb. 19, 1962; *WP,* Nov. 19, 1997; Elliston, *Psywar on Cuba,* 83, 87–90.

247 **Kennedy wanted to see him:** JHG OH, JFKL; JFK Appointment Book, Feb. 5, 1962, JFK-MPF-PAB, JFKL; *Columbus Evening Dispatch,* Feb. 5, 1962.

248 **At 9:40 the two men sat down:** JHG OH, JFKL.

248 **Now the doors were opened:** *WP,* Feb. 6, 1962; *NYT,* Feb. 6, 1962.

248 **The next morning NASA announced:** *NYT,* Feb. 6, 1962; French and Burgess, *Into That Silent Sea,* 91; JFK, 1962: *Public Papers,* 1962: 125; JHG to JFK, Feb. 9, 1962, WHCF, Box 655, JFKL.

249 **February was living up:** *AW,* Feb. 19, 1962; JFK, *Public Papers,* 1962: 139–40.

249 **"We've had bad luck":** *NYT,* Feb. 14, 15, and 16, 1962; *WP,* Feb. 16, 1962; *TNO,* 419–20. Powers put fifty-fifty odds on launching MA-6 on February 20 (French and Burgess, *Into That Silent Sea,* 92).

250 **The press, meanwhile:** *TNO,* 421; *Newsweek,* Feb. 5, 1962; *AW,* Feb. 19, 1962; Kraft, *Flight,* 156; *Space Digest,* Jan. 1962; "Eyewitness: Flight Plan for Orbit," CBS News, Jan. 19, 1962, Box 98-331/31 Cronkite Papers, Briscoe Center.

250 **After the tenth postponement:** *NYT,* Feb. 17, 1962; French and Burgess, *Into That Silent Sea,* 91–92; JHG, "Pilot's Flight Report," Box 66, Folder 3, JGA; Voas OH, JSC; *WP,* Feb. 16, 1962.

251 **"How refreshing":** *Space Digest,* Mar. 1962; *Zanesville Times Recorder,* Feb. 19, 1962.

251 **Word at the Cape:** *WP,* Feb. 20, 1962; French and Burgess, *Into That Silent Sea,* 92; *Life,* Mar. 2, 1962; *TNO,* 422.

252 **The pad crew:** French and Burgess, *Into That Silent Sea,* 93–94; Kraft, *Flight,* 156.

CHAPTER 15: GODSPEED, JOHN GLENN

253 **At a quarter past five:** Glenn and Taylor, *Memoir,* 340–41; Van Riper, *Glenn,* 169–70; French and Burgess, *Into That Silent Sea,* 103; Reingold to Parker, "Glenn Cover, Take One," Feb. 21, 1962, Box 28, Folder 542, HL; NBC News, Feb. 20, 1962, accessed at https://www.youtube.com/watch?v=t0RwiDK7yJA&list=ULVg9QX2-W6fA&index=143; Fields et al., eds., *Post-Launch Memorandum Report for Mercury-Atlas No. 6 (MA-6),* Mar. 5, 1962, NASA, 10-2, Kraft Papers, Box 5, Folder 1, Virginia Tech; *Results of the First United States Manned Orbital Space Flight,* 52; Barbour, *Footprints on the Moon,* 45, 47.

254 **To Glenn, it all seemed routine:** JHG OH, JGA; Glenn and Taylor, *Memoir,* 341, 343; Williams, "Friendship 7," "Go!," 1; NBC News; NASA film, accessed at https://www.youtube.com/watch?v=KDzSRMTDnyE&t=1s.

254 **workmen in white caps:** French and Burgess, *Into That Silent Sea,* 106–08; Wendt OH, JSC; Dick Auerbach and Jim Kitchell, "Project Mercury Handbook, MA-6," n.d. [Jan.–Feb. 1962], NASA HRC, File 16349; *Results of the First United States Manned Orbital Space Flight,* 7–8, 155; "Earth Path Indicator," accessed at https://airandspace.si.edu/collection-objects/earth-path-indicator-mercury-4.

255 **Above his knees:** Rocco Petrone, a launch engineer, claimed that Glenn was so upset by the picture of the naked woman on January 27 that the flight might have been scrubbed for that reason. In truth, the morning after the scrub, Bibby found a note taped to the lamp above her drawing board. It was from Glenn. He wrote that he had loved the drawing and was going to get it framed and hang it at home, in his den. "I have no idea what Annie thought about that!" Bibby said, years later (Bibby, "Story Time"; McGlynn, "Breaking Through the Glass Gantry").

255 **Small problems:** *Results of the First United States Manned Orbital Space Flight,*

70, 119; French and Burgess, *Into That Silent Sea,* 107–09; Glenn and Taylor, *Memoir,* 342; JHG, *Debriefed,* 19.

256 **Now the nation was waking up:** NBC News; Barbour, *Footprints on the Moon,* 48; Sidey to Parker, "Waiting for Glenn," Feb. 21, 1962, Box 28, Folder 542, HL.

256 **Annie Glenn had awakened:** *Life,* Mar. 2, 1962; Wainwright, *The Great American Magazine,* 278.

257 **They filled the time:** *Life,* Mar. 2, 1962.

257 **"Looks like the weather":** "Procedures Log," Feb. 20, 1962, p. 5, accessed online; *TNO,* 423; French and Burgess, *Into That Silent Sea,* 111; "Glenn Cover, Take One"; Williams, "Go!," 2; Glenn and Taylor, *Memoir,* 343; *Results of the First United States Manned Orbital Space Flight,* 120.

257 **The count resumed at T-45:** *Life,* Mar. 2, 1962; Glenn and Taylor, *Memoir,* 342–43; Wainwright, *The Great American Magazine,* 270; JHG to "Troops," n.d. [Jan. 1962], Box 47.1, Folder 23, JGA.

258 **Now she called Lyn and Dave:** *Life,* Mar. 2, 1962; Glenn and Taylor, *Memoir.* "We were all really uptight," David Glenn recalled of that moment. "You didn't know what you were going to be seeing on live TV. . . . You wonder if he's going to turn into a ball of fire" (author int., David Glenn, Mar. 9, 2020).

258 **"What must be the thoughts":** *Life,* Mar. 2, 1962; NBC News, accessed at https://www.youtube.com/watch?v=Vg9QX2-W6fA&list=ULt0RwiDK7yJA&index=141. At T-22 there was a twenty-five-minute hold when a valve stuck on a liquid oxygen pump; and with six and a half minutes left in the count, there was a two-minute hold when the Bermuda tracking station had a power outage (*TNO,* 423, 426).

259 **The call and response:** *NYT,* Oct. 28, 1998; NASA film; French and Burgess, *Into That Silent Sea,* 116, 143; Carpenter et al., *We Seven,* 291; Carpenter and Stoever, *For Spacious Skies,* 226; Carpenter OH, JSC; Stoever e-mail to author, June 11, 2020.

260 **Three engines erupted:** Shelton to Parker, "Glenn Cover, Take Two," Feb. 21, 1962, Box 28, Folder 542, HL; Shelton, *Countdown,* 57; NASA film; Carpenter et al., *We Seven,* 291–92.

260 **On the beach, people jumped:** NASA film; Barbour, *Footprints on the Moon,* 48; *NYT,* Feb. 21, 1962.

261 **The mood at the White House:** JFK Appointment Book, Feb. 1962, JFK-MPF-PAB, JFKL; "Waiting for Glenn"; transcript, "The Flight of John Glenn," CBS News Extra, Feb. 20, 1962, Box 98-331/31, Cronkite Papers, Briscoe Center; Wainwright, *The Great American Magazine,* 279; *Life,* Mar. 2, 1962.

261 **"We got a pitch!":** "The Flight of John Glenn"; *Life,* Mar. 2, 1962; Carpenter et al., *We Seven,* 293; Voas, draft (with JHG revisions), "John Glenn's Three Orbits of the Earth," 2; *Results of the First United States Manned Orbital Space Flight,* 120, 149.

261 **The rocket was also speeding up:** French and Burgess, *Into That Silent Sea,* 119; JHG OH, JGA; Voas, draft, "John Glenn's Three Orbits," 3–4; *Results*

of the First United States Manned Orbital Space Flight, 120; Carpenter et al., *We Seven,* 294.

262 **nearing the end of powered flight:** Carpenter et al., *We Seven,* 295; Voas, draft, "John Glenn's Three Orbits," 5–6; JHG, *Debriefed,* 43, 75; JHG OH, JGA; *Postlaunch Memorandum Report for Mercury-Atlas No. 6 (MA-6),* Mar. 5, 1962, accessed at https://osdn.net/projects/sfnet_mscorbaddon/downloads/Research/Mercury%20MA-6/MA6_PostLaunchReport.pdf/, 7–22; *Results of the First United States Manned Orbital Space Flight,* 120–21, 150.

263 **Glenn had a substantial list:** Carpenter et al., *We Seven,* 295; image of flight plan accessed at https://natedsanders.com/LotDetail.aspx?inventoryid=43935; author int., Voas, Feb. 14, 2020; "John Glenn Notebook," accessed at https://airandspace.si.edu/multimedia-gallery/web11481-2010hjpg?id=2668; JHG, *Debriefed,* 19.

264 **But in these first moments:** Carpenter et al., *We Seven,* 296–97; author int., Voas, Feb. 14, 2020; *Astronaut's Handbook,* Dec. 4, 1959, Box 58, Folder 2, JGA; *NYT,* Feb. 21, 1962; JHG, *Debriefed,* 31–33.

264 **In a few minutes' time:** *Results of the First United States Manned Orbital Space Flight,* 124, 152; JHG, *Debriefed,* 38–39; Carpenter et al., *We Seven,* 296–97; Glenn and Taylor, *Memoir,* 347–48; Dana, "The Astronaut," accessed at https://www.youtube.com/watch?v=PVxfJYw59cM; Buckbee with Schirra, *The Real Space Cowboys,* 77.

265 **"This is *Friendship* 7, still on ASCS":** *Results of the First United States Manned Orbital Space Flight,* 124, 153–55; Carpenter et al., *We Seven,* 298; Voas, draft, "John Glenn's Three Orbits," 13, 15; JHG OH, JGA.

266 **As Africa rolled behind him:** Carpenter et al., *We Seven,* 300–01; *Results of the First United States Manned Orbital Space Flight,* 127, 129; Glenn and Taylor, *Memoir,* 349–50; JHG, *Debriefed,* 41.

267 **On the East Coast:** *Life,* Mar. 2, 1962; Makemson, *Media, NASA, and America's Quest for the Moon,* 80–81, 83; NBC News. NBC reported that night that sixty million people had watched the coverage of Glenn's flight (*NYT,* Feb. 21, 1962).

267 **In New York, the crowd:** *NYT,* Feb. 21, 1962; *WP,* Feb. 21, 1962; *WES,* Feb. 21, 1962; NBC News.

268 **Life stood still:** "Waiting for Glenn"; JFK Appointment Book, Feb. 20, 1962, JFK-MPF-PAB, JFKL; Tazewell T. Shepard to JHG, Feb. 14, 1962, WHCF, Box 655, JFKL.

268 **Down the hall, in the press office:** "Waiting for Glenn"; "NASA's Mercury Control Center at the Cape," accessed at https://www.youtube.com/watch?v=2QPFhyyAv8Q; "NASA Edge: Mercury Mission Control Room," accessed at https://www.youtube.com/watch?v=jHfdTwmDs9w; Kraft, "Tracking the First U.S. Man in Orbit by Radio and Radar," n.d. [Feb. 1962], p. 2, accessed online; Kranz OH, JSC; Williams, "Friendship 7," "Go!," 3.

CHAPTER 16: A REAL FIREBALL

270 **The night, for Glenn:** Voas, draft, "John Glenn's Three Orbits," 19; Carpenter et al., *We Seven,* 301–02; JHG, *Debriefed,* 41.

270 **Over his headset:** Though NASA's transcript reads, "kinda passes rapidly," the recording makes clear that Cooper said "time passes rapidly" (*Results of the First United States Manned Orbital Space Flight,* 158; audio file, "Air-Ground Communications of the MA-6 Flight, Part 2," accessed at https://science.ksc.nasa.gov/history/mercury/ma-6/sounds/ma-6-transcript-audio-2.html).

271 **As he passed above Perth:** *Results of the First United States Manned Orbital Space Flight,* 159; Cooper OH, JSC.

271 **He flew eastward:** *Results of the First United States Manned Orbital Space Flight,* 131, 161; Carpenter et al., *We Seven,* 303–04.

272 **"Roger, *Friendship 7*":** *Results of the First United States Manned Orbital Space Flight,* 161–62. At first, Glenn thought the fireflies might be some of the 350 million copper filaments, each thinner than a human hair, that the Air Force, in October 1961, had tried to scatter in orbit to create a vast cloud of tiny antennas that could—it was hoped—boost long-range radio broadcasts on Earth. The payload never deployed from its satellite and was thought to be lost ("Press Conference, Flight of MA-6, Feb. 23, 1962," NASA News Release No. 62-41A, p. 3, accessed online; *NYT,* Oct. 22, 1961; *Wired,* Aug. 13, 2013). The fireflies, as Carpenter discovered on MA-7, were flakes of frost from the exterior of the capsule; by banging on the inside of the hatch, Carpenter was able to make clouds of the particles appear outside it (*TNO,* 453). "Both Carpenter and Glenn were ridiculed by peers for their fascination with the phenomenon," Kris Stoever recalled (Stoever e-mail to author, May 12, 2020).

272 **But a few seconds later:** Carpenter et al., *We Seven,* 304–05; JHG notes, n.d. [Nov. or Dec. 1961], "Mercury Program, 1961," Box 65, Folder 21, JGA; JHG OH, JGA. For the ASCS problems on MA-5, see chapter 12.

273 **Except that Enos:** *Results of the First United States Manned Orbital Space Flight,* 163–64; "Glenn Cover, Take Two"; JHG notes.

274 **The "little difficulty" was a warning light:** French and Burgess, *Into That Silent Sea,* 146; Williams, "Friendship 7," "Go!," 5–7; Kranz, *Failure,* 68–69; Shelton and Reingold to Parker, "Glenn Cover, Trouble Aloft," Feb. 22, 1962, Box 28, Folder 543, HL; author int., Roberts, Dec. 17, 2018. In an interview the next day, Yardley told Shelton and Reingold of *Time* that the warning light first flashed at T-plus 95 minutes, only seconds before Shepard asked Glenn to stand by for the call with JFK.

274 **What the signal said:** *Results of the First United States Manned Orbital Space Flight,* 28–29; "Glenn Cover, Trouble Aloft"; "Project Mercury Handbook, MA-6"; NASA Memorandum, Sept. 11, 1961, Box 64, Folder 42, JGA; JHG OH, JGA; Carpenter et al., *We Seven,* 305–06.

274 **This landing bag, or "impact skirt":** Adding a landing bag to the capsule had presented various problems from the start, but never, at any stage in the cycle of redesign and retesting, had the heat shield deployed of its own accord (McDon-

nell Aircraft Corp., "Summary of Compliance," Sept. 12, 1960, Box 67, Folder 10, JGA; JHG notes, 1960, Box 69, Folder 44, JGA; Ad Hoc Report, Apr. 12, 1961, JFKL; Hammack and Heberlig, NASA Space Task Group, "The Mercury-Redstone Program," Paper No. 2238-61, *Space Flight Report to the Nation,* Oct. 9–15, 1961, 25–26).

275 **The debate, hushed but heated:** "Glenn Cover, Trouble Aloft"; Williams, "Go!," 7; *WP,* Feb. 23, 1962; Kranz, *Failure,* 68–70; author int., von Ehrenfried.

275 **Decades later, it would remain unclear:** Williams, "Go!," 8; Kranz, *Failure,* 70; *NYT,* Feb. 22, 1962; Arabian OH, JSC.

276 **The source of this idea:** Faget OH, JSC; author int., Roberts, Dec. 17, 2018; "Glenn Cover, Trouble Aloft."

276 **There were related fears:** "Project Mercury Familiarization Manual," SEDR 104–3, Nov. 1961 (revised May 1962), Sect. 8, 2-10, accessed at https://osdn.net/projects/sfnet_mscorbaddon/downloads/Research/Mercury/MercuryFamiliarizationManual20May1962.pdf/; "Glenn Cover, Trouble Aloft"; author int., Roberts, Dec. 17, 2018; Kraft, *Flight,* 159; Arabian OH, JSC. Roberts recalls that the burning fragments were a greater worry than the aerodynamics (author int., Roberts, Dec. 17, 2018).

277 **Then, as if on cue:** "Network Status Monitor Report to Flight Director, MA-6, Test Date, Feb. 20, 1962," accessed online; Williams, "Go!," 8. In his memoir, Kranz writes that the data from the remote sites heightened the uncertainty because "half of the CapCom reports indicated they were seeing Segment 51, and the others were not" (Kranz, *Failure,* 69–70). But official reports from the global network indicate that on Glenn's second pass, all but one tracking station that answered the query (a few did not) picked up a strong signal (above 80 percent) that the landing bag had deployed. The exception was Muchea, which did not detect the signal on Glenn's second orbit, but on his third put it at 85 percent ("Network Status Monitor Report").

277 **Glenn himself had said nothing:** Kraft, *Flight,* 158; Kranz, *Failure,* 71.

277 **"Even if a plane":** Carpenter et al., *We Seven,* 35; Voas OH, JSC.

278 **The control center's approach:** Author int., von Ehrenfried; French and Burgess, *Into That Silent Sea,* 132; *Results of the First United States Manned Orbital Space Flight,* 168, 170; Carpenter et al., *We Seven,* 307.

278 **Seven minutes later:** *Results of the First United States Manned Orbital Space Flight,* 171; Glenn and Taylor, *Memoir,* 355; Carpenter et al., *We Seven,* 307.

279 **Kraft had been right:** "Transcript of MA-6 Press Conference, Feb. 20, 1962," NASA News Release No. 62-41, p. 10, accessed online; *Results of the First United States Manned Orbital Space Flight,* 167–68, 170; Carpenter et al., *We Seven,* 306–08; *TNO,* 430.

280 **As dawn approached:** "Press Conference, Flight of MA-6, Feb. 23, 1962," NASA News Release No. 62-41A, pp. 3–4, accessed at https://www.scribd.com/document/54248326/Press-Conference-Flight-of-MA-6; *Results of the First United States Manned Orbital Space Flight,* 173–75, 177; Glenn and Taylor, *Memoir,* 356; JHG OH, JGA; Carpenter et al., *We Seven,* 308. According to Kranz,

this was a slip by the Canton Island CapCom, Rodney F. Higgins, who was not supposed to let Glenn in on their concerns—even partially (Kranz, *Failure,* 72). Yet "if there was a failure in this quite successful mission," Higgins wrote in his report two weeks later, "it was the failure of the Cape to keep us advised of what was going on." Higgins had been instructed to ask Glenn about the landing bag, but no one had told him why (Higgins, "MA-6 Remote Site Trip Report," Mar. 6, 1962, accessed at https://osdn.net/projects/sfnet_mscorbaddon/downloads/Research/Mercury%20MA-6/MA6_NeworkReport.pdf/).

281 **In the control center, the mood:** Kranz, *Failure,* 71–74; author int., von Ehrenfried; Roberts, e-mail to author, Mar. 11, 2020; "Procedures Log," 12.

281 **One way or another:** Williams, "Go!," 9–10, 15; Stoever and Rene Carpenter, e-mail to author, Mar. 10, 2020; Kraft, *Flight,* 2.

282 **With ten minutes left:** "Glenn Cover, Trouble Aloft"; Kraft, "Report on Test 5460 (MA-6)," Feb. 27, 1962, 5, 8, accessed online; Williams, "Go!," 14.

282 **But even that:** Kranz, *Failure,* 74; "Test 5460, Composite Message Summary," 14, 17, accessed at https://osdn.net/projects/sfnet_mscorbaddon/downloads/Research/Mercury%20MA-6/MA6_Memos.pdf/; *Results of the First United States Manned Orbital Space Flight,* 186; Glenn and Taylor, *Memoir,* 358–59; Carpenter et al., *We Seven,* 313. "The astronaut was somewhat hesitant," Kraft noted in his report on the flight, "but did as directed" (Kraft, "Report").

283 **Kraft had told the Hawaii station:** Kraft, "Report"; Williams, "Go!," 15; Kranz, *Failure,* 74.

284 **Then another fragment:** *Results of the First United States Manned Orbital Space Flight,* 188; Carpenter et al., *We Seven,* 313–15; Glenn and Taylor, *Memoir,* 359–60.

284 **By the time Glenn flew over Texas:** Kranz, *Failure,* 75; *Results of the First United States Manned Orbital Space Flight,* 189; Carpenter et al., *We Seven,* 316. Williams, in his memoir, reflected that "there is no fear in his voice . . . yet to me there is within it the note of resignation of a professional who has sensed death may be only moments away" (Williams, "Go!," 16).

285 **On North Harrison Street:** *Life,* Mar. 2 and 9, 1962; author int., David Glenn; *WP,* Feb. 22, 1962. Lyn and her boyfriend were among the fifteen nominees for class president; each homeroom made a nomination (*NYT,* Feb. 21, 1962).

286 **During the four hours:** Author int., David Glenn; *Life*; "Glenn Cover, Take Two"; Makemson, *Media, NASA, and America's Quest for the Moon,* 82.

286 **Across the country, people froze:** *WP,* Feb. 21, 1962; "Waiting for Glenn," Feb. 21, 1962, Box 28, Folder 542, HL; "The Flight of John Glenn," CBS News Extra, Feb. 20, 1962, Box 98-331/31, Cronkite Papers, Briscoe Center.

287 **The last voice:** *Results of the First United States Manned Orbital Space Flight,* 190. At the moment Shepard's transmission cut out, he was starting to say that Glenn should jettison the retropack as soon as he hit 1 or 1.5 g's, at which point the aerodynamic pressure on the capsule would likely be strong enough to keep the heat shield in place; but Glenn never heard the command (Carpenter et al., *We Seven,* 318).

287 **The blunt end hit:** "Project Mercury Handbook, MA-6." The ablative heat shield was made of glass fibers and resin: the resin boiled off, while the fibers provided stability (*Results of the First United States Manned Orbital Space Flight,* 7). Glenn and Taylor, *Memoir,* 363; JHG OH, JGA.

288 **Outside the window:** JHG, *Debriefed,* 9–10, 15; Glenn and Taylor, *Memoir,* 362–63; *Results of the First United States Manned Orbital Space Flight,* 190. "This," Glenn wrote later, "was a bad moment." He "kept wondering: 'Is that it? Do I feel it?'" (Carpenter et al., *We Seven,* 318; Glenn and Taylor, *Memoir,* 363).

288 **Shepard, too, called:** *Results of the First United States Manned Orbital Space Flight,* 75, 190; Carpenter et al., *We Seven,* 319; Glenn and Taylor, *Memoir,* 363–64; "Glenn Cover, Trouble Aloft"; Williams, "Go!," 17; French and Burgess, *Into That Silent Sea,* 139; Thompson, *Light This Candle,* 274–75; *NYT,* Feb. 22, 1962. The radio blackout lasted four minutes and twenty-two seconds. "I thought we'd lost him," Shepard said to one of the controllers (*Postlaunch Memorandum Report,* 8-5; "Glenn Cover, Trouble Aloft").

289 **"Oh, pretty good":** *Results of the First United States Manned Orbital Space Flight,* 190; Carpenter et al., *We Seven,* 319.

289 **He was twelve miles high:** *Results of the First United States Manned Orbital Space Flight,* 190–91; Carpenter et al., *We Seven,* 319–21; *TNO,* 432–33; JHG, *Debriefed,* 15–16; "Project Mercury Familiarization Manual," 9–3, 9–4; Glenn and Taylor, *Memoir,* 364.

CHAPTER 17: THE BIG LIFT

291 ***Friendship* 7 slammed into the ocean:** *Results of the First United States Manned Orbital Space Flight,* 135, 192; JHG, *Debriefed,* 16–17; *Life,* Mar. 9, 1962; Reingold to Parker, "Glenn Cover, Take Three," Feb. 21, 1962, Box 28, Folder 542, HL; "Transcript of MA-6 Press Conference, Feb. 20, 1962," 12.

291 **Lookouts on the destroyer:** *Postlaunch Memorandum Report,* 9-1; "Transcript," 9; JHG, *Debriefed,* 17; *Results of the First United States Manned Orbital Space Flight,* 135; *Life,* Mar. 9, 1962; *NYT,* Feb. 21, 1962; author int., Lyn Glenn; Kranz, *Failure,* 76; *John Glenn: A Hero's Life, 1921–2016, Time* Commemorative Edition (Middletown, DE, 2017), 33. Despite strict orders against it, crewmen also painted inscriptions and put their decal on the capsule. These would be removed by NASA (*WES,* Feb. 22, 1962).

292 **They reacted at first:** *WES, WP,* and *NYT,* Feb. 21, 1962; Carpenter et al., *We Seven,* 323; Williams, "Friendship 7," "Go!," 19.

293 **On the South Grounds:** "Waiting for Glenn"; *WP* and *NYT,* Feb. 21, 1962.

293 **"I know that I express":** JFK, *Public Papers,* 1962: 150.

293 **Returning to the Oval Office:** JFK Appointment Book, Feb. 20, 1962; Reeves, *President Kennedy,* 287–88; "Waiting for Glenn"; JFK, *Public Papers,* 1962: 150; Sidey, *John F. Kennedy,* 239. In January, JFK had not been planning to meet Glenn in Florida; he instructed LBJ to go in his place (JFK to LBJ, Jan. 20, 1962, WHCF, Box 655, JFKL).

294 **Glenn, on the *Noa*:** JHG, *Debriefed,* 9; *TNO,* 434; JHG OH, JGA.

294 **With Glenn back on Earth:** Powers OH, NASA HRC; *Life,* Mar. 2, 1962; Glenn and Taylor, *Memoir,* 366–67.

295 **It was time:** CBS News Extra, Feb. 20, 1962, Box 98-331/31, Cronkite Papers, Briscoe Center; NBC News Special Report, Feb. 20, 1962, accessed at https://www.youtube.com/watch?v=rDikrP7AfWg&list=ULrDikrP7AfWg&index=144; Makemson, *Media, NASA, and America's Quest for the Moon,* 83.

295 **The networks moved on:** *Bergen-Rockland (NJ) Record,* Feb. 21, 1962; NBC News Special Report; "The Flight of John Glenn"; Barbour, *Footprints on the Moon,* 48.

296 **NASA, too, held a press conference:** "Transcript of MA-6 Press Conference," 4–5.

296 **As evening approached:** *TNO,* 434; *Life,* Mar. 9, 1962; Carpenter et al., *We Seven,* 323; "Space Triumph!," Universal-International Newsreel, n.d. [Feb. 1962], accessed at c-span.org/video/?303736-1/john-glenns-friendship-7-mercury-space-flight-1962; Voas OH, JSC.

297 **On the *Randolph*:** *TNO,* 434; Carpenter et al., *We Seven,* 323; JHG notes, n.d. [Feb. 1962], Box 65, Folder 29, JGA; Shelton to Parker, "Glenn Cover, Reaction Add," Feb. 23, 1962, Box 28, Folder 543, HL.

297 **As the Navy pilot prepared:** Carpenter et al., *We Seven,* 323; JHG diary, n.d. [after Apr. 22, 1944], Box 13, Folder 33, JGA.

298 **And now, having flown:** Carpenter et al., *We Seven,* 325–26; Glenn and Taylor, *Memoir,* 369.

298 **For lack of a parade:** *WP,* Feb. 21 and 22, 1962; *WES,* Feb. 21, 1962; Robert Z. Pearlman, "Search Continues for Secret Stamp," Feb. 20, 2012, Space.com, accessed at https://www.space.com/14629-secret-stamp-john-glenn-historic-spaceflight.html; "Giori Innovation," n.d., Smithsonian National Postal Museum, accessed at https://postalmuseum.si.edu/exhibition/stamps-take-flight-creating-america%E2%80%99s-stamps-classic-engraving/giori-innovation. To maintain the element of surprise—and as a hedge against the possibility that Glenn would not make it back alive—officials took pains to keep the process secret. All correspondence was verbal; designers worked late at night, on the weekend, or at home; models and die proofs were carried by special messenger; the stamps were printed in a sealed room; and boxes were sent to post offices without explanation and labeled DO NOT OPEN. When the sealed-off room began to arouse suspicion among employees, the Bureau of Engraving and Printing even spread a false rumor that it was printing test runs of multicolored money (*WP,* Feb. 21, 1962).

298 **The occasion was marked:** *Life,* Mar. 9, 1962; *WES,* Feb. 21, 1962; *NYT,* Feb. 21, 1962; "Glenn Cover, Reaction Add"; JFK Appointment Book, Feb. 20, 1962; Sidey to Parker, "Add Waiting for Glenn," Feb. 21, 1962, Box 28, Folder 542, HL.

299 **In a special report:** NBC News Special Report.

299 **At the White House dinner:** "Toasts at the Dinner Given by the President," Feb. 20, 1962, POF, Series 3, Speech Files, JFKPOF-037-017, JFKL; *Life,*

Mar. 2, 1962; *NYT,* Feb. 25, 1962; *Miami News,* Feb. 21, 1962. By coincidence, before the flight had been scheduled for February 20, Charles and Anne Morrow Lindbergh were put on a draft guest list for the White House dinner; for reasons unknown, they did not make the final cut (Letitia Baldrige to JFK and Jacqueline Kennedy, Feb. 6, 1962, JFK Presidential Papers, White House Staff Files of Sanford L. Fox, Series 2, JFKWHSFSLF-012-003, JFKL).

300 **"The nation would be deluding itself":** *WES,* Feb. 21 and 25, 1962; "Transcript of MA-6 Press Conference," 8; author int., Voas, Apr. 26 and 30, 2020; Wolfe, *The Right Stuff,* 60; *NYT,* Feb. 23 and 25, 1962.

300 **And the world took note:** *WP,* Feb. 22, 1962; *NYT,* Feb. 25, 1962; USIA, *18th Review of Operations, Jan. 1–June 30, 1962,* 8; Dino A. Brugioni, "The Tyuratam Enigma," *Air Force Magazine,* Feb. 23, 2009, accessed at https://www.airforcemag.com/article/0384tyuratam/.

301 **At the Vatican:** *WES,* Feb. 21 and 22, 1962; *WP,* Feb. 22, 1962; *NYT,* Feb. 15 and 25, 1962; Sidey to Parker, "Nation's Lead," Feb. 22, 1962, Box 28, Folder 543, HL.

301 **In the Soviet Union:** "The President's Intelligence Checklist," Feb. 21, 1962, accessed at https://www.cia.gov/library/readingroom/docs/DOC_0005992205.pdf; *NYT,* Feb. 21, 22, and 23, 1962.

302 **"I regard it":** JFK, *Public Papers,* 1962: 151–52, 157–58; *NYT,* Feb. 22, 1962; *WES,* Feb. 23, 1962; Bundy to Webb, Feb. 23, 1962, NSF, Box 334, JFKL. Kennedy was right to be skeptical: also on February 21, in a separate letter, Khrushchev complained bitterly about JFK's refusal to join him for a summit on disarmament, warning that the standoff would end in apocalypse. "It is not for nothing," he wrote, "that they say that once in ten years even an unloaded gun goes off" (Khrushchev to JFK, Feb. 21, 1962, Document 37, *Foreign Relations of the United States, 1961–1963,* vol. 6, *Kennedy-Khrushchev Exchanges,* accessed at https://history.state.gov/historicaldocuments/frus1961-63v06/d37).

303 **But the space race:** *NYT,* Feb. 21 and 25, 1962; "Glenn Cover, Reaction Add."

303 **Lyndon Johnson talked:** Shelton to Parker, "Glenn Cover, Reaction Add," Feb. 23, 1962, Box 28, Folder 543, HL; JHG OH, JGA.

303 **"You are a real can-do man":** "Glenn Cover, Reaction Add"; *Life,* Mar. 2, 1962; *NYT,* Feb. 24, 1962.

304 **Around 8:35 a.m.:** *WES,* Feb. 23, 1962; *WP,* Feb. 24, 1962.

304 **The vice president's plane:** *WES,* Feb. 23, 1962; Shelton to Parker, "Glenn Cover, the Reception," Feb. 23, 1962, Box 28, Folder 543, HL; French and Burgess, *Into That Silent Sea,* 176–77; *NYT,* Feb. 24, 1962.

305 **"Security," Sidey wrote:** Sidey to Parker, "Meeting with Glenn (Nation)," Feb. 23, 1962, Box 28, Folder 543, HL; *Christian Science Monitor,* Feb. 23, 1962; "Glenn Cover, the Reception"; *WES,* Feb. 23, 1962; *NYT,* Feb. 24, 1962.

305 **On the Skid Strip:** Bisney and Pickering, *Space-Age Presidency,* 36; Reingold to Parker, "Glenn Cover, the Glenn Report," Feb. 23, 1962, Box 28, Folder 543, HL; "Meeting with Glenn (Nation)."

306 **The motorcade pulled away:** "Meeting with Glenn (Nation)"; *WES,* Feb. 22,

1962; *Postlaunch Memorandum Report,* 10-1. *Friendship* 7 "looked as if it had really been cooked," Bill Shelton of *Time* observed. "It was tarnished . . . to multi-colored iridescence" ("Glenn Cover, the Reception").

306 **By now the group:** Bisney and Pickering, *Space-Age Presidency,* 38–42; Kraft, *Flight,* 162–63; JHG OH, JGA; *WP,* Feb. 24, 1962; *NYT,* Feb. 24, 1962; "Meeting with Glenn (Nation)"; "Eyewitness to History: John Glenn's Homecoming," CBS News, Feb. 23, 1962, accessed at https://www.youtube.com/watch?v=-Hw8WZbCAyY.

307 **Finally, now, they arrived:** "Glenn Cover, the Reception"; *NYT,* Feb. 24, 1962; "Eyewitness to History"; JFK, *Public Papers,* 1962: 159.

307 **After bestowing the medal:** *NYT,* Feb. 24, 1962; JFK, *Public Papers,* 1962: 159–60.

308 **the president stood in front:** "Meeting with Glenn"; "Flight of John Glenn"; "Eyewitness to History"; author int., Lyn Glenn; *NYT,* Feb. 24, 1962.

309 **The ceremony conferred:** JFK, *Public Papers,* 1962: 159–60; *NYT,* Feb. 24, 1962; *WP,* Feb. 24, 1962; "Press Conference, Flight of MA-6." "The heat-shield deploy switch between stringers 2 and 3 had a very loose rotary stem," McDonnell engineers determined on February 22. "The switch would make and break electrical contact when the rotary stem was moved up and down" (*Postlaunch Memorandum Report,* 10-1).

309 **"I can't pin":** "Press Conference," 10, 15; *WES,* Feb. 22, 1962.

310 **It was a pointed comment:** Glenn and Taylor, *Memoir,* 367; Carpenter and Stoever, *For Spacious Skies,* 227; author int., Voas, Feb. 14, 2020; Kraft, *Flight,* 160–61; Kranz, *Failure,* 77; French and Burgess, *Into That Silent Sea,* 149; Carpenter et al., *We Seven,* 308–09.

310 **The "death peril":** "Press Conference," 13; author int., Schepp; author int., Roberts, Dec. 17, 2018.

311 **He showed Glenn:** Author int., Schepp; author int., Roberts, Dec. 17, 2018, Apr. 27 and 28, 2020; *Postlaunch Memorandum Report,* 11-1. "The ASCS crew got a good laugh when we read the Mission Critique," Roberts recalled years later (Roberts e-mail to author, May 16, 2020).

312 **The last time there had been a parade:** Author int., Lyn Glenn; *WES,* Feb. 26, 1962; author int., Voas, Apr. 23, 2020.

312 **Though nearly a week:** JHG OH, JFKL; Glenn and Taylor, *Memoir,* 373; Phillips to Bundy, "Highlights of Current NASA Activities," Apr. 10, 1962, NSF, Box 282, JFKL. Just before the Washington parade, Glenn visited the White House and signed a special globe owned by the American Geographical Society; it bore the signatures of fifty-six other explorers. Glenn added his own just above Lindbergh's, in the mid-Atlantic (Bisney and Pickering, *Space-Age Presidency,* 48–49).

313 **Just after one o'clock:** *Life,* Mar. 9, 1962; *Space News Roundup,* Manned Spacecraft Center, Mar. 7, 1962, NASA HRC; Wolfe, *The Right Stuff,* 277–78; *Cong. Record,* 87th Cong., 2nd Sess., Vol. 108, Part 3 (Feb. 26–Mar. 15, 1962): 2902. Glenn's seventeen-minute speech was interrupted by applause twenty-five times (*WES,* Feb. 26, 1962).

313 ***"Friendship* 7 is just a beginning":** *Cong. Record,* 2902-2903; *Life,* Mar. 9, 1962; *Time,* Mar. 9, 1962.

314 **New York, on March 1:** "Schedule of Events, New York City, Mar. 1, 1962," Feb. 28, 1962, Box 47.1, Folder 57, JGA; UPI, Feb. 28 and Mar. 1, 1962, Box 47.1, Folder 57, JGA; *Newsweek,* Mar. 12, 1962; Wolfe, *The Right Stuff,* 278–79; *Space News Roundup.*

314 **Two days later:** *Dover (OH) Daily Reporter,* Mar. 5, 1962; "Schedule, New Concord," Mar. 3, 1962, and pamphlet, *Welcome to Muskingum College,* Box 47.1, Folder 55, JGA; Kettlewell, *Our Town,* 116–19.

EPILOGUE: ESCAPE VELOCITY

316 **"Is the moon":** *NYT,* Feb. 25, 1962; Vivien Davis Reid to Annie Glenn, Jan. 24, 1962, Box 19, Folder 8, JGA; JHG OH, JGA; Elven, *The Book of Family Crests,* 1: 34–35. The Glenn family crest, a variation on the theme, bore the motto *Ad astra:* "To the stars" (*NYT,* Feb. 25, 1962).

316 **"This stuff":** "Press Conference, Flight of MA-6," 9; "Glenn Cover, Reaction Add"; JHG OH, JFKL; JHG OH, JGA; Franklyn W. Phillips to Bundy, "Highlights of Current NASA Activities," Apr. 10, 1962, NSF, Box 282, JFKL; USIA, *18th Review of Operations, Jan. 1–June 30, 1962,* 20–21; author int., Aldrich.

317 **Glenn, meanwhile, went back:** JHG OH, JGA; Voas OH, JSC; author int., Voas, Apr. 26, 2020; Hiden T. Cox to Webb, Mar. 22, 1962, and Webb to Kenneth O'Donnell, Mar. 23, 1962, Glenn File, NASA HRC; Lincoln to Webb, Mar. 6, 1961, POF, Box 84, JFKL; Glenn and Taylor, *Memoir,* 384.

318 **Glenn tried to stay on task:** Author int., Voas, Apr. 26, 2020; JHG letter, n.d. [Sept. 1962], Box 48, Folder 37, JGA; Powers OH, NASA HRC; Carpenter and Stoever, *For Spacious Skies,* 326.

318 **Schirra told CBS News:** *Boston Globe,* Sept. 14 and 15, 1962; "Statement of Dr. Gilruth" and "Statement of John Glenn," Sept. 14, 1962, Box 48, Folder 37, JGA; JHG letter; *Atlanta Journal,* Sept. 26, 1962; *Shreveport Times*, Sept. 15, 1962.

318 **In the summer of 1963:** JHG OH, JGA; Glenn and Taylor, *Memoir,* 392–93; author int., Lyn Glenn; *NYT,* July 20, 1963, and Nov. 13, 1983.

319 **Glenn asked for time:** Author int., Voas, Apr. 26, 2020; JHG OH, RFK Project, JFKL; JHG OH, JGA; Carpenter and Stoever, *For Spacious Skies,* 358n; JHG OH, NASA HRC. Over the years, Glenn sometimes trafficked in the JFK rumor himself, possibly as a salve for his own wounded ego, though he conceded late in life that he had "never known whether it was true" (Glenn and Taylor, *Memoir,* 394).

319 **Late in 1963 he drafted:** JHG notes, n.d. [1963], Box 47, Folder 9, JGA.

320 **In the early afternoon:** Author int., Voas, Apr. 26, 2020; Voas email to author, May 24, 2020.

320 **For Glenn, John Kennedy's death:** Author int., Lyn Glenn; Voas e-mail; JHG OH, RFK Project, JFKL.

320 **In 1958, John Foster Dulles:** *NYT,* Jan. 17, 1958; McDougall, . . . *the Heavens and the Earth,* 295–96.

320 **Before the decade was out:** Siddiqi, *Challenge to Apollo,* 385, 403–04, 408, 511, 513, 516; McDougall, . . . *the Heavens and the Earth,* 248, 285; Taubman, *Khrushchev,* 615–21.

321 **Not long after Glenn's flight:** *Report to the Congress from the President of the United States: United States Aeronautics and Space Activities 1962*, 57, accessed at https://history.nasa.gov/presrep1962.pdf; McDougall, . . . *the Heavens and the Earth,* 351; Schlesinger, *A Thousand Days,* 919–21; JFK, *Public Papers,* 1963: 695.

321 **On February 20, 1997:** JHG, Remarks in New Concord, Feb. 20, 1997, accessed at https://ohiomemory.org/digital/collection/p267401coll36/id/11551/.

322 **His path to office:** Warren H. Leimbach, "Re: Lt. Col. John Glenn," Mar. 5, 1964, and John J. Costanzi, "Clinical Record," June 22, 1964, Box 18, Folder 57, JGA; Glenn and Taylor, *Memoir,* 400–04, 409; *Nashville Tennessean,* May 10, 1964.

322 **He tried again:** Richard F. Fenno Jr. notes, June 23–25, 29, and July 22, 1982, Box 8, Folder 6, Fenno Papers, University of Rochester; Glenn and Taylor, *Memoir,* 427–48, 433–36; *NYT,* Nov. 13, 1983; *WP,* Feb. 20, 1975, and June 18, 1982.

322 **In the Senate he stood apart:** Fenno notes, July 21, 1983, and July 15, 1984; *NYT,* Oct. 11, 1981, and Nov. 13, 1983; author int., Lyn Glenn; *New York,* Jan. 31, 1983. In college, Glenn had written a paper titled "Problems of the Nomination for the President," in which he argued that candidates should take strong stands on the issues and not run on "personality traits or attractiveness" (Shelton to Parker, "Astronaut John Glenn Cover—Bioperse—Take I—Early Days," Jan. 19, 1962, Box 28, Folder 535, HL).

323 **One night in the 1980s:** Author int., Lyn Glenn; Lyn Glenn e-mail to author, May 14, 2020.

324 **In 1995, reviewing materials:** Glenn and Taylor, *Memoir,* 473–77; *Time,* Aug. 17, 1998; *Vanity Fair,* Oct. 1998.

324 **The scientific agenda:** Author int., Parazynski; Helen Thorpe, "Can John Glenn Do It Again?," *Texas Monthly,* Oct. 1998; Rinker Buck, "The Ripe Stuff," *Vanity Fair,* Oct. 1998; *WP,* Oct. 29, 1998; author int., Lyn Glenn; author int., David Glenn; author int., McCurry and Schneiders.

325 **Yet there was no denying him now:** *NYT,* July 29 and Oct. 30, 1998; *WP,* Oct. 29, 1998; author int., David Glenn; author int., Parazynski.

BIBLIOGRAPHY

MANUSCRIPT COLLECTIONS AND PERSONAL PAPERS

Walter Cronkite Papers, Dolph Briscoe Center for American History, University of Texas, Austin, TX

Richard F. Fenno Jr. Papers, University of Rochester, Rochester, NY

Robert R. Gilruth Papers, Virginia Tech, Blacksburg, VA

John Fitzgerald Kennedy Papers, John F. Kennedy Presidential Library, Boston, MA

National Security Files

Pre-Presidential Papers, Senate Files

President's Office Files

White House Central Files

Christopher C. Kraft Papers, Virginia Tech, Blacksburg, VA

NASA Headquarters Historical Reference Collection, Washington, DC

Non-Senate Papers, John H. Glenn Archives, Ohio State University, Columbus, OH

Don A. Schanche Papers, Hargrett Rare Book and Manuscript Library, University of Georgia, Athens, GA

Theodore C. Sorensen Papers, John F. Kennedy Presidential Library, Boston, MA

Time Inc. Dispatches, Houghton Library, Harvard University, Cambridge, MA

Jerome Wiesner Papers, Massachusetts Institute of Technology, Cambridge, MA

PERSONAL INTERVIEWS

Arnold Aldrich, telephone interview, December 20, 2018

David Glenn, telephone interview, March 9, 2020

Lyn Glenn, telephone interview, May 14, 2020

Mike McCurry and Greg Schneiders, Washington, DC, February 12, 2019

Don Morway, telephone interview, October 17, 2019

Scott Parazynski, telephone interview, May 30, 2020

Jerry Roberts, telephone interviews, December 17, 2018, and April 27 and 28, 2020

Robert Schepp, telephone interview, January 28, 2019

Kris Carpenter Stoever, telephone interviews, June 1 and July 5, 2018, and June 24, 2019

Robert Voas, Bethesda, Maryland, February 14, 2020, and telephone interviews, February 19, April 23, 26, and 30, and May 12 and 21, 2020

Manfred "Dutch" von Ehrenfried, telephone interview, August 29, 2020

ORAL HISTORY TRANSCRIPTS

American Institute of Aeronautics and Astronautics ("Space Stories")

- Maxime Faget
- Paul Purser
- Joseph Guy Thibodaux Jr.

John F. Kennedy Oral History Collection, John F. Kennedy Presidential Library, Boston, MA

- Hugh Dryden
- John H. Glenn Jr.
- Theodore C. Sorensen

John H. Glenn Archives, Ohio State University, Columbus, OH

- Annie Glenn
- John H. Glenn Jr.

Johnson Space Center, Houston, TX

- Donald D. Arabian
- Scott Carpenter
- Gordon Cooper
- Charles J. Donlan
- Maxime Faget
- Paul Haney
- Richard S. Johnston
- Kenneth Kleinknecht
- Christopher C. Kraft
- Eugene Kranz
- C. Frederick Matthews
- Dee O'Hara
- Walter M. Schirra Jr.
- Alan B. Shepard Jr.
- Robert F. Thompson
- Robert Voas
- Manfred "Dutch" von Ehrenfried
- Guenter F. Wendt

NASA Historical Reference Collection, Washington, DC

- Charles J. Donlan
- Maxime Faget
- Robert R. Gilruth

John H. Glenn Jr.
Paul Haney
Christopher C. Kraft
George M. Low
Charles W. Mathews
John A. Powers
Julian Scheer
Abe Silverstein
Donald K. Slayton
Robert F. Thompson
Jerome Wiesner
Walter C. Williams

National Air and Space Museum, Washington, DC
Robert R. Gilruth

Robert F. Kennedy Oral History Collection, John F. Kennedy Presidential Library, Boston, MA
John H. Glenn Jr.

BOOKS

Aeronautical and Astronautical Events of 1961: Report of the National Aeronautics and Space Administration to the Committee on Science and Astronautics. U.S. House of Representatives, 87th Cong., 2nd Sess. Washington, DC: GPO, 1962.

Astronautical and Aeronautical Events of 1962: Report of the National Aeronautics and Space Administration to the Committee on Science and Astronautics. U.S. House of Representatives, 88th Cong., 1st Sess. Washington, DC: GPO, 1963.

Atkinson, Joseph D., Jr., and Jay M. Shafritz. *The Real Stuff: A History of NASA's Astronaut Recruitment Program.* New York: Praeger, 1985.

Baker, David. *NASA Mercury: 1956 to 1963 (All Models).* Somerset, UK: Haynes, 2017.

Barbour, John. *Footprints on the Moon.* New York: Associated Press, 1969.

Berg, A. Scott. *Lindbergh.* New York: G. P. Putnam's Sons, 1998.

Beschloss, Michael. *The Crisis Years: Kennedy and Khrushchev, 1960–1963.* New York: Edward Burlingame, 1991.

Bisney, John, and J. L. Pickering. *The Space-Age Presidency of John F. Kennedy: A Rare Photographic History.* Albuquerque: University of New Mexico Press, 2019.

Bizony, Piers. *The Man Who Ran the Moon: James E. Webb, NASA, and the Secret History of Project Apollo.* New York: Thunder's Mouth, 2006.

Brinkley, Douglas. *American Moonshot: John F. Kennedy and the Great Space Race.* New York: Harper, 2019.

Buckbee, Ed, with Wally Schirra. *The Real Space Cowboys.* Burlington, ON: Apogee, 2005.

Burgess, Colin. Friendship 7*: The Epic Orbital Flight of John H. Glenn, Jr.* Chichester, UK: Springer, 2015.

Burrows, William E. *This New Ocean: The Story of the First Space Age*. New York: Random House, 1998.

Cadbury, Deborah. *Space Race: The Epic Battle Between America and the Soviet Union for Dominion of Space*. New York: HarperCollins, 2006.

Caidin, Martin. *The Astronauts*. New York: E. P. Dutton, 1960.

———. *Man into Space*. New York: Pyramid, 1961.

Caro, Robert A. *The Years of Lyndon Johnson: Master of the Senate*. New York: Knopf, 2002.

Carpenter, M. Scott, L. Gordon Cooper Jr., John H. Glenn Jr., Virgil I. Grissom, Walter M. Schirra Jr., Alan B. Shepard Jr., and Donald K. Slayton. *We Seven*. New York: Simon & Schuster, 1962.

Carpenter, Scott, and Kris Stoever. *For Spacious Skies: The Uncommon Journey of a Mercury Astronaut*. New York: New American Library, 2004.

Chertok, Boris. *Rockets and People: Hot Days of the Cold War*. Vol. 3, NASA SP-2009–4110. Washington, DC: GPO, 2009.

Cooper, Gordon, with Bruce Henderson. *Leap of Faith: An Astronaut's Journey into the Unknown*. New York: HarperCollins, 2000.

Dallek, Robert. *Camelot's Court: Inside the Kennedy White House*. New York: HarperCollins, 2013.

———. *An Unfinished Life: John F. Kennedy, 1917–1963*. Boston: Little, Brown, 2003.

Dick, Steven J., ed. *Historical Studies in the Societal Impact of Spaceflight*. NASA SP-2015-4803. Washington, DC: NASA, 2015.

Elliston, Jon, ed. *Psywar on Cuba: The Declassified History of U.S. Anti-Castro Propaganda*. Melbourne: Ocean Press, 1999.

Elven, John Peter. *The Book of Family Crests*. London: Henry Washbourne, 1840.

Final Report of the Committee on Commerce, U.S. Senate, Subcommittee of the Subcommittee on Communications, 87th Cong., 1st Sess. *Part III: The Joint Appearances of Senator John F. Kennedy and Vice President Richard M. Nixon and Other 1960 Campaign Presentations*. Washington, DC: GPO, 1961.

French, Francis, and Colin Burgess. *Into That Silent Sea: Trailblazers of the Space Era, 1961–1965*. Lincoln: University of Nebraska Press, 2007.

Fursenko, Aleksandr, and Timothy Naftali. *"One Hell of a Gamble": Khrushchev, Castro, and Kennedy, 1958–1964*. New York: W. W. Norton, 1997.

Gaddis, John Lewis. *The Cold War: A New History*. New York: Penguin, 2005.

Gavin, Lieutenant General James M. *War and Peace in the Space Age*. New York: Harper & Brothers, 1958.

Gerovitch, Slava. *Soviet Space Mythologies: Public Images, Private Memories, and the Making of a Cultural Identity*. Pittsburgh, PA: University of Pittsburgh Press, 2015.

Glenn, John. *John Glenn, Jr.—Debriefed: First Thoughts and Pilot's Debriefing of His Historic* Friendship 7 *Flight*. Edited by Scott Sacknoff. Bethesda, MD: SpaceHistory101.com Press, 2015.

Glenn, John, and Nick Taylor. *John Glenn: A Memoir*. New York: Bantam, 1999.

Glennan, T. Keith. *The Birth of NASA: The Diary of T. Keith Glennan*. NASA SP-4105. Washington, DC: GPO, 1993.

Goldman, Eric F. *The Crucial Decade—and After: America, 1945–1960*. New York: Vintage, 1961.

Goodwin, Richard N. *Remembering America: A Voice from the Sixties*. Boston: Little, Brown, 1988.

Green, Constance McLaughlin, and Milton Lomask. *Vanguard: A History*. NASA SP-4202. Washington, DC: NASA, 1970.

Grey, Jerry, and Vivian Grey, eds. *Space Flight Report to the Nation*. New York: Basic Books, 1962.

Grimwood, James M. *Project Mercury: A Chronology*. NASA SP-4001. Washington, DC: NASA, 1963.

Halberstam, David. *The Coldest Winter: America and the Korean War*. New York: Hyperion, 2007.

———. *The Fifties*. New York: Villard, 1993.

Hansen, James R. *Spaceflight Revolution: NASA Langley Research Center from Sputnik to Apollo*. NASA SP-4308. Washington, DC: GPO, 1995.

Hearing Before the Committee on Science and Astronautics. U.S. House of Representatives, 87th Cong., 1st Sess., on Discussion of Russian Man-in-Space Shot, April 13, 1961.

Hearing Before the Committee on Science and Astronautics. U.S. House of Representatives, 87th Cong., 1st Sess., on H.R. 6169, April 12, 1961.

Hearings Before the Committee on Aeronautical and Space Sciences. U.S. Senate, 87th Cong., 1st Sess., on H.R. 6874, June 7, 8, and 12, 1961.

Hearings Before the Committee on Science and Astronautics, and Subcommittees Nos. 1, 3, and 4. U.S. House of Representatives, 87th Cong., 1st Sess., on H.R. 3238 and H.R. 6029 (superseded by H.R. 6874), 1962 NASA Authorization, March 13, 14, 22, and 23, and April 10, 11, 14, and 17, 1961.

Hearings Before the Subcommittee of the Committee on Appropriations. U.S. Senate, 87th Cong., 1st Sess., on H.R. 7445, June 21, 1961.

Herken, Gregg. *Cardinal Choices: Presidential Science Advising from the Atomic Bomb to SDI*. Stanford, CA: Stanford University Press, 2000.

Herman, Arthur. *Douglas MacArthur: American Warrior*. New York: Random House, 2016.

Hersch, Matthew H. *Inventing the American Astronaut*. New York: Palgrave Macmillan, 2012.

January 1962 Economic Report of the President, Hearings of the Joint Economic Committee. 87th Cong., 2nd Sess., January 26, 1962.

Kempe, Frederick. *Berlin 1961: Kennedy, Khrushchev, and the Most Dangerous Place on Earth*. New York: G. P. Putnam's Sons, 2011.

Kennedy, John F. *The Strategy of Peace*. Edited by Allan Nevins. New York: Harper & Brothers, 1960.

———. *Why England Slept*. New York: Wilfred Funk, 1940.

Kettlewell, Ken. *Our Town: New Concord, Ohio, the Birthplace of John Glenn*. Lima, OH: Express Press, 2001.

Kraft, Christopher. *Flight: My Life in Mission Control*. New York: Dutton, 2001.

Kranz, Gene. *Failure Is Not an Option: Mission Control from Mercury to* Apollo 13 *and Beyond*. New York: Simon & Schuster, 2000.

Lamb, Lawrence E. *Inside the Space Race: A Space Surgeon's Diary*. Austin, TX: Synergy Books, 2006.

Lambright, W. Henry. *Powering Apollo: James E. Webb of NASA.* Baltimore, MD: Johns Hopkins University Press, 1995.

Launius, Roger D. *Apollo's Legacy: Perspectives on the Moon Landings.* Washington, DC: Smithsonian Books, 2019.

Leopold, George. *Calculated Risk: The Supersonic Life and Times of Gus Grissom.* West Lafayette, IN: Purdue University Press, 2016.

Logsdon, John M., ed. *Exploring the Unknown.* Vol. 1, *Organizing for Exploration.* Washington, DC: NASA, 1995.

———, ed., with Roger D. Launius. *Exploring the Unknown.* Vol. 7, *Human Spaceflight: Projects Mercury, Gemini, and Apollo.* Washington, DC: NASA, 2008.

———. *John F. Kennedy and the Race to the Moon.* New York: Palgrave Macmillan, 2010.

Makemson, Harlen. *Media, NASA, and America's Quest for the Moon.* New York: Peter Lang, 2009.

McDougall, Walter A. . . . *the Heavens and the Earth: A Political History of the Space Age.* New York: Basic Books, 1985.

Mercury Project Summary Including Results of the Fourth Manned Orbital Flight, May 15 and 16, 1963. NASA SP-45. Washington, DC: NASA, 1963.

Mieczkowski, Yanek. *Eisenhower's Sputnik Moment: The Race for Space and World Prestige.* Ithaca, NY: Cornell University Press, 2013.

Murray, Charles, and Catherine Bly-Cox. *Apollo: Race to the Moon.* New York: Simon & Schuster, 1989.

Neufeld, Michael J. *Von Braun: Dreamer of Space, Engineer of War.* New York: Knopf, 2008.

O'Donnell, Kenneth P., and David Powers. *"Johnny, We Hardly Knew Ye": Memories of John Fitzgerald Kennedy.* Boston: Little, Brown, 1972.

Official Book of the Fair. Chicago: A Century of Progress, Inc., 1932.

Patterson, James T. *Grand Expectations: The United States, 1945–1974.* New York: Oxford University Press, 1996.

Proceedings of a Conference on the Results of the First U.S. Manned Suborbital Space Flight, June 6, 1961. Washington, DC: GPO, 1961.

Proceedings of First National Conference on the Peaceful Uses of Space, Tulsa, Oklahoma, May 26–27, 1961. NASA SP-8. Washington, DC: GPO, 1962.

Project Mercury: Man-in-Space Program of the National Aeronautics and Space Administration. Report of the Committee on Aeronautical and Space Sciences, U.S. Senate, Dec. 1, 1959. Washington, DC: GPO, 1959.

P.S. I Listened to Your Heart Beat: Letters to John Glenn. Houston, TX: World Book Encyclopedia Service/ Doubleday & Co., 1964.

Public Papers of the Presidents of the United States: Dwight D. Eisenhower, 1953–1960. 8 vols. Washington, DC: GPO, 1954–61.

Public Papers of the Presidents of the United States: John F. Kennedy, 1961–1963. 3 vols. Washington, DC: GPO, 1962–64.

Reeves, Richard. *President Kennedy: Profile of Power.* New York: Simon & Schuster, 1993.

The Report of the President's Commission on National Goals. New York: American Assembly, Columbia University, 1960.

Results of the First United States Manned Orbital Space Flight, February 20, 1962. Washington, DC: NASA, 1962.

Rosenblith, Walter A., ed. *Jerry Wiesner: Scientist, Statesman, Humanist.* Cambridge, MA: MIT Press, 2003.

Rosholt, Robert L. *An Administrative History of NASA, 1958–1963.* NASA SP-4101. Washington, DC: GPO, 1966.

Schirra, Wally, with Richard N. Billings. *Schirra's Space.* Boston: Quinlan Press, 1988.

Schlesinger, Arthur M., Jr. *Robert Kennedy and His Times.* New York: Ballantine, 1978.

———. *A Thousand Days: John F. Kennedy in the White House.* Boston: Houghton Mifflin, 1965.

Seamans, Robert C., Jr. *Aiming at Targets: The Autobiography of Robert C. Seamans, Jr.* NASA SP-4106. Washington, DC: GPO, 1996.

Shelton, William Roy. *Countdown: The Story of Cape Canaveral.* Boston: Little, Brown, 1960.

Shepard, Alan, and Deke Slayton with Jay Barbree. *Moon Shot: The Inside Story of America's Apollo Moon Landings.* New York: Open Road, 2011.

Shesol, Jeff. *Mutual Contempt: Lyndon Johnson, Robert Kennedy, and the Feud That Defined a Decade.* New York: W. W. Norton, 1997.

Siddiqi, Asif A. *Challenge to Apollo: The Soviet Union and the Space Race, 1945–1974.* NASA SP-2000–4408. Washington, DC: GPO, 2000.

———. *The Red Rockets' Glare: Spaceflight and the Soviet Imagination, 1857–1957.* New York: Cambridge University Press, 2010.

Sidey, Hugh. *John F. Kennedy, President.* New York: Atheneum, 1964.

Slayton, Donald K. "Deke," with Michael Cassutt. *Deke! U.S. Manned Space: From Mercury to the Shuttle.* New York: Forge, 1995.

Sorensen, Theodore C. *Counselor: A Life at the Edge of History.* New York: Harper, 2008.

———. *Kennedy.* New York: Harper & Row, 1965.

Stone, Robert, and Alan Andres. *Chasing the Moon: The People, the Politics, and the Promise That Launched America into the Space Age.* New York: Ballantine, 2019.

Swenson, Loyd, Jr., James M. Grimwood, and Charles C. Alexander. *This New Ocean: A History of Project Mercury.* NASA SP-4201. Washington, DC: GPO, 1966.

Taubman, William. *Khrushchev: The Man and His Era.* New York: W. W. Norton, 2003.

Thompson, Neal. *Light This Candle: The Life and Times of Alan Shepard, America's First Spaceman.* New York: Crown, 2004.

The Training of Astronauts. Washington, DC: National Academy of Sciences / National Research Council, 1961.

Van Riper, Frank. *Glenn: The Astronaut Who Would Be President.* New York: Empire, 1983.

Wainwright, Loudon. *The Great American Magazine: An Inside History of* Life. New York: Ballantine, 1986.

Walker, Martin. *The Cold War: A History.* New York: Henry Holt, 1993.

Weitekamp, Margaret A. *Right Stuff, Wrong Sex: America's First Women in Space Program*. Baltimore, MD: Johns Hopkins University Press, 2004.

Wiesner, Jerome B. *Where Science and Politics Meet*. New York: McGraw-Hill, 1965.

Wilson, Sloan. *The Man in the Gray Flannel Suit*. New York: Simon & Schuster, 1955.

Wolfe, Tom. *The Right Stuff*. New York: Farrar, Straus and Giroux, 1979.

ARTICLES

Almond, Gabriel A. "Public Opinion and the Development of Space Technology." *Public Opinion Quarterly* 24, no. 4 (Winter 1960): 553–72.

Erskine, Hazel Gaudet, ed. "The Polls: Defense, Peace, and Space." *Public Opinion Quarterly* 25, no. 3 (Autumn 1961): 478–89.

———. "The Quarter's Polls." *Public Opinion Quarterly* 25, no. 4 (Winter 1961): 657–65.

Launius, Roger D. "Public Opinion Polls and Perceptions of U.S. Human Spaceflight." *Space Policy* 19 (2003): 163–75.

Voas, Robert B. "John Glenn's Three Orbits in *Friendship 7*." *National Geographic* 121, no. 6 (June 1962): 792–827.

Wallrich, Bill. "Superstition and the Air Force." *Western Folklore* 19, no. 1 (January 1960): 11–16.

UNPUBLISHED MANUSCRIPTS

Levasseur, Jennifer. "Pictures by Proxy: Images of Exploration and the First Decade of Astronaut Photography at NASA." PhD diss., George Mason University, 2014.

Williams, Walter C. "Go!" Unpublished manuscript, 1967, NASA HRC.

ILLUSTRATION CREDITS

1. NASA
2. John Glenn Archives, The Ohio State University
3. John Glenn Archives, The Ohio State University
4. John Glenn Archives, The Ohio State University
5. John Glenn Archives, The Ohio State University
6. Library of Congress, U.S. News & World Report Magazine Photograph Collection
7. Paul Schutzer/The LIFE Picture Collection via Getty Images
8. Flip Schulke/Corbis Premium Historical via Getty Images
9. W/B/Camera Press/Redux
10. NASA
11. ITAR-TASS News Agency/ Alamy Stock Photo
12. AP Photo/Murray L. Becker
13. AP Photo/WJS
14. Bill Taub Photos
15. NASA
16. NASA
17. Michael Rougier/The LIFE Picture Collection via Getty Images
18. John Glenn Archives, The Ohio State University
19. Edward Hausner/The New York Times/Redux
20. NASA
21. NASA
22. NASA
23. NASA
24. Michael Rougier/The LIFE Picture Collection via Getty Images
25. NASA
26. AP Photo
27. AP Photo
28. AP Photo
29. AP Photo
30. AP Photo
31. John Glenn Archives, The Ohio State University

INDEX

Page numbers after 333 refer to endnotes.